Cold Fusion

Studies in Slavic Literature, Culture, and Society
General Editor: Thomas Epstein

Volume 1
Vladimir Odoevsky and Romantic Poetics: Collected Essays
Neil Cornwell

Volume 2
Women and Russian Culture: Projections and Self-Perceptions
Edited by Rosalind Marsh

Volume 3
Russian Postmodernism: New Perspectives on Post-Soviet Culture
Mikhail Epstein, Alexander Genis, and
Slobodanka Vladiv-Glover

Volume 4
Sight and Sound Entwined: Studies of the New Russian Poetry
Gerald Janecek

Volume 5
Cold Fusion: Aspects of the German Cultural Presence in Russia
Edited by Gennady Barabtarlo

Volume 6
Russian Literature and Its Demons
Edited by Pamela Davidson

COLD FUSION

Aspects of the German Cultural Presence in Russia

Edited by

Gennady Barabtarlo

Berghahn Books
NEW YORK • OXFORD

Published in 2000 by
Berghahn Books
www.berghahnbooks.com

Library of Congress Cataloging-in-Publication Data

Cold fusion : aspects of the German cultural presence in Russia /
edited by Gennady Barabtarlo.
p. cm. — (Studies in Slavic literature, culture, and society : v. 5)
Includes bibliographical references and index.
ISBN 1-57181-188-5 (alk. paper)
1. Literature, Comparative—Russian and German. 2. Literature, Comparative—German and Russian. 3. Russia—Civilization—German influences. I. Barabtarlo, Gennady, 1949– . II. Series.
PG2981.G3C65 2000
891.709—dc21 99-40047
CIP

British Library Cataloguing in Publication Data
A catalogue record for this book is available from
the British Library.

Printed in the United States on acid-free paper

In memory of
Konstantin Bogatyrev

CONTENTS

ACKNOWLEDGMENTS

I am indebted to the Mid-American University International organization and its then chairman Dr. John Heyl, and to the German Academic Exchange Service (DAAD) and its officer Barbara Motyka, as well as to the College of Arts and Science of the University of Missouri, for their beneficent support of the symposium held at Columbia in February and March 1997 from which this volume derives. I am thankful to Mme. Sirotinski for her kind permission to publish a translation of Shalamov's short story.

I must include a special, if inadequate, word of thanks to Mrs. Robert Arnold, the administrative assistant in the Department of German and Russian Studies at the University of Missouri, for her invaluable help in organizing and running the symposium.

I should like to extend here expressions of gratitude to the authors of this anthology—both those from among the original symposium participants and those specially invited to contribute—for their expedient and patient compliance with the various technical and temporal requirements of the enterprise. In this regard, I am particularly indebted to Miss Maria Toshchakoff-Ostashoff, who painstakingly read the essays and called attention to many faults in want of correction, and who compiled the name index.

Neva embankment, with view of Kunstkammer and Academy of Sciences. Photograph by William Brumfield.

INTRODUCTION

Orthoclastic Cleaving

Gennady Barabtarlo

Ptolemy calls the Baltic by its old name, the Slavonic Gulf, because Slavs inhabited its shores.[1] A much more recent author attempts a *philological solution* to the problem of placing the ancient Slavs on the European map: he argues, brilliantly, that their aboriginal habitat must have lain in a place where yew and ivy grew (because the Russian *tis* and *pliushch* are common to all Slavs and are of Slavonic derivation) but beech did not, since *buk* comes from the German *Buche.* "This gives the parallels Ösel-Kurland-Kovno-Vilno-Grodno-Kamenets Podolsk-Kishinev and Königsberg-Danube Mouths."[2] Recent archaeological discoveries rather astonishingly place it much farther west, into Ostholstein (the southeastern region of Schleswig-Holstein), where two stages of Slav occupation have been proposed: first, several fortified settlements in the late seventh and eighth centuries, and then an expansion across the entire region, with many fortresses and even some trading centers. And quite recently, German archeologists Klaus Goldmann and Günter Wersmusch found the long-sought city of Vineta *underwater,* at the mouth of the Oder in eastern Germany. This appears to be the very city of which monk Adam of Bremen said centuries ago that it was the "biggest city in all of Europe," populated by "Slavs, Saxons, Greeks, and Barbarians," all Orthodox Christians.[3] Suddenly, the German and Slavic historical abodes appear to have been not merely contiguous but also overlapping in large areas centuries before coming to a cultural engagement.

A fusion requires contact, often a clash, usually under considerable pressure; the process releases a great deal of energy in one form or another. The German-Russian contacts were variegated yet curiously biased; their clashes were tremendous; and the German presence on the Russian scene during the three-century span between the reigns of the first and the last Romanov, while variously affecting diverse aspects of Russian culture, led to an interesting assimilation of the active side of the fusion process. The evolution of influence and the reaction to it, the phenomena of cultural consumption and subsumption taking turns, as it were, and the intellectual energy thus released, are the chief subjects of the essays comprising this book. As far as we are aware, this collective study takes up this particular tangle of problems for the first time in the English language.

I.

It is possible to argue that, manifold as the German presence in Russia was, its thrust and pathos left especially deep traces in the realm of verbal expression in a manner both provident and lasting. There is a well-known set of cultural novelties related to literary expression that the Germans introduced in Russia. The German Muscovite Wilhelm Mons, in love with Peter I's wife Catherine (and beheaded in 1724), wrote some of the very first secular verse in the Russian language, even if in German transliteration. These love poems had "a quaint intensity that makes us believe he was something of a poet."[4] At about the same time, two other Germans living in Moscow, Pastor Glück and Magister Pauss, experimented with rendering Lutheran prosodic hymns into Russian, trying to preserve the meter and avoid words borrowed from Western languages. The great Russian poetry, especially in its predominant iambic rhythms, owes its organization and indeed its origin to the equisyllabic and tonic prosody that Lomonosov adapted from the metrical system fashioned in Germany by Opitz. This was no historical accident; it is a significant, and rarely acknowledged, fact that of all Western European languages German is phonetically and rhythmically the closest to Russian; in terms of prosody that closeness is quite organic, which explains why Russian renditions of German poetry have been so successful and blend so naturally into the Russian canon (Zhukovski, Lermontov, and Tiutchev stand out among many less illustrious examples).

The first Russian grammars were all compiled either by Germans or in German; at least three such treatises appeared before Lomonosov's *Rossiiskaia grammatika* (1755), which was indebted to its predecessors.[5] The very first one, by Heinrich W. Ludolf, a German from Erfurt and an œcumenist, was published in Latin at Oxford in 1696; it was a sensitive and astonishingly advanced description by a polyglot (with a missionary purpose in mind) of what was then a *dialectus vulgaris*, not suitable for any literary purpose excepting the law code. The other two were in German, with Latin mediation: one by Elias Kopijewitz ("Iliushka Fedorov"), printed in Stoltzenberg (near Danzig) in 1706, the other, an anonymous supplement to the 1730 edition of the excellent *German-Latin-Russian Lexicon* published by the Russian Academy of Sciences. This establishment was so heavily staffed by Germans that the thoroughly frustrated Lomonosov, one of the few Russians there, fought the German party without success until his death.[6] In 1672, another dweller of Moscow's Germantown, Pastor Gregory, was asked to assemble an amateur company to stage plays from a German traveling repertory, translated into a particularly awkward brand of Slavonic, at the court of Czar Alexis. The incredible duration of these first "Russian" dramas (some dragged for eight hours on end) was matched only by their utter silliness, and one does not know whether to marvel more at the patience or the curiosity that kept the royal spectators alive, let alone awake, during the performance.

Russified and even native Germans gave Russia the majority of its very best dictionaries and superb reference books, often unsurpassed in their kind: Vostokov's (Ostenek's) two-volume Church Slavonic dictionary, Michelson's compendium of Russian idiomatic expressions, Vasmer's classic etymological dictionary, Hilferding's establishing editions of Russian folklore, Unbegaun's explicatory *Russian Surnames* (as well as his good and concise grammar and versification treatises), and above all Dal's peerless interpretative lexicon (Dal' was half-German and half-Dane). And it is fitting that the standard (and best by far) German-Russian dictionary by Pawlowski and Semberg, which went through four editions between the 1880s and 1911, is as solid as it is quaint, for it lists Russian words and forms (such as *resti* or *chetshi*) that are not available in the language but could be constructed theoretically (as could "vailable" and "structed" in English).

Ludolf—who, despite his remarkable linguistic erudition and a good ear, could not spare his first Russian grammar of a foreigner's solecisms, e.g., *shtekan* for *stakan* (glass), *baren* for *paren'* (servant boy), a misconception of the verbal aspect, and so on—embarked on his project and spent eighteen months in Moscow mainly because he wished to facilitate the unification of all churches into one Universal Church. Most Germans who settled in Russia, however, had no such ambition and kept their religious tenets to themselves, which made their presence possible and even relatively welcome in a country whose historical foundation and existence rested on the uncontaminated purity of the Orthodox faith as transferred from the pious Greeks. Any form of missionary advances would have been much less tolerable than the deliberate spreading of a contagious deadly disease, but if isolated and contained within the community of the carriers, foreign creeds were not repelled. Until the mid-nineteenth century, Protestantism remained exclusively a religion of foreigners (with the exception of the Baltic provinces and later Finland), with almost no attempts to proselytize, because these foreigners generally pursued practical interests in Russia. Ludolf was a remarkable exception.

The zeal came later. But that earlier restraint made Protestant foreigners welcome in Russia, and their cultural presence was more influential than that of immigrants from Roman Catholic countries—with the famous exception of pre-revolutionary France in the second half of the eighteenth century; but then, the *anti*-Catholic tendencies engendered by the intellectual reign of Voltaire and the Encyclopedists were imported as well.[7] The first Lutheran *kirche* was built in Moscow in 1576 (for the captive Livonians), whereas the first Roman Catholic church was allowed only a century later (in 1684). Eric Amburger shows the unimpeded propagation of the German cultural presence in Russia until 1914, when everything came to an unexpected end.[8]

The first Germans came to Russia, along with their pastors, during the reign of Ivan IV ("the Terrible") as technicians, physicians, and manufacturers. A German colony settled in Moscow outside the city walls, over the Jahooza River. That German Freetown (Liberty, "*sloboda*") had enormous influence on young Peter I. The famous Russian historian Sergei Soloviev said that Moscow Germantown was a stepping stone to the appearance and prominence of St. Petersburg in the sense in which the ancient city of Vladimir was to Moscow.[9] Protestant German communities spread from the Russian provinces to the Urals and Siberia. Since they fostered industry, the Germans enjoyed certain advantages accorded by the state, especially during and after Peter I's reign. They gained commanding positions in many important spheres. Not only the Academy of Sciences but also the state bureaucracy employed a great number of Germans. For example, the Ministry of Finance under Nicholas I had Germans in about 70 percent of its upper- and middle-range executive positions. The military had many Germans in its upper ranks. In a characteristic, although probably untrue, contemporary story, General Ermolov—the hero of the Napoleonic war, an ardent Russian patriot, and a bold-tongued man (other sources of this anecdote name Suvorov)—when asked by Emperor Alexander (or, respectively, Emperor Paul) to choose his own reward before retirement, declined any and, the Emperor being insistent, replied with a deep bow: "Your Majesty may deign to promote me to Germanhood"—"How do you mean?"—"Why, in that case I will, with Your Majesty's leave, have anything I could wish."[10] Indeed, from the first third of the eighteenth century and throughout the nineteenth, many Germans had permanent and hereditary residentship at the Russian court, and, in the eighteenth century, on the Russian throne. Catherine II published a most advantageous offer in Western Europe inviting foreigners to come to Russia and enjoy freedom from all taxation for thirty years—and twenty-seven thousand Germans came and settled 104 agricultural colonies along the banks of the Volga River and elsewhere.

Since the Turks had been driven out of Bessarabia early in the nineteenth century, many Germans replaced them there. Later, with the emergence of new sects in Germany, which were harassed severely under Friedrich I (Würtemberg), German colonies sprang up near the shores of the Black and Azov seas, and eventually in the Caucasus and Georgia. In the Volyn' and Galicia regions, the number of Germans increased from 1,200 in 1832 to 5,800 after the reforms of the 1860s, and then to an astounding 75,000 in 1888 (later, however, the droughts in the Volga area forced many to move to Canada, the United States, and Brazil). About 30,000 Germans lived in Moscow alone before the first German war of this century (7,000 of them were German subjects)—the second largest ethnic group of Muscovites after the Russians.[11]

When Dorpat University (later Iuriev, now Tartu) was founded in 1802, it was the only university in the Empire outside Moscow. All subjects except Russian literature were taught in German for almost the entire century; the president of the University was also appointed as a curator of all the form schools in the whole Baltic region. Confessional matters were also well established and structured. Local Lutheran consistories were subordinated to the General Evangelical Consistory of All Russia, established in 1832, whose charter was drawn by Professor Neimann, a Dorpat professor of law. There was even a special "Subsidy Fund for the Evangelical Lutheran Communities in Russia," instituted in 1859, which lent material support to enterprises ranging from the construction of German churches and schools to salaries and pensions for pastors and teachers.

Dorpat University was one of the chief channels of propagation of critical rationalism in religion (it once issued a prayerbook that contained no mention of Jesus Christ). On the other hand, and in part as a reaction to the academic rationalism and modernism of the Protestant universities, chiliastic and mystical trends gained ground in some parts of Germany early in the nineteenth century, particularly during and after the Napoleonic wars. Anticipation of the end of the world, which the state's reactive persecution only reinforced, drove some dissidents to America, while many others went eastward, to the hospitable Russia of Alexander I.

Eventually, the appearance of new sects, especially *Stunden* and Baptist, made tremendous and devastating inroads in the Lutheran and Calvinist communities, and, in the southern provinces, spread to the Russian population—even though Russian laws continued to punish severely any propaganda of non-Orthodox tenets or practices.

The history of the German cultural presence in Russia ended with the end of Russia proper, as the outcome of a tremendous confrontation between the two countries. The war with Germany led to a mass deportation of Russian Germans from the areas adjacent to the front lines into the hinterlands, thus uprooting solid, established communities. The historical thread of our theme is snipped when it runs into a symmetrical conclusion: in April 1917 the Germans, who had been pouring funds into the coffers of the Russian radical Marxist party for several years, delivered Lenin to St. Petersburg in a sealed coach as a catalyst of the general catastrophe and a fateful civil war that ensued; in November 1920, at the end of that war, a Russified German (Baron von Wrangel), Russia's last commander-in-chief, expatriated the Russian armies from the Crimea, the last patch of free Russia, to Constantinople, under the onslaught of the encroaching Reds.

The Bolsheviks extirpated and scattered the German settlement in Russia altogether, even if not in one blow; the "autonomous" German region in the lower Volga lingered and was industrialized, producing trams and trolleys for the entire country in the Z.I.U. ("Plant named after Uritski," head of the Petrograd Cheka) in "Engels-City." World War II put an end to that tri-centennial chapter of history, as great numbers of people of German origin—real, suspected, or simply with the misfortune of bearing German-sounding names—were exterminated or banished to languish in the deserts of Central Asia,[12] from which some of them, destitute, strangely changed, and speaking a peculiar mixture of dialects, were able to "repatriate" to West Germany in the late 1970s and 1980s, leaving the land of their fathers and returning to that of their forefathers.[13]

II.

This volume owes its concept and many of its constituent articles to a unique international symposium held at the University of Missouri in late February and early March 1997.[14] The Department of German and

Russian Studies was exploring possibilities for collaborative research that would cross-stitch the two disciplines. The thought seemed meaningful and, oddly enough, without known precedent at American universities. Fledgling Russian programs were placed in the capacious nests of German departments in America even before World War II. For instance, Samuel Hazzard Cross, the founder of Russian studies at Harvard, began as an instructor in German, and later became assistant, associate, and then full professor of Slavic languages, while at the same time remaining chairman of the Department of Germanic Languages and Literatures within which the Russian program lurked. This cohabitation became fashionable in the 1950s, out of administrative expediency and perhaps because of a vague notion of geopolitical proximity of the two areas. There are now numerous such bifocal programs throughout the country, but few have ever achieved or even attempted a reciprocity in research. Does the cultural history of the two civilizations, especially during the three-hundred-year span of their interaction, make some such cross-disciplinary coordination fruitful and desirable?

Two interesting and perhaps interrelated facts become immediately salient. First, as is amply evident even from the sketch above, one can speak at length of the German cultural presence in Russia, yet hardly anything significant can be adduced in the reverse direction (see, however, Alexander Dolinin's essay in this volume for a curious, and characteristic, exception). The Germans made many and diverse attempts first to carve a cultural space for themselves in the vast Russian Empire, then to develop or influence various sides of its life—military, financial, industrial, intellectual—and in the process became Russified almost to the exact degree to which that influence succeeded. After all, any cultural penetration requires and effects an ethnic self-sacrifice, as the active ethnic identity is diminished or even lost in the process. Enduring foreign influence (in a homogeneous nation) is scarcely possible over a great cultural distance or over existing ethnic partitions; a fusion is in order. However, the cultural, and therefore religious, differences between the two nations were so important as to make this fusion riddled with complications. These difficulties and their issues gave pet themes to almost every major Russian writer and scores of minor ones since at least the mid-eighteenth century, from Fonvizin to Nabokov (both, curiously enough, were in part of German extraction).

The second peculiarity of the situation concerns the seldom discussed fact that it is much easier to find a Slavic scholar well informed of German cultural history in general and of the history of German-Russian relations in particular than a German one versed in the Russian counterpart. Most contributors to this collection are indeed Slavists, and this is neither an accident nor a result of the compiler's bias but rather a spontaneous reflection of that state of affairs. The *presence* here described was not a trade of cultural goods and vices but a unidirectional, West to East exportation, with largely assimilative consequences for the exporters.

While obvious historical and political aspects of the Russo-German relationship over the past three or even four centuries have received due attention from scholars, the range of these remarkably varied, important, and peculiar cultural relations has not been fully assessed as such, certainly not in English.[15]

These relations were indeed varied in the sense that they engaged every acknowledged sphere of intellectual and common life: religion and thought, education and ideology, sciences and their applications, arts and letters, custom and language. The articles collected in this book touch upon most of these.

They were, and remain, important not only because these two civilizations have imparted, separately and summarily, a powerful tractive force to the course of world affairs in the past centuries, but also because one has significantly affected the other in a great variety of ways. The articles that follow explore these ways and point up their direct or oblique cultural significance.

And they were peculiar inasmuch as they were always marked by the singular paradox already mentioned: the German cultural presence in Russia resulted either in a more or less complete assimilation that gave us the phenomenon of a *Russified German*, a frequent and familiar figure in Russian society, as well as on the literary scene—a complete and even eager assimilation, sometimes—or else that presence could result in a pronounced mutual repulsion and a durable notion of the other nation's cultural and even ethnic inferiority. Examples are well known and numerous in every century since the seventeenth. This curious paradox has been noted in passing or put in focus in several of the essays, and can be derived from almost all of them.

As mentioned above, this particular set of problems has not yet been, to our best knowledge, defined, let alone addressed in a systematic fashion (except within the great "Wuppertal Project" in Germany[16]), and certainly not in the Anglophone world. This volume brings to light the scope of the German presence on the Russian cultural scene as a sequence of significant cultural episodes, arranged more or less chronologically and thematically strung together to form a pattern of fitful yet fruitful and lasting relations. Specific areas considered by the authors include philosophy, literature of every genre, literary history and theory, music, and architecture.

This range owes its breadth to the special expertise of the contributors—distinguished American, European, and Israeli scholars, all specialists in Russian or German literature, in English or in comparative studies, in the fine arts or in music history—eminent professors, authors of a great number of monographs, essays, and articles. This wide range of collected scholarship has not, however, resulted in an amorphous spread of disjointed compilation; rather, it sets out to explore various sides of one thematic body that gave this book its title.

III.

The symposium mentioned above was meant to be held in October 1996, which proved to be impracticable. It was dedicated—as is this book—to the memory of Konstantin Bogatyrev, a poet, a brilliant translator of German poetry and prose into Russian, an erudite man of letters, and, above all, an intrepid man. Had the event happened in 1996 as planned, it would have marked the twentieth anniversary of Bogatyrev's assassination. The fifty-one-year span of his life was tragically and perhaps significantly divided into two halves: he was arrested by the Soviet political police at not quite twenty-six years of age and subsequently sentenced to twenty-five years of imprisonment, and even though he was released, with millions of others, after enduring five years in some of the most atrocious dungeons and concentration camps, he was murdered by the KGB soon after his original sentence was to have been completed.

Bogatyrev was a man of rare genius, which became incontestably and almost instantly clear to everyone who came to know him, including such vastly different people as Pasternak, Akhmatova, Böll, Jakobson, and Sakharov. His international renown was limited and the circumstances of his life and death have not been widely known. The brief sketch that follows cannot do justice to his achievement but will at least serve to explain to the general reader of this book the meaning of its dedication.[17]

Bogatyrev was not a political figure; he engaged in no outward political activity and was not an "officially recognized" dissident, although he was close to, and publicly protested the persecution of, many famous ones, among them Sakharov, Daniel, and Kopelev. Thus it is not easy to explain to those who are not intimately familiar with the hideous intricacies of the Soviet regime why it chose Bogatyrev, of all people, for a model murder. Of course, he was staunchly and fiercely anti-Soviet, but so was a good part of the Moscow and Leningrad intelligentsia. During this more or less herbivorous period (the late 1960s to 1970s), such views, neither concealed nor carried into action, usually led only to a stunted career and relative poverty, far more seldom to prison terms, and only very rarely to state-ordered assassinations. It was, some have said, an act of pure terror—to send a bone-chilling message of dread to other anti-Soviet intellectuals. This is no place to dwell on a discussion of this significant and politically consequential episode in the history of the advanced years of the Soviet state. Suffice it to say that this gentle, enormously gifted, and highly uncommon man was one of the very few Russians who were genuinely free-minded and whose utter contempt for the regime, which was at once horrible and risible, knew no stages of gradual disappointment in Marxism, or Leninism, or any blend of these execrable ideological prescriptions, as was typical of many an intellectual in his circle. His contempt and rejection were unmitigated and much harder to hide.

Bogatyrev was born in Prague on 12 March 1925. His father, a famous scholar, was a prominent specialist in Slavic folklore and a co-founder of the Moscow Linguistic Society, along with Roman Jakobson, Prince Nikolay Trubetskoy, Gregory Vinokur, and others. Like millions of other Russians from all walks of life, he had emigrated after the Civil War and the collapse of Russia. Unlike most of the émigrés, however, the Bogatyrev family returned to the Soviet Union in the mid-1930s—first Mrs. Bogatyrev with the young Konstantin, followed by Professor Bogatyrev. Amazingly, they were not harassed then, amidst the bloodbath of the Soviet pre-war scene, and Peter Bogatyrev was even allowed to teach Slavistics at Moscow University. His son enlisted at the end of the war; its finale and aftermath found him in Berlin. By that time he had developed such a strong anti-Soviet and anti-communist mindset that he seriously considered escaping to the West Zone (a feat that could be accomplished at the time rather comfortably by means of city transportation), and only the thought of his parents in Moscow held him back.

It was through his father that Bogatyrev made the acquaintance of Boris Pasternak, whom he admired above all poets, and Pasternak steered him toward translating Rilke (Bogatyrev's command of literary German was superb from his childhood). Bogatyrev entered Moscow University late in the 1940s, and in February of 1951 he was arrested and sucked into the state security extermination labyrinth on the absurd charge of having plotted to assassinate Stalin. In actual fact, Bogatyrev had once dropped a disparaging remark about the tyrant to a fellow student, and the latter reported him to one of the local student watch groups. Bogatyrev spent half a year of the preliminary interrogations in the most dreadful "Sukhanovka" prison not far from Moscow, of which almost nothing is known because very few have survived, or retained enough sanity to recount the experience. Solzhenitsyn has little to say about it in his monumental *Gulag Archipelago.* American diplomat Dolgun was interrogated there after the war, and reported that he was beaten by the deputy minister for state security. Dolgun was stretched across boards, and Comrade Riumin used a flexible rubber club, trying to hit the sciatic nerve, an indescribable torture during which atrocious pain soon leads to clinical madness.

Bogatyrev did not like to dwell on his experience in that place. He was sentenced to execution by firing squad and spent some time on death row, but on the eve of the thirty-fifth anniversary of the Bolshevik coup d'état, Stalin deigned to strike a few names from the top of the (alphabetically arranged) execution list, and Bogatyrev's name was close enough to the top. His sentence was commuted to twenty-five years of hard labor in extermination camps, which was his own age at the time. His parents were exiled to Voronezh. He spent the next four-and-a-half years in one of the worst penal labor dumping grounds, in the Vorkuta mining region, beyond the Arctic Circle. Several times he was almost killed by predatory criminals placed there by the authorities to terrorize political prisoners. One tremendous consolation for him was his correspondence with the fearless Pasternak, who sent him books, among them Pasternak's translation of *Faust* (1953), with a cheering inscription ("Be brave, take courage, things will change soon"); right below it, a lilac stamp says "Allowed for personal use. Lieutenant Fillin."

He was released in 1956, but never "rehabilitated" officially. He became Pasternak's close friend, and the poet entrusted Bogatyrev with delicate commissions when he was hounded and reviled by the state for *Dr. Zhivago* and the Nobel Prize that it had won him. Bogatyrev was admitted to the Union of Soviet writers on the recommendation of Anna Akhmatova, and resumed his studies at the university. His dissertation was devoted to Thomas Mann's treatment of time flow in *The Magic Mountain.* He also translated and wrote commentaries to several of Mann's novels included in the ten-volume collection published in Moscow in 1961. He published translations of German and Austrian poets ranging from Hölderlin to Trakl to Kästner, all of whom had one feature in common: the prodigious technical difficulty that their verse presented to a translator.

But his main and most glorious literary achievement was his work on Rilke's poetry, begun at Pasternak's bidding. Rilke had tremendous formative influence on Pasternak himself (his father, the great artist Leonid Pasternak, had known Rilke since the latter's visit to Russia at the beginning of this century, through Tolstoy, who was a good friend of the Pasternak family). That influence was so profound that Pasternak deliberately declined to translate Rilke himself, out of an acute and peculiar sense of reverence—except for the two poems he rendered in Russian for illustrative purposes, to include in his second autobiography. But he encouraged Bogatyrev to try it, and Bogatyrev's work is in a sense a tribute to his favorite poet and friend.

Everyone who has read Rilke knows that he is, artistically speaking, untranslatable, because his resilient, capacious, and branching metaphors are packed tightly in a well-ordered classical rhythmic form and often hang on long rhyming chains strung together, moreover, by frequent enjambments. Numerous attempts to render Rilke into English have been variously unsatisfactory. The Russian language as a medium appears in principle to be better suited for the task because Russian verse, with its relatively stable prosodic system adapted from the German, and with its consonant-appointed rhyming preferences, is rhythmically and prosodically much closer to German. In general, as was noted above, Russian translations of German poetry are among the very best accomplishments of this sort in Russian literature; in fact, some have become a solid component of it. Notwithstanding several excellent versions of some of Rilke's best-known poems by such superb translators as Witkowsky, Bisk, Mikushevich, or Levik, Bogatyrev is, after a fashion, the only true translator of Rilke into Russian.

What makes Bogatyrev's work stand out so obviously? Aside from sheer technical virtuosity and an unusually keen sense of idiomatic Russian, he profited from the fact that Rilke belonged to a relatively narrow circle of poets who did not write sundry poems gathered in chronological collections. Rather, he wrote *books of poems*, which were compositionally and thematically arranged in a manner similar to chapters in a good novel. Indeed, it is possible to say that all of Rilke's books of poetry are novels in verse. But the poems that make up his books appear sharply uneven when measured by the inspirational charge that generated and informed them, if only because the best of them contained that charge in such concentration and of such alpine purity that was impossible to sustain for long or to match on demand. When his enormously powerful poetic engine was disengaged, Rilke was quite capable of writing, in a coasting mode as it were, series of intricately crafted poems that lacked intense inspiration, were repetitive in style and trope, and produced not a shiver of recognition but a disbalancing effect akin to seasickness. It is easy to see why most translators avoid rendering an entire book, a complex narration in verse chapters, with all its summits and valleys and pitfalls, and instead translate only excerpts—usually the same poems over and over again. It is, however, one thing to render for the tenth time the "Panther," or the "Archaic Torso of Apollo," or the "Spanish Danceuse," and quite another to have the stamina, the genius, the erudition, and the patience to translate an entire book of Rilke—which, arguably, is the only way to translate him correctly. To make a splendid version of one of Rilke's numerous little masterpieces is a tour de force, but then any talented translator is almost assured of success because he starts his climb already at the great altitude of the original and is helped along by the uplifting force of its metaphors. But to translate a complete book of Rilke's poetry is a heroic deed, an artistic exploit of uncommon proportion, because it requires not only passion but also patience and a sense of coordination. And this is especially true if the book in question is *Neue Gedichte*, both parts of it.

Bogatyrev possessed the requisite genius, stamina, diligence, and erudition to translate into Russian both volumes of this most difficult book, and to translate them faithfully, retaining moreover the meter and the rhyme scheme of Rilke's original. This fantastic achievement can be credited in almost equal parts to Bogatyrev's own poetic talent and to his intimate knowledge of Pasternak's poetic language, which in turn owed much to Rilke's. Bogatyrev would so often and so happily come upon certain images, vertiginous in complexity yet natural in diction, precisely because they were organically connected to Pasternak's images and tropes, which he knew by heart in both the usual and literal senses of the phrase. The quality of this accomplishment defies credulity.

Bogatyrev celebrated his fiftieth birthday in March 1975 and the formal end of his prison term in February of the year following. In April of that year, he was ambushed on the staircase of his apartment building by two secret police operatives and expertly hit in the back of his head with a heavy object, with the intention to crack the skull. His old mother found him in a pool of blood, unconscious outside the door to their apartment. He did not die then and there, but in a hospital two months later, apparently without regaining consciousness. His doctor later confessed that she had been paid a visit by KGB agents who issued a strong and emphatic prediction that Bogatyrev would not survive. When he died on 18 June 1976, there was a chilling sense that he had been on death row for all those twenty-five years, and the connection between 1976 and 1951, not immediately grasped by some, suddenly presented itself with a shudder of realization. His funeral drew a huge crowd of thousands of Moscow's intelligentsia to the seventeenth-century Church of the Transfiguration in Peredelkino, a village near Moscow turned in later times into a country retreat for Soviet writers; he was buried there in the churchyard cemetery, not far from Pasternak's grave.

Bogatyrev's translations of *Neue Gedichte* came out a few months thereafter. "He did not live to see his last book," as the hackneyed phrase might have it. But of course for any good translator of poetry—and for any good poet, for that matter—the lingering moment of inspired composition is much more acutely felt and is much more vibrant and exalting than the sight of the ultimate product of that inspiration, printed and bound. It is the difference between rapture and mere satisfaction: only the first is otherworldly.

"What is my debt to Russia? She made me what I have become; inwardly, I came out of her depth, all my underground sources are there!"[18]

One can say that while most good Russian translations allow Rilke's poetry to flash, here and there, with a rich and natural Russian sheen, only Bogatyrev's work makes Rilke's Russification—potent, organic, and begging for a shaping expression—complete. At the artistic level, nothing emblematizes the theme of this book better than this elaborate, sublime cooperation.

Notes

1. *Geographikë Syntaxis,* III, p. 19.
2. W. J. Entwistle, "The Chronology of Slavonic," in *The Transactions of the Philological Society,* 1944 (Hertford, 1945), p. 38.
3. Tony Paterson, "Sonar Ship Homes in on Atlantis of North," *Sunday Telegraph* (London), 26 September 1999, p. 39.
4. Dmitri Mirsky, *A History of Russian Literature from Its Beginnings to 1900,* ed. Francis J. Whitfield (New York, 1958), p. 34.
5. See B. O. Unbegaun, "Russian Grammars before Lomonosov," *Oxford Slavonic Papers* VIII (1958), pp. 98–116. Another one, by Michael Groening, a Swede from Finland and Königsberg, was published in 1750 in Stockholm. This one sported an appendix with examples of Russian expository prose—a collection of sundry anecdotes culled and translated from various unknown sources, mostly French chapbooks, under the title "Jocular and Other Stories." The following riotous example of gruesome wit, however, could scarcely pass as French:

> A thief, upon seeing a Merchant entering a Comedy, followed him aspiring to appropriate his handsome gold buttons, which he had on his velvet caftan; and in order better to succeed in that endeavour, situated himself behind him. At the end of the first act, he proceeded to cut the caftan with the view to obtaining the buttons [*nachal rezat' kaftan dlia poluchenia pugovits*]. Whereupon the Merchant, having noticed the deceit, produced [*vinul,* a misspelling reflecting a much mocked mispronunciation] a penknife from his pocket and applied himself so deftly as to succeed in cutting off the thief's ear, upon which the latter cried out: "My ear! my ear!" The Merchant also cried out: "my buttons, my buttons." "Here, have them back," the thief told the Merchant, to which he replied in kind, "and here is your ear" (no. 37, p. 300).

6. A contemporary wit trenchantly called the institution the "dreifüssige Akademie," at the expense of the threesome of the Fusses, the German mathematician and his two sons, who had found there a comfortable perch, as did at the time many other unemployed German scholars.
7. According to one utterly paradoxical theory that was advanced recently, the origin of Russian Orthodoxy is not Slavic, and furthermore not even Byzantine, as has been unswervingly and universally held. Rather, its foundation was laid by "the Germanic peoples." The proponent of this rather astonishing view traces a bold historical line from the ninth-century Princes Askold and Dier and Teutonic knights in ancient Novgorod to the present "Patriarch Alexii, Baron von Rediger," and complains that Russian chauvinism prevents Orthodox Germans in Russia from having services in their native language. See Monk Ambrose (von Sievers), "'Early will I seek Thee' (Germans in the Russian Orthodoxy)," in *Religion and Democracy: Towards the Freedom of Conscience,* vol. 2 (Moscow, 1993), pp. 373–84. Russian hagiography does preserve, however, a true early instance of an amazing conversion. Blessed Isidore was born in Germany of wealthy parents, came to Russia in the first part of the fifteenth century and was received from the Roman Church into the Orthodox. He settled in the city of Rostov and assumed the hardship of a voluntary ascetic deed known as "a Fool-for-Christ-Sake" (*Iurodivyi Khrista radi*), living in a hut and wandering about town in rugs, begging for alms and prophesying. He died in 1474 and was glorified as a saint; the Russian church celebrates his day on 14 (27) June.
8. See Erik Amburger's *Geschichte des Protestantismus in Russland* (Stuttgart, 1961). See also Protopriest Michael Pomazanski: *O zhizni, o vere, o Tserkvi.* (1946 1976), issue 2 (Jordanville, 1976), which has an excellent synopsis of Amburger's book, on which I rely here.
9. See the most recent book on the history of that peculiar community by V. Kovrigina, *Nemetskaia sloboda Moskvy i eyo zhiteli v kontse XVII—pervoi chetverti XVIII vv.* [Moscow's Germantown and Its Inhabitants from the End of the Seventeenth to the First Quarter of the Eighteenth Centuries] (Moscow, 1998).
10. "In order to be liked by the sovereign [Alexander I] in the second half of his reign, one had to be a foreigner or have a German surname. Our generals who served the country so well in 1812—Raevski, Ermolov and others—were ignored or under suspicion." A. M. Muraviev, in *Vospominania i rasskazy*

deiatelei tainykh obshchestv 1820-kh godov (Moscow, 1931), vol. 1, p. 119. A curious retrospective reference to the general situation—and to Ermolov's stance in particular—can be seen in Hertzen's and Ogarev's oppositional émigré newspaper *Kolokol:* "Whence this sad multitude of Germans in the Russian service which keeps growing and not at all diminishing since the times of A. P. Ermolov's wisecracks?" (1 January 1867, 231–32ff.). I thank Professor Alexander Ospovat of the University of California at Los Angeles for pointing out these sources.

11. There have been recent signs of special interest in the subject of the German presence in Moscow—see *Nemtsy Moskvy. Istoricheskii vklad v kul'turu stolitsy, Sbornik dokladov mezhdunarodnoi konferentsii* [Muscovite Germans: Their Historical Contribution to the Culture of the Capital City. A Collection of Papers Read at the International Conference] (Moscow: Obshchestvennaia akademiia nauk rossiiskikh nemtsev [Community Academy of Sciences of the Russian Germans!], 1997). And a special exhibition "Germans in Moscow: Four Centuries with Russia" ran at the Moscow Historical Museum in winter of 1999, sponsored in part by Bayer of aspirin fame—the firm had opened a branch in Moscow in 1883.
12. Some 51,000 managed to get to Germany with the retreating troops. The numbers of those deported to Central Asia and Siberia are impossible even to estimate with reasonable accuracy, but it ranges from 650,000 (according to Amburger 1961) to 1,500,000 and more. Of course, very many died on the way to their destination or during the first year.
13. An especially fascinating and poignant "repatriation" case, that of Wilhelm Draugel, which hinged on a linguistic test, is reported in Professor I. N. Gorelov's pioneering article "Imprinting and Speech as a Basic Complex of Socialization," published (in Russian) in *Politicheskii diskurs v Rossii. Materialy rabochego soveshchania 30 marta 1997 goda,* Insititut Iazykoznaniia Rossiiskoi Akademii Nauk (Moscow, 1997), pp. 13–18.
14. The event was subsidized by the Mid-American University grant and by a grant from the German Academic Exchange Service (DAAD); we are most grateful to these generous institutions.
15. But there have been several recent publications in Russian. Here are a few that caught our attention:

    Nemtsy v Rossii. Problemy kul'turnogo vzaimodeistviia [Germans in Russia. Problems of Cultural Interaction] (St. Petersburg, 1998).

    Konstantin Shchepetov, *Nemtsy glazami russkikh* [Germans as Viewed by Russians] (Moscow, 1995).

    Kentavr pered Sfinksom: Germano-Rossiiskii Dialog [The Centaur Facing the Sphinx: A German-Russian Dialogue] (Moscow, 1995).

    Nemtsy v Rossii: Istoriko-Kul'turnye Aspekty [Germans in Russia: Historical and Cultural Aspects] (Moscow, 1994).

    L. Malinovski, *Nemtsy v Rossii i na Altae* [Germans in Russia and in the Altai Region] (Barnaul, 1995).

There is even a periodical on the history and culture of Germans in Russia entitled *Phoenix* and published, irregularly since 1993, out of Alma-Ata, Kazakhstan, where many Germans were exiled during and after World War II.

16. *Russen und Rußland aus deutscher Sicht und Deutsche und Deutschland aus russischer Sicht von den Anfängen bis zum 20.Jahrhundert: Wuppertaler Projekt zur Erforschung der Geschichte deutsch-russischer Fremdenbilder* (Munich, 1985–1992). Six volumes, in two series, have appeared so far.
17. But see *Poet-perevodchik Konstantin Bogatyrev*, ed. Wolfgang Kasack, Jefim Etkind, and Lew Kopelew (Munich, 1982), and its German version, *Ein Leben nach dem Todesurteil* (Bornheim, 1982).
18. Rilke to Leopold Schlötzer, 21 January 1920.

Chapter 1

A SWEDISH COUNTESS AND A RUSSIAN EMPRESS

Ch. F. Gellert as a Source of Inspiration for Enlightened Writing in Russia

Annelore Engel-Braunschmidt

There is no definite boundary in literary history between enlightenment and sentimentalism; nevertheless, writers of the late eighteenth and early nineteenth centuries had a predilection for one or the other direction, depending on their characters and their intentions.[1] One source that offered rich material for translation, adaptation, or inspiration was the work of the prolific German writer Christian Fürchtegott Gellert (1715–1769), whose fables, comedies, treatises and one novel were widely read throughout Germany and its neighbors. Whereas Catherine II (1729–1796), the Russian empress of German origin, preferred enlightenment and reason in her reference to Gellert, Nikolai Karamzin (1766–1826) felt inspired by sentimentalism and morals.[2] Karamzin, the main representative of Russian sentimentalism, not only has maintained his place in the history of Russian literature, but also has given the reader—with stories like *Poor Liza* or his *Letters of a Russian Traveller*—texts that may be read with delight even today. By contrast, Catherine's "literary" heritage, apart from her memoirs, has only recently begun to be discovered. Much effort is presently being made to establish the empress as a writer[3] because of a specific contemporary interest in gender studies and, of course, the two-hundredth anniversary of the empress's death in 1996. Her works, in fact, offer material for literary criticism and have already won her a place among the literature of eighteenth-century Russia. Self-determined as she was in her deeds, Catherine, the focus of this essay, was equally independent in her thoughts, literary or otherwise, being extremely well read and thus able to draw on her knowledge in every respect.

The main evidence of Catherine's knowledge of Gellert is her comedy *O vremya!* (O time!), written in Russian and modeled on Gellert's play *Die Betschwester* (The Devotee [of religion]; 1745), though the empress did not reveal her authorship,[4] nor did she mention her model, either. *O vremya!* (1772)[5] was to be the first of her comedies, and Gellert's drama, famous at its time in Germany, served to get her started with a new literary genre, with "another craft" as she said.[6] When she launched into literature, Catherine had been empress for almost ten years and by then had arrived at a transition from legislative operations to executive ones. Her son Paul (1754–1801), heir to the throne, had come of age and was about to be married—enough reason to become concerned with the individual and his place in society. Catherine's use of the original German play was so radical that *O vremya!* must be considered an original, too; its remodeling definitely offers more insight into Catherine's educational and literary purposes than would a direct translation.[7]

At the same time that Catherine adapted *Die Betschwester*, she also started writing her memoirs (in French), and although there is no proof[8] that she had read Gellert's *Leben der Schwedischen Gräfinn von G**** (The Life of the Swedish Countess of G***; 1747/48), it is hard to believe she had not. Having "no proof" does not mean that there are no facts. For instance, there is no "proof" in Catherine's memoirs that another Catherine played a decisive part in her life, although Ekaterina Dashkova contributed to some extent to Catherine the Great's ascent to the throne. Ekaterina Dashkova simply does not exist in the memoirs.[9] Catherine did a lot of reading during those eighteen years of waiting for the throne, most often arming herself with books[10] to escape boredom at court. According to her memoirs, she loved novels[11]—at least until she began to study more serious works on history and philosophy.

Between 1750 and 1770, Gellert was praised as a national poet of all German countries—"[s]eine Vorlesungen über Moral und Poesie zogen in den fünfziger und sechziger Jahren die akademische und adelige Jugend aus ganz Europa nach Leipzig"[12]—but with his death in 1769 his fame began to fade. In Russia, however, he became popular around 1765, first, through translations of his fables[13] and, as early (or late) as 1766, through his novel. These and other translations of his works produced an afterlife in Russia that lasted much longer than in Germany. The German writer's renown throughout Europe may have been one incentive for Catherine's choice; another may have been the model of autobiography "as" novel, which Catherine could exploit for her own purposes.

When Catherine left Germany at the age of fourteen, she also left behind her mother tongue and switched to Russian, although she remained fluent in French. Only fourteen of the sixty-seven years of her life were spent in Germany,[14] compared to fifty-three in Russia (forty-three of these in power). Motivated by a strong determination to be as Russian as possible, she maintained a powerful dissimilarity with her husband, with his inclination to everything Holsatian and non-Russian.[15] That Catherine's mother passed on any information about Gellert to her daughter seems doubtful. Although Johanna Elisabeth von Anhalt-Zerbst kept up a correspondence with Gellert, even inviting him as a guest,[16] and although she was also a foster mother of Gellert's young pupil Friedrich Karl von Bose (from whom she could have learned how eager his teacher was to answer questions about morality when asked for advice and help), there are no direct clues about Catherine's contact with Gellert, since she was not free to correspond with her mother in her early Russian years. If Catherine succeeded in writing to her mother or received any notes from her, family matters would be discussed rather than literature.[17]

When in 1766 Ivan F. Rumyantsev presented the public with the first Russian translation of Gellert's *Leben der Schwedischen Gräfinn G****, this work had already seen five editions in Germany and by that time had been translated into French (1754), English (three translations), Italian, and Polish.[18] Catherine, an enemy of everything boring and traditional as well as a friend of things new and original, tackled Russian comedy and Russian opera.[19] She did not write verse, nor did she write novels; yet her autobiography can be considered a novel, her only one, in the same way that Gellert wrote only one autobiographical novel. The (real) empress uses her own life for literary purposes as much as Gellert makes his (fictitious) countess use hers. In both novels, sincerity and authenticity are strategies,[20] but whereas Gellert casts virtue and Divine Providence as the governing principles in life, Catherine gives priority to virtue and worldly power. One might be surprised to find the sincere Catherine skilled in the art of dissembling, yet this seems to be another literary device of "sincerity," as she frankly tells her readers what she actually had in mind when pretending: at the age of two, Catherine says, she was placed under the tutelage of Mlle. Cardel, a slightly false woman who "avoit grand soin que je parusse devant mon père et ma mère telle que je puisse leur plaire et elle aussi. Cela fit que je devins dissimulée pour mon âge."[21] Catherine knew when to speak and, even better, when to keep silent. She describes herself as a woman who knew how to curb her temper and win everybody's confidence through discretion; she spent much time studying people's thoughts and intentions. In order to become familiar with the world and its people, she would even listen to gossip.[22]

*Zhizn' Shvedskiya grafini G**** (The Life of the Swedish Countess G***)[23] was Rumyantsev's first translation; accordingly, in his foreword to the book he gave an explanation with a *captatio benevolentiae.* As *kapitan*, he taught fortification at a Petersburg land-military school for the offspring of nobility and, at the same time, headed the institute's printing press.[24] His *Shvedskaya grafinya*, despite having origins as noble as Gellert's countess, does not tell her life in the same plain and reasonable words as her German counterpart. Considering all talkativeness to be superfluous, Gellert's countess in the German original comes to terms with the facts of life as with the women's narrative conventions of her time right from the beginning: "Vielleicht würde ich bey der Erzählung meines Geschlechts eben so beredt und geschwätzig als andere seyn, wenn ich denn anders viel zu sagen wüßte. Meine Aeltern sind mir in den zartesten Jahren gestorben ..."[25] This dry (*sukhovatyi*) style[26] is sentimentalized in Russian: "Mozhet stat'sya, chtob ya rasskazyvaya moi priklyucheniya, staralas' chitatelei moikh o znatnosti familii nashei uverit'; no roditeli moi ostavili menya v samom mladenchestve ..."[27] Sentimentalization in Rumyantsev is thus achieved by leaving out Gellert's ironic allusion to women's eloquence. In the German edition, the countess criticizes the literary superfluousness so common to eighteenth-century female writing; Rumyantsev retains this stereotype of female narration. Also, the noble origin of the countess, which Rumyantsev points out (*znatnost*), seems to have obligated him to avoid the sober and colloquial style of the German original, a discourse that did not correspond with the decorum of courtly life.

Furthermore, while reason governs feeling in Gellert's writing, this is not to the taste of his translator. Gellert writes:

> Er [mein Vetter] lieh mir seinen Verstand, mein Herz recht in Ordnung zu bringen, und lenkte meine Begierde zu gefallen nach und nach von solchen Dingen, die das Auge einnehmen, auf diejenigen, welche die Hoheit der Seele ausmachen. Er sah, daß ich wußte, wie schön ich war; um desto mehr lehrte er mich den wahren Wert eines Menschen kennen, und an solchen Eigenschaften einen Geschmack finden, die mehr durch einen geheimen Beyfall der Vernunft und des Gewissens, als durch eine allgemeine Bewunderung belohnet werden.[28]

In the Russian version, the "order of the heart" is completely left out, and the "secret approval of reason and conscience" finds itself replaced by "a clear conscience":

> On [moi dyadya] upotrebil ves' svoi razum ko vpereniyu v menya poznaniya o tekh veshchakh, koi vysokost' dukha sostavlyayut, i usmatrivaya, chto ya o krasote moei byla ne bezyzvestna, staralsya nauchit' menya poznaniyu istinnoi tseny chelovecheskoi, i takim svoistvam, kotorye bolee chistoyu sovest'yu, nezheli obshchim udivleniem, nagrazhdaemy byvayut.[29]

Rumyantsev misinterprets the original by omitting or changing key words that highlight or at least insinuate the dual perspective of head and heart. If we agree with Hill that "What is most important … is Gellert's treatment of the basic dimensions of existence of the personalities he presents, emotionality and rationality,"[30] Rumyantsev's omissions, as well as his far more affected language, enhance the sentimentalist, and hence superficial, impression. He chooses to read sentimentalism into Gellert.

With Catherine it is rather the other way around: she reads reason rather than emotion into Gellert. The empress may have been particularly attracted by Gellert's novel because it takes on the form of a woman's autobiography. She might, furthermore, have been influenced by the elements of political reason that were turning up more frequently with the development of a bourgeois literary, as well as social, culture (elements such as the elimination of class distinction can, of course, also be found in Rousseau's *Nouvelle Héloïse* [1761] and other works of eighteenth-century literature with which Catherine was acquainted). Gellert presents a variety of women who are magnanimous and modest, and, if not already endowed with a most noble heart, at least apt for an education toward it. This is partly what Catherine was aiming at when she produced her first dramatic work in which she referred to her own founding of the Smol'nyi Institute for the education of (noble) maidens.[31] In her play she depicts backward Muscovite women of the old generation indignantly discussing a Petersburg institution where young ladies are being educated, and thus criticizes a certain Russian reluctance to gain knowledge and enlightenment.

Gellert's novel offered even stronger material for contrasting the primitive and the educated; for example, in Part Two, we come across the description of some amoral and barbarian Russians in Siberia, passages that somehow seem to correspond to the *Voyage en Sibérie* by l'Abbé Chappe d'Auteroche (1771). The Siberian theme, though, is a double-edged sword. Barbarism is certainly a bad quality, but it depended on whether it was noticed by the empress or from abroad. Although barbarism was not a major topic in Gellert's writing, it was in Chappe d'Auteroche's, and he received his answer from the empress in her famous *Antidote.* In her memoirs, when Catherine recounts how her mother had tried to arouse her fears of the foreign country's instability, she depicts herself as a courageous and confident German princess of fourteen who had set her mind on Russia and decided that, with God's help, she would be able to manage.[32] Catherine's mother warned her about the bad state of affairs in Russia, regarding the government rather than the people in general; the inhabitants of distant Siberia were almost inconceivable. Gellert presents a Siberia of the early eighteenth century peopled with barbarians in order to increase the number of obstacles his characters have to fight.[33] Among them, however, there is a Polish Jew with a good heart, who is rescued by the Swedish count and, in return, becomes the prisoner's benefactor and liberator: "Dieser Mann ist auf die edelste Art dankbar gewesen, und hat mir bewiesen, daß es auch unter dem Volke gute Herzen giebt, das sie am wenigsten zu haben scheint"[34]—"virtue rewarded" on a larger scale.

However, where Gellert's countess demonstrates that virtue is considered "nie als eine beschwerliche Bürde, sondern als eine angenehme Gefährtinn …, die uns die Reise durch die Welt erleichtern hilft"[35] and Rousseau's Héloïse presents virtue as a continuous effort to conquer passion, lending spiritual value to the renunciation of earthly pleasure, Catherine avoids all matters of virtue and, rather, demonstrates her aptness for deceit on her dangerous journey through courtly life in pursuit of power: "[A]joutez à cela un pressentiment innée de mon état futur, qui ne laissoit pas que de me donner du courage à soutenir tout ce que j'avois à souffrir et à essuyer de désagréments journellement de plus d'un coté."[36]

With her literary taste and love of entertainment, Catherine certainly had more affinities with eighteenth-century adventurism than with the later novels of the *Goethezeit,*[37] and felt quite at home with the extreme (literary) situations described by Gellert, for example "[d]iscoveries of accidental incest and accidental bigamy, captivity in Siberia, deaths by murder and suicide, from being struck by lightning, from privation and torture.…"[38] If the empress achieved something new in her literary career, she certainly did not do so deliberately; yet she was daring enough to experiment, and maybe her works would have been more influential if more of them had been known to the public. The monarch-playwright was in a powerful position to use the stage to display power, on the one hand,[39] and to enlighten her audience, on the other, especially since she believed so strongly in the possibilities and advantages of education. Moreover, by stylizing herself in an autobiography, Catherine was able to apply strategies meant for and appealing to readers other than Voltaire, with whom she kept up a long-term correspondence on political, social, and cultural matters. As a writer of autobiography, she was able to add, eliminate, correct, and finally pass on to future generations what she deemed suitable for public knowledge about herself as *mater patriae*. By the time that Catherine wrote herself into her drama *O vremya!*, Gellert had already offered a model of a woman uniting in herself the double role of father and mother by having his novel's heroine brought up as a man in the morning and as a woman in the afternoon. In *O vremya!* we meet only potential fathers, who still have to grow into their roles. By contrast, the mothers shown in the play are horrible creatures who dominate their children with ignorance, wickedness, cant, and superstition, and who have to be replaced by a younger and better generation. Since the mothers in the play are bad characters and the fathers are either absent or too young to be responsible, there is only one person to show the way to a better future: this person is the author of the drama, Catherine herself, who thus inaugurates herself as the ideal father and mother.

Matchmaking and married life are important motifs in Catherine's works. Both Gellert's countess and Catherine's "I" get married at the age of sixteen "without having loved before," but there is one difference. While the countess, as well as other women in the novel, enjoys three sorts of voluptuousness (*Wollust*)[40]—religion, sex, and virtue—Catherine, in her memoirs, does not view religion as a higher power to be obeyed as a guiding principle in life, but rather as pure ritual and a means to become firmly rooted in the life of the Russian people. She keeps more or less silent about sex, clearly realizing that it would not be suitable for her to break the rules. Still, she shows an inclination toward disobedience and the unconventional when, in her memoirs, she portrays herself galloping on her cushion as a girl while Mlle Cardel thinks she is fast asleep.[41] Catherine does not make the erotic subtext explicit, yet this interpretation is implied in another context. The grown-up woman in her memoirs remembers the very liberal Countess Bentink, whom she had admired as a girl of fourteen, a woman who "montoit comme un écuyer,"[42] lived separated from her husband, and brought up her illegitimate child. Although Catherine would not follow Countess Bentink's example, she was secretly attracted by such an unconventional way of life.[43] Like Countess Bentink, Catherine would indulge in horseback riding later on in life to match her vitality and compensate for the lack of it in her husband. Thus, whereas Gellert's countess rejoices at the fulfillment of her marriage ("Unser Beylager wurde ohne Gepränge, mit einem Worte, sehr stille, aber gewiß sehr vergnügt vollzogen"),[44] Catherine has to confess: "Frau Kruse versuchte am nächsten Morgen uns junge Eheleute auszufragen. Ihre Hoffnungen erwiesen sich als trügerisch. Und in dem Zustand verblieben die Dinge während der folgenden neun Jahre ohne die geringste Änderung."[45]

The countess finds her innocence and guileless capacity "to please" rewarded by marriage and thus gains complete happiness.[46] Catherine is also eager to please, which becomes evident in her memoirs when she mentions three things she had never lost sight of once she became firmly rooted in Russia: "(1) Plaire au Grand Duc, (2) plaire a [*sic*] l'Impératrice, and (3) plaire à la nation."[47] However, she failed to please her (rather peculiar) husband and was disliked by her continuously jealous aunt Elizabeth I; the only field she actually succeeded in (according to her own words) was the nation. Her memoirs' happy ending is not marriage, but the ascension to the throne. This was the goal of a woman who took her own fate in hand instead of acquiescing to Providence. Catherine focuses her narrative on this goal (the Russian State), weaving in the motif of the crown that had been promised to her as early as 1742.[48] In Catherine's memoirs *gosudarstvo rossiiskoe* (i.e., the crown of the Russian Empire) occupies the place that is usually taken by husbands in novels and comedies.

Eliminating class distinction, as mentioned above, should not be misunderstood in a social sense.[49] Gellert does not deal with social change but with the moral perfection of the individual. And Catherine, for her part, was far from depriving the aristocracy of their privileges and fighting for social equality. It is true that in Gellert's novel the count does not marry Caroline because she is "ein Frauenzimmer vom bürgerlichen Stande" (that she is not a noble woman and thus not his equal). Comparing herself to Caroline, however, the countess remarks: "Ich sah bey nahe keinen Vorzug, den ich vor ihr hatte, als daß ich adelich gebohren war. Und wie geringe ist dieser Vorzug, wenn man ihn vernünftig betrachtet!"[50] Reason can find no privilege in noble birth alone, for in virtue and beauty the two women are equal. However, if the Swedish court had not objected to marriage between members of different classes (a count and a middle-class woman), Caroline would not have had the wonderful opportunity to prove her magnanimity by sacrificing her love to the count's career. Here, class distinction and the court's disapproval can be read as a social motivation for what is actually supposed to be the decision of a virtuous character. A similar sacrifice is offered in Gellert's *Betschwester* by Lorchen, who renounces Simon's love in order to make him marry her friend Christianchen[51]—an act of selflessness unacceptable to Catherine. By implicitly discussing and refusing Gellert's ideal of friendship, Catherine reveals her sympathy for secular values, especially when contrasting friendship with principle.[52]

Women of the nobility had to be either in the same position as the empress or a man in order to be noticed.[53] As both father and mother of the country, the empress had no choice but to work: Catherine detested idleness and, consequently, got up early herself, kept discipline throughout the day, and would never be found leisurely whiling away her time.[54] Thus, she could also reproach her predecessor Elizabeth for not understanding affairs of the state because the latter never attempted to comprehend state matters or studied any documents. In her memoirs, Catherine praises her own valet I. I. Shuvalov for his brains and diligence, in total contrast to the idleness and inertia all around her.[55] "A room of one's own" is the desired realm in which Gellert's countess and the Russian empress wish to discover themselves, a room where they can read and write (and where Catherine, while still Grand Duchess, could escape from spies). Still, the empress had little in common with the kind of work with which Gellert surrounds his countess. Commerce (set in the bourgeois Netherlands) and the financial transactions that the countess and her husband deal with at the end of the novel are middle-class business and far removed from aristocratic Russia. There are even monetary calculations included in Gellert's novel, and money is frequently used as an argument. Gellert creates mid-eighteenth-century characters who are aristocrats by birth but become merchants. In this way, Gellert re-evaluates nobility: aristocrats become more bourgeois in a positive sense, being self-made aristocrats in terms of money and education. The "countess by name only"[56] in Gellert's writing holds a particular influence on Catherine, the "workaholic," the minister of women's education, and the ideal representative of the vital aristocracy. To transform Russian society in this way would have pleased Catherine as an enlightened ruler; however, economically, the middle classes at that time had no chance to prosper.[57]

The enlightened ruler had to strip Gellert of sentimentalist content and discourse in order better to express her reformative (if not literary) intentions and visions. Catherine's writings carefully avoid everything heartfelt, moments of great agitation, and tears—the hallmarks of sentimentalism. The domain of culture served Catherine well to underscore her image as both an enlightened ruler and an independent and well-educated woman. Writing was to become her daily preoccupation, and Russian literature, previously poor in production, could now be presented to the West: Catherine, who was not at all modest, had her comedy *O vremya!* translated into French and immediately sent off to Voltaire. Her memoirs can be considered as Gellert's *Schwedische Gräfinn* in the language of Voltaire. Just as enlightenment and sentimentalism can be drawn from Gellert's works, prompting M. Yu. Koreneva to argue that Gellert's works fully *answered* the needs and tastes of the Russian *reader*,[58] so too can a different perspective be advanced: they *created* the needs of the Russian *writer*, at least as far as Catherine and Karamzin are concerned.

Notes

1. The women writers' penchant for certain genres needs special attention. See Frank Göpfert, "Catherine II et les femmes-écrivains de son temps," in *Catherine et l'Europe*, ed. Anita Davidenkoff (Paris, 1997) [= Collection historique de l'Institut d'études slaves, vol. XXXVIII], pp. 237–45.
2. At the end of the eighteenth century, Karamzin moved into literary history with Gellert's virtue and sentiment on his lips when, in his *Letters of a Russian Traveller*, he recalled his professor from school days, who would teach morality according to Gellert's lectures and sometimes call out with great warmth: "My dear ones, become some day as Gellert taught you, and you will be happy!" Karamzin's professor was Johann Matthias Schaden, a German and a student of Tübingen University, who died in the same year as Catherine. In 1756, Schaden was invited to Moscow University where, later on, he was to become director of the famous boarding school attached to the university that Karamzin attended in 1779–83. Cf. Hans Rothe, *N. M. Karamzins europäische Reise. Der Beginn des russischen Romans. Philologische Untersuchungen* (Bad Homburg v.d.H. und Zürich, 1986) [= Bausteine zur Geschichte der Literatur bei den Slaven, vol. 1], p. 46. Gellert's fable *Incle and Yarico* had wrung tears from Karamzin's eyes, and *The Green Donkey* made him laugh out loud. Moreover, in his *Letters*, Karamzin describes how he posed in literary exaltation at the monuments to Gellert in Lipsick. Having visited Oeser's monument in Wendler Park, he passed on to the Church of St. John's, where he admired the allegory of Religion offering Virtue a medal with a portrait of Gellert crowned with laurel on it. Later that same memorable day, he would sit in the window of his room and indulge in sentiment, reading elegies and odes on the death of the beloved poet: "I shall read, feel, and, maybe, shed tears. I dedicate today's night to the memory of the virtuous man. *Here* he lived and *here* he taught virtue." Nikolai M. Karamzin, *Izbrannye sochineniya*, vol. 1, ed. P. N. Berkov and G. P. Makagonenko (Moscow and Leningrad, 1964), p. 287f. (my translation). It should not be overlooked, though, that Karamzin destroyed the peaceful pastoral in his famous story *Poor Liza*. His "arcadia was sceptical." Tanya

Page: Karamzin, in *Handbook of Russian Literature*, ed. Victor Terras (New Haven and London, 1985), p. 215.

3. See the collection of essays: *Katharina-Lesebuch. Literarisches aus der Feder der russischen Zarin Katharina II*, ed. Frank Göpfert (S.l., 1996). See also the collection of essays: *Katharina II. Eine russische Schriftstellerin*, ed. M. Fajnštejn and F. Göpfert (Wilhelmshorst, 1996) [= Frauen Literatur Geschichte, vol. 5]; Yurii V. Stennik, "Rol' Ekateriny II v razvitii russkoi literatury XVIII veka," *Russkaya literatura* 4 (1996), pp. 3–20; and *Russkaya kul'tura poslednei treti XVIII veka—vremeni Ekateriny vtoroj. Sbornik statei* (Moscow, 1997).
4. Although the empress assumed the "mask of anonymity" (an expression used by L. D. O'Malley for what is usually called *transparent incognito*), Catherine's audience in the court theatre, where *O Vremya!* was performed three times in 1772, knew who the author was. See Lurana D. O'Malley, "The Monarch and The Mystic: Catherine the Great's Strategy of Audience Enlightenment in *The Siberian Shaman*," *SEEJ* 41:2 (1997), pp. 224–42, here 224.
5. This is the date Catherine herself noted, although the play may have been conceived as early as 1770 and was probably written in 1771. See E. L. Kostina, "K istorii rannikh komedii Ekateriny II," in *XVIII vek. Sbornik 18* (St. Petersburg, 1993), pp. 300–306.
6. "Drugoe remeslo," as Catherine called it in her journal *Vsyakaya vsyachina*. See Kostina, "K istorii rannikh komedii," p. 301.
7. See Annelore Engel-Braunschmidt, "Modernisierung durch Literatur: Ch. F. Gellerts 'Betschwester' und Katharinas 'O Zeit!'," in *Rußland zur Zeit Katharinas II.: Absolutismus, Aufklärung, Pragmatismus*, ed. Eckhard Hübner et al. (Cologne 1998) [= Beiträge zur Geschichte Osteuropas, vol. 26], pp. 235–51. The first identification of Gellert as a model for Catherine's play is made by A. Chebyshev, "Istochniki komedii imp. Ekateriny *O vremya!*," *Russkaya starina* 2 (1907), pp. 390–409. And see a similar remark made by Dragutin Prohaska in his "Die Vorlage zur Komödie 'O vremja!' von Katharina II.," *Archiv für Slavische Philologie* 27 (1905), pp. 563–77.
8. "Nicht bekannt ist …, ob die Kaiserin [Gellert's] Roman *Leben der schwedischen Gräfin von G**** kannte.…" Claus Scharf, *Katharina II. Deutschland und die Deutschen* (Mainz, 1995) [= Veröffentlichungen des Instituts für Europäische Geschichte Mainz, vol. 153], p. 183.
9. See Georg Witte, "Zwei Katharinen: Autobiographische Skizzen im Rußland des 18. Jahrhunderts," in *Maskeraden. Geschlechterdifferenz in der literarischen* Inszenierung, ed. E. Bettinger and J. Funk (Berlin, 1995) [= Geschlechterdifferenz & Literatur, vol. 3], pp. 98–114.
10. "[J]e m'armois dans ma chambre d'un livre." *Sochineniya Imperatritsy Ekateriny II na osnovanii podlinnykh rukopisei*, ed. A.N. Pypin, vol. 12: *Avtobiograficheskie zapiski*, 1907 (reprint, Osnabrück, 1967), p. 75. Hereafter quoted as *Sochineniya.*
11. "Je ne lisais alors que des romans; ceux-ci ne laissaient pas que de m'échauffer l'imagination, chose dont je n'avais aucunement besoin …" (*Sochineniya*, p. 92f.). Although she preferred French books, Voltaire in particular, she would also read German literature, such as A. v. Thümmel or Ch. F. Nicolai. See the notes to Erich Boehme's German edition and translation of the memoirs: *Memoiren*

der Kaiserin Katharina II. Nach den von der Kaiserlich Russischen Akademie der Wissenschaften veröffentlichten Manuskripten, vol. 1 (Leipzig, 1913), p. 165. Hereafter quoted as *Memoiren.*

12. "During the fifties and sixties, his lectures on morality and poetry drew the academic and noble youth from all over Europe to Lipsick." Bernd Witte, "Der Roman als moralische Anstalt. Gellerts 'Leben der Schwedischen Gräfin von G ...' und die Literatur des 18. Jahrhunderts," *Germanisch-romanische* Monatsschrift 30 (1980), pp. 150–68, here p. 150; my translation.
13. See Yurii D. Levin, ed., *Istoriya russkoi perevodnoi khudozhestvennoi literatury. Drevnyaya Rus'. XVIII vek*, vol. 1: *Dramaturgiya. Poeziya* (St. Petersburg, 1995), pp. 124–30.
14. "[I]mperatritsa Rossiiskaya, nemka po proiskhozhdeniyu," as L. G. Kislyagina proudly pointed out. See L. G. Kislyagina, "Ekaterina II v povsednevnom bytu," in *Russkaya kul'tura poslednej treti XVIII veka – vremeni Ekateriny vtoroi. Sbornik statej* (Moscow, 1997), pp. 22–25, here p. 24.
15. If we follow Catherine's characterization, Peter did not like Praskovya Nikitichna for two reasons: "Son plus grand défaut étoit qu'elle étoit Russe, et le second qu'elle remplaçoit mad. Krouse, Holsteinoise d'extraction" (*Sochineniya*, p. 133).
16. See Scharf, *Katharina II*, p. 183.
17. For the Ministry of Foreign Affairs' control over Catherine's correspondence since 1747, see Ekaterina, *Sochineniya*, p. 102f.
18. See Christian Fürchtegott Gellert, *Gesammelte Schriften. Kritische, kommentierte Ausgabe*, vol. 4: *Roman, Briefsteller. Leben der Schwedischen Gräfinn von G***. Gedanken von einem guten deutschen Briefe. Briefe, nebst einer praktischen Abhandlung von dem guten Geschmacke in Briefen*, ed. Bernd Witte (Berlin and New York, 1989), pp. 233 and 240. Quotations from *Leben der Schwedischen Gräfinn von G**** will be from this edition (cited as *Gesammelte Schriften*). Neither Witte nor Fechner mention Rumyantsev's translation; see Christian Fürchtegott Gellert, *Leben der schwedischen Gräfin von G****, ed. Jörg-Ulrich Fechner (Stuttgart, 1985) [= RUB 8536], p. 162. Yet, whereas Fechner does not know any Russian translation at all, Witte notes a later one: *Zizn'* [*sic*] *svedskoj* [*sic*] *grafini fon G****. Tambov 1792; see Gellert, *Gesammelte Schriften*, p. 241.
19. See M. N. Shcherbakova, "Das Genre 'Musik zum Drama' im Schaffen Katharinas II.," *Balagan* 2 (1996), pp. 3–11, and "Musykal'naya stsenografiya 'istoricheskikh predstavlenii' Ekateriny II: fenomen avtorskoj idei v kontekste khudozhestvennoi real'nosti 1780/1790-kh godov," in M. Fajnštejn and F. Göpfert, eds., *Katharina II*, pp. 49–60
20. The epigraph to Part I of Catherine's memoirs reads: "Dedié à mon amie la comtesse Bruce, née comtesse Roumenzof, à laquelle je puis tout dire sans que cela tire à consequence," (*Sochineniya*, p. 5). And later on: "Je suis obligée de rendre justice à la verité, et de dire les choses comme elles étoient" (Ibid., p. 167). Compare with Gellert's countess: "Ich fürchte, wenn ich meine Tugenden und Schwachheiten noch so aufrichtig bestimmte ..." (*Gesammelte Schriften*, p. 4). See also the count's letter: "... bedenken Sie, ob dieser Fehler nicht eine Wirkung meiner Aufrichtigkeit seyn kann" (Ibid., p. 5 and passim).

21. "… took great care that I, and she as well, should appear before my father and mother so as to please them. This made me become deceitful despite my age" (*Sochineniya*, p. 5f; my translation).
22. Ibid., pp. 161f.
23. *Chast' pervaya. Sochinennaya gospodinom professorom Gellertom. S nemetskogo perevel kapitan Rumyantsov* (St. Petersburg, 1766); henceforth cited as Gellert, *Zhizn*'). Printing the second part of the novel caused some problems because of the passages on Swedish war prisoners in Siberia, yet there were copies at hand. See Yurii D. Levin, ed., *Istoriya russkoi perevodnoj khudozhestvennoi literatury. Drevnyaya Rus'. XVIII vek.*, vol. 2: *Proza* (St. Petersburg, 1996), p. 196. The second part finally came out in 1768.
24. See Levin, *Istoriya,* vol. 1, p. 196.
25. "Maybe I would be as eloquent and talkative as other women in the tale about my family, if only I had much to tell. My parents died when I was still of a most delicate age …" (Gellert, *Gesammelte Schriften*, p. 3; my translation).
26. Levin, *Istoriya,* vol. 1, p. 196.
27. "Maybe in telling my adventures I would try to convince the reader of my family's nobility; but my parents left me behind when I was still a baby …" (Gellert, *Zhizn'*, p. A 3; the text is given in modernized spelling; my translation).
28. "He [my uncle] lent me his reason so that I might put my heart in order, and by and by he directed my desire to please from those things that please the eye to those that constitute the highness of the soul. He saw that I knew how beautiful I was; and all the more so he taught me to know a man's true value and to relish in those qualities that are rewarded by the secret approval of reason and conscience rather than by common admiration" (Gellert, *Gesammelte Schriften*, pp. 3f.; my translation).
29. "He [my uncle] used all his reason to implant into me the comprehension of those things which constitute the highness of the spirit, and realizing that I was not ignorant of my beauty he tried to teach me the comprehension of man's true value and of those qualities that are rewarded by a clear conscience rather than by common admiration" (Gellert, *Zhizn'*, pp. A 4f.; my translation).
30. David Hill, "Die Schwedische Gräfin: Notes on Early Bourgeois Realism," *Neophilologus* 65 (1981), pp. 574–88, here p. 578.
31. On Catherine's educational policy, see Isabel de Madariaga, *Catherine the Great: A Short History* (New Haven and London, 1990), and Evelyne Enderlein, "CATHERINE II pédagogue (De l'éducation en général et celle des filles en particulier)," in M. Fajnštejn and F. Göpfert, eds., *Katharina II*, pp. 81–102.
32. "[Ma mère] me dit qu'il y avoit beaucoup aussi à risquer, vu le peu de stabilité des choses de ce pays; je lui répondis, que le bon Dieu pourvoyeroit à la stabilité, si Sa volonté étoit que cela fut; que je me sentois assez de courage pour m'y exposer, que le coeur me disoit, que tout iroit bien" (*Sochineniya*, p. 32).
33. Regarding Gellert's possible sources for this compilation, see Mechthild Keller, "Literarische Würze: Russisches bei Gellert und Münchhausen," in *Russen und Rußland aus deutscher Sicht. 18. Jahrhundert. Aufklärung*, ed. Mechthild Keller (Munich, 1987) [= West-östliche Spiegelungen, Reihe A, vol. 2], pp. 494–98.

34. "This man was grateful in a most noble way and proved to me that one can also find a good heart among a people who seems to have one least of all" (Gellert, *Gesammelte Schriften*, p. 51; my translation).
35. "… never as a heavy burden but as a pleasant travel companion who gives us ease on our journey through the world" (ibid., p. 4; my translation).
36. "Add to this an inborn presentiment of my future state of being which left me with nothing but the courage to endure all I had to suffer and to overcome the daily disagreements from more than one side" (Ekaterina, *Sochineniya*, p. 174; my translation).
37. See Wolfgang Martens, "Über Weltbild und Gattungstradition bei Gellert," in *Festschrift für Detlev W. Schumann zum 70. Geburtstag*, ed. Albert R. Schmitt (Munich, 1970), pp. 74–82, esp. p. 81.
38. Hill, "Die Schwedische Gräfin," p. 575.
39. On the impact of the court on Russian nobility, see Richard Wortmann, *Scenarios of Power. Myth and Ceremony in Russian Monarchy*, vol. 1: *From Peter the Great to the Death of Nicholas I* (Princeton, 1995), passim.
40. Another kind of *Wollust* as expressed in Gellert's essay "Von den Annehmlichkeiten des Mißvergnügens" was translated into Russian only in 1781. See Peter Brang, "A. M. Kutuzov als Vermittler des westeuropäischen Sentimentalismus in Rußland," *Zeitschrift für slavische Philologie* 30 (1962), pp. 44–57, here p. 51.
41. Ekaterina, *Sochineniya*, pp. 17f.
42. "… mounted a horse like an equerry" (ibid., p. 26; my translation).
43. See Scharf, *Katharina II*, p. 110.
44. "Our nuptials were performed without pomp, in one word, very quietly but certainly very merrily" (Gellert, *Gesammelte Schriften*, p. 7; my translation).
45. "The next morning, Mrs. Kruse tried to cross-examine us, the young couple. Her hopes proved delusive. And things remained in this state during the following nine years without the slightest change" (Ekaterina, *Memoiren*, p. 120; my translation). The passage is suppressed in the Academy edition, cf. Ekaterina, *Sochineniya*, p. 69.
46. "Ich hatte ihm gefallen, ohne mich darum zu bestreben … Unsere Ehe selbst war nichts, als Liebe, und unser Leben nichts, als Vergnügen" (Gellert 1989, pp. 4 and 11).
47. *Sochineniya*, p. 57.
48. For the prophesy of the crown, see *Sochineniya*, p. 13. Examples of the crown as Catherine's real-life leitmotif, including moments of extraordinary willpower in the pursuit of her aim, would be: "Cette couronne alors se mit à trotter dans ma tête et y a beaucoup trotté du depuis" (15); "… dans mon petit particulier je me destinois à lui [le Grand Duc], et cela parcequ'il étoit de tous les partis, qu'on proposoit, le plus considérable" (25); "[Ma mère] détourna l'esprit de mon père du voyage de Russie et … ce fut moi qui les y déterminois, tous les deux" (31); "… mais au vrai je crois que la couronne de Russie me plaisoit plus que sa personne" (45); "Le prince évêque de Lubec tint la couronne sur la tête du Grand Duc et le grand veneur c-te Alexis Rosoumofski la tint sur la mienne. C'est lui encore qui à mon couronnement a porté la couronne aussi" (68); "… l'espérance ou la perspective non de la couronne céleste, mais bien de la

couronne terrestre me soutenoit l'esprit et le courage" (105). See also G. Witte, "Zwei Katharinen," p. 107.

49. As seems to be the case in Levin, *Istoriya,* vol. 1, p. 195: "Gellert takzhe podnimaet temu soslovnykh pregrad, prepyatstvuyushchikh braku."
50. "I saw almost no privilege as compared to her, apart from being of noble birth. And what an insignificant privilege this is when you look upon it reasonably" (Gellert, *Gesammelte Schriften,* pp. 9f.; my translation).
51. Lorchen: "Genug, ich opfere die Liebe der Freundschaft auf, mein Herz mag dawider sagen was es will." Christian F. Gellert, *Lustspiele,* Faksimiledruck nach der Ausgabe von 1747 (Stuttgart, 1966), p. 224.
52. "Babet se tut et me bouda un peu, mais je ne lui dis rien, et je vit que cela la fâchoit. J'en souffrois, mais mes principes furent plus forts que mon amitié dans ce moment-là" (Ekaterina, *Sochineniya,* p. 33).
53. Catherine about M. Ya. Dolgorukova: "Je n'ai point connu jamais de femme qui rassembla un plus grand nombre de mérites, et si elle avoit été un homme, ç'auroit été un homme dont on auroit parlé avec éloge" (*Sochineniya,* p. 163).
54. See de Madariaga, *Catherine the Great,* pp. 203–18.
55. *Sochineniya,* pp. 106f.; on the general idleness at court, see ibid., p. 126.
56. See John Van Cleve, "A Countess by Name Only: Gellert's *Schwedische Gräfin,*" *The Germanic Review* 55 (1980), pp. 152–55.
57. The proto-bourgeois elements in Gellert have been pointed out frequently. See Van Cleve, "A Countess by Name Only," Hill, "Die Schwedische Gräfin," and Sybille Späth, "Väter und Töchter oder die Lehre von der ehelichen Liebe in Gellerts Lustspielen," in *Leben und Werk des Christian Fürchtegott Gellert,* ed. Bernd Witte (Munich, 1990), pp. 51–65.
58. "Imenno ėta obrashchennost' k serdtsu kak nel'zya luchshe sootvetstvovala dukhu vremeni, kogda sfera chuvstv, protivopostavlyaemaya sfere razuma, stala priobretat' vse bol'shee znachenie."—"Pri ėtom perevody iz Gellerta … vpolne sootvetstvovali obshchemu napravleniyu razvitiya russkoi basni 1770-kh gg" [For it was exactly this turning to the heart that corresponded to the spirit of the time as nothing else did when the sphere of emotion, as opposed to the sphere of reason, began to gain importance. — Nevertheless, translations from Gellert suited well the general tendency in the development of the Russian fable in the 1770s] (Levin, *Istoriya,* vol. 2, pp. 125 and 129; my translation).

Chapter 2

From King-as-Poet to Poet-as-King

Derzhavin between Friedrich of Prussia and Catherine II

Anna Lisa Crone

Derzhavin's self-conception as a poet by the end of his life was something akin to that of a "king of poetic language," an independent "monarch or president of the republic of letters." The purpose of this chapter is to assess the German influence on this conception of the morally and aesthetically independent poet as a kind of "king," since Derzhavin developed this idea under the influence of two German monarchs—his own Russian empress, Catherine II, and Friedrich the Great, the king-poet whose odes he translated in the 1770s and whose view of the ideal poet exerted a lasting influence upon him.

Derzhavin learned German as a young child, and it was through German that he assimilated European culture, including ideas and trends that had originated in France. In the German colonial settlement of Schafhausen in the midst of the Pugachev Rebellion (1774), Derzhavin first read the German translations of the French-language odes of Friedrich the Great.[1]

Khodasevich wrote of Derzhavin's encounter with these anonymously published odes: "When he looked into the mirror of Friedrich's poetry, Derzhavin for the first time saw himself."[2] What he saw in that mirror, in my opinion, was the image of the poet and poetic service that he had been groping toward and that he would make his own from 1774 on. Derzhavin's use of the material in Friedrich's volume would differ over time and appeared in at least three stages. The first, comprising the period from about 1774 to 1782, includes the translation of Friedrich's odes from German, which, as Etkind has shown, influenced Derzhavin's poetic style.[3] These years also witnessed Derzhavin's adoption of Friedrich's notion of the king as the first servant of the people and of the poet as the first servant of Truth and thereby of his sovereign. In the second stage (1783–88), Derzhavin attempts to force this "German" conception of the poet upon his own sovereign, herself of German birth; this attempt reached its apogee in "The Vision of the Mirza" and the preparatory "Sketch" for that poem. In the third stage (1788–1816), Derzhavin maintains and elaborates upon Friedrich's high evaluation of the poet, but separates this concept from association with Catherine II or any subsequent tsar. Derzhavin's exaltation of Poetry in the service of Truth and his conceptualization of himself as the kingly leader of this higher sphere, expressed well before the advent of romanticism, were profoundly influenced by Friedrich the Great's ideas on poetry.

Derzhavin's Reaction to Friedrich II's Odes: The Chitagalai Cycle

In 1774 Derzhavin borrowed from his German friend Wilhelmi Friedrich's odes in the German edition *Vermischte Gedichte* (Diverse Poems), which had appeared simultaneously with the French original *Poésies Diverses* in 1760. As a young poet, Derzhavin was particular about his choice of models; the work of a great sovereign who was also a poet was a properly authoritive model for imitation. It is most probable that Derzhavin knew who had written the poems, though Friedrich had been Russia's enemy in the Seven Years' War.[4] Ronald Vroon considers most important the fact that the volume contains Friedrich's "The Art of War," a work whose authorship was generally known and which had been rendered into Russian prose by Derzhavin's friend and superior General Bibikov.[5] Moreover, Derzhavin directly alludes to Friedrich in his original ode to Bibikov, and there are numerous other broad hints at the authorship in the volume itself. For example, in the mystifying "Avant-propos" the author is called a "Grand Prince"; the volume is published in Berlin, and the places alluded to in the odes themselves are Prussian (such as the Elbe River). One of the odes celebrates the re-establishment of the Academy of Sciences, which had been left in disarray by Friedrich's father; the ode to the French poet Gresset invites the latter to Berlin; one ode is entitled "To the Prussians"; the ode "To Maupertuis" was dedicated to the French mathematician who was a member of Friedrich's intimate circle and head of the Prussian Academy of Sciences. The final ode of the collection, "To Voltaire," displays an intimate friendship with the elderly French philosopher. Catherine's own court was of a very Voltairean bent; information about Voltaire's connection with Friedrich was in the air, and there was only one Grand Prince alive, in Berlin, who was known to be a poet and a friend of Voltaire and Maupertuis.

Since the sometimes strained relationship between Catherine II and Friedrich II could have deterred Derzhavin from translating the works of an unfriendly monarch,[6] the fact that Derzhavin published his collection much more "anonymously" than had Friedrich suggests that he did indeed know the identity of its author. Vroon treats Derzhavin's *Ody, sochinennye pri gore Chitagalae, 1774-ogo goda* (Odes, composed near Mount Chitagalai, in 1774) as one of the first autobiographically driven cycles in Russian poetry. He suggests that the personal significance of this early confrontation with the verse of the king-poet was so great that Derzhavin could not resist his attraction to poems that so resonated with his current life.[7] Here we shall examine Friedrich's prescriptions to men and to poets and Derzhavin's attempt to follow them.

Derzhavin chose to translate four odes on general, so-called philosophical subjects: I. "Ode to Flattery," III. "Ode to Calumny," IV. "On Steadfastness," and IX. "To Maupertuis" (on the ephemerality of human life).[8] Vroon emphasizes the colossal breach of decorum by the young Derzhavin in the ordering of the poems in his vanity publication:

> In the distribution of texts there is a notable breach of literary etiquette: translations precede original poems, prose precedes poetry, philosophical odes precede panegyric ones. No less striking is the breach of political and societal etiquette: the ode to General Bibikov is placed before the ode to Catherine II. The ode to Catherine closes this collection while, according to the rules of good tone, it should have opened it. All these exceptions to tradition show that for Derzhavin such an "unorthodox" placement of poems had a principle behind it; the collection had a new meaning, exceeding its literal content.[9]

One must agree with Vroon's assessment of Derzhavin's extreme license and independence in dispensing with the traditional placement of poems and even materials. Vroon attributes it to the highly personal nature of the collection for the poet and the fact that there is a kind of development between the first and eighth poems. But the Chitagalai collection is more striking in what it reveals about Derzhavin's early sense of self-importance as a poet and the ideas behind his growing independence. He not only translated the four philosophical odes that were close to his own interests, but also created four original odes in response to Friedrich's, ipso facto entering into a dialogue with the monarch-poet and, in a sense, placing himself on Friedrich's level. He certainly knew from *Vermischte Gedichte* of Friedrich's high regard for Voltaire and Gresset, whom the king addressed as his equals or betters. But the leap from this knowledge to Derzhavin's sense that he, a veritable unknown at the time, could enter so freely into a dialogue with the king-poet by presenting his answering odes is quite another matter. Despite the fact that in later years Derzhavin was dismissive of his earliest verse, including the translations and original odes in this collection—"Ode to Greatness," "On Nobility," "Ode on the Death of General Bibikov," and "Ode on the Birthday of Catherine the Great," composed in 1774 during the Pugachev Rebellion—at the time this choice was indicative of the maverick, upstart nature of Derzhavin's character and talent that was often to exasperate highly placed persons such as Catherine, Potemkin, and Tsar Pavel, who wished Derzhavin would act like a more docile court poet.

Vroon sees the translated and original odes as related, based on the general themes struck in each pairing.[10] His view that themes such as calumny and the death of "the great" were on the poet's mind at the time and reflective of his own life is valuable and correct, yet only partially explains the significance of this collection. There is another way to view the interrelation of the translations and the original odes: the originals, on the one hand, can be seen as a *theory* of the ideal functioning of the poet (and ideal kingship), and the translations, on the other, can be seen as *poetic practice*, demonstrating Derzhavin's understanding and acceptance or rejection of the German king's recommendations. Friedrich's views on poetry were the catalyst that helped Derzhavin's own views crystallize.

The pervasive subjects of the four odes Derzhavin selected to translate are the true greatness of the enlightened monarch, "first servant of his people," and, more important for Derzhavin, the kind of poet and poetry such a monarch needs to support him in his rule, to foster his increasing greatness. The first three odes, especially "On Steadfastness," emphasize in very stoic terms the adversities of life, which are numerous and inevitable. The theme of the ephemerality of human life dominates the fourth, "To Maupertuis," and is later developed in Derzhavin's original poem "On the Death of Prince Meshcherskii" (1779). The latter, one of Derzhavin's most celebrated odes, is, as Etkind has shown, the closest Derzhavinian "twin" of Friedrich's ode to Maupertuis.[11] In the

Prussian case, the ideal king and the poet were the same person. Despite Catherine's talent as a writer in the Russian language, she was not a poet, and Derzhavin came to conceive of his role as that of a Friedrich-style poet for a Friedrich-like monarch.

The first four of Friedrich's odes selected for translation are very classicist in form and abstract in their subject matter, which is full of allegory and allusions to Greco-Roman history and mythology. Friedrich emphasizes that the "good" monarch is always open to honest, constructive criticism. True to her reputation, Catherine was more open to criticism than many earlier Russian tsars. Derzhavin addresses her in his "Ode to Felitsa": "And they say, verily, That one can always tell you the truth."[12] The second four odes in the collection, Derzhavin's originals, demonstrate both his conformity to Friedrich's abstract positions and his departures therefrom. On close reading of Friedrich's eleven odes, it becomes clear that Derzhavin chose those which are least connected with concrete persons and events. The issue of being a great poet for a great king dominates the first three translated odes, "Ode to Flattery," "Ode to Calumny," and "On Steadfastness." The notion of poetry as national service was a mainstay of Derzhavin's poetic creed until the end of his life. Friedrich's entire cycle of eleven odes is rather more concerned with the optimal functioning of the poet than with ideal kingship. In "To Gresset" it is made clear that the good poet not only inspires kings to greatness but that the poet's recording of the sovereigns' deeds is necessary for both posterity and the contemporary world to see them as they truly are. Thus, Friedrich says to Gresset, whom he considers a powerful poet, "You made the heroes what they are."[13]

Derzhavin's initial placement of "Ode to Flattery" (fourth in Friedrich's collection) is important. Abject and excessive flattery was a problem that Derzhavin and others perceived in the Russian panegyric in the 1770s. In her dissertation on Derzhavin's odic practice, Jane Gary Harris discusses at length the ways in which the official court genre was sorely compromised by obsequiousness and insincerity.[14] Derzhavin, beginning his earliest extant ode to Catherine (1767), strikes the note of truthful praise for his monarch ("khvalit' nelestno," to praise without flattery).[15] Service to the Holy Truth is a major theme in Friedrich's verse. Flattery is more important to Derzhavin because it compromises his most favored genre (the ode), whereas Friedrich devotes more time to calumny. Derzhavin places "Ode to Calumny" second; thus his first two translated odes oppose dishonesty in poetry, whether it be laudatory or disparaging. Derzhavin saw flattery as most deleterious to the monarch and the state in Russia. Calumny, on the other hand, was not often directed toward the

Russian monarch in verse—publicly, at any rate. Yet Friedrich's ode emphasizes that even kings commonly bear the brunt of calumny:

> And you, venomous viper,
> Wounded this debonair ruler,
> A Prince, born to serve the public welfare.[16]

In Friedrich's view, as presented in the ode, the objects of calumny are usually "great men," and the motivation for slander is usually envy. Derzhavin repeats this often as he becomes increasingly concerned with envy directed toward himself. In the original "Ode to Greatness," Derzhavin goes so far as to say that the truly great are such *because* of their inevitable triumph over adversity, including slander. Friedrich goes further to say that for a poet to engage in calumny is to "prostitute his talent."[17] The third translated ode, "On Steadfastness," defines true nobility and heroism as stalwart firmness both in hard times and in prosperity, and contends that character is forged in adversity, another position found in the works of Horace and Cicero that Derzhavin was to espouse throughout his life. "My happiness lies within me," Friedrich writes of his constancy. He also emphasizes serenity: "Nothing will destroy my tranquility." Derzhavin's "My tranquility lies within me" in the closing lines of his "Ode to Fortune" (1788) echoes this sentiment. This theme re-emerges in "On Moderation" (1792), in both odes referring not to some other "great" figure, but rather to Derzhavin himself as a man and a poet.

The pessimistic yet militant mood of the first two translated odes, with their emphasis on two poisons—flattery and calumny—soars in the rampant catastrophe of "On Steadfastness." This mood is followed in "To Maupertuis" with philosophical pessimism: the fleeting quality of human life, time's destruction of all human endeavor, and the absence of any form of immortality for man. At the very end of Derzhavin's life, this same pessimistic note is struck in the unfinished slate ode: "The river of time in its flow,/Carries away all the deeds of men."[18] However, Derzhavin's position was inconsistent, since we know that he believed that certain exceptional great men, even some of his contemporaries, such as Potemkin, would live on in eternity.[19] In "The Waterfall" (1791–94) Derzhavin repeats this pessimistic theme, only to counter it immediately.[20] Had he lived to complete his final ode, he might well have reversed his position in it as well. Friedrich also wavers on this question, as we readily see in the ode to Voltaire in which he contradicts his denial of immortality that appeared in "To Maupertuis." But Derzhavin does not translate those odes of Friedrich's that contradict the largely negative, stoical conclusions. Friedrich's legacy to Derzhavin is a picture of great men tempered like precious metal by adversity and all the forces that conspire to corrupt virtue and prevent great deeds.

The second, original half of the Chitagalai collection indeed answers and attempts to present the kind of poetry called for by Friedrich. These odes exemplify Derzhavin's rather free interpretation of the role of the poet outlined in Friedrich's "theories." Derzhavin's original odes are panegyrics with a "philosophical" bent. Unlike the more theoretical and abstract odes he chose from Friedrich, Derzhavin composes two odes, "Ode to Greatness" and "To Nobility," with "abstract" titles, but they contain large topical sections related to living persons and current occasions, the *ad hominem* concretization of the abstract that was to become his hallmark. Derzhavin Russifies his odes, writing of great heroes on the current Russian scene—the tsaritsa in v and viii, and Pyotr A. Rumyantsev, the great hero of the recent Turkish war, in ode vi. In vii, Derzhavin praises the recently deceased General Bibikov, his own military superior, in a lamenting eulogy that is clearly the most heartfelt of the poems. The final ode (viii), marking the occasion of Catherine's birthday, depicts her with some specificity of detail, but as an ideal Enlightenment monarch in accordance with Friedrich's rather standard prescriptions.

What is remarkable about the view of Friedrich we derive from his odes and from Derzhavin's selection is the high valuation the monarch placed on poetry and poets. Vroon pairs the final ode to Catherine with the first one in Derzhavin's collection, "Ode to Flattery," to emphasize that the birthday ode is an example in practice of truthful praise. But Derzhavin, never wanting to be viewed as a flatterer, underscored his sincerity in several metapoetic comments in "Ode to Greatness."

> Should someone think I was a flatterer,
> I do not praise a goddess here, the joy of hearts,
> Because I will leave that sounding, luxuriant
> Honor for posterity.[21]

This sincerity is echoed again in the odes to Bibikov[22] and to Catherine.[23] "Sincerity formulae" are frequent in his poetry, for example, the repeated statement "I have sung a righteous hymn here." In these four poems Derzhavin identifies truly virtuous Russians, indicating the adversities with which they struggled. The funereal ode to Bibikov, paired by Vroon with "To Maupertuis," can be seen thematically as the negation of its partner.[24]

Derzhavin appears to draw the conclusion that the best poet, often not a king, should be a true, honest friend to his king, like the Cesarion described in "Ode to Flattery":

> Cesarion, faithful friend,
> I find in you the model
> Of the primary of all virtues.
> Reveal to us without harshness
> Both our errors and faults:
> Thus is the gold which the fire prepares
> Purified and separated
> From lead and the baser metals.[25]

It should be noted that these last words appear twice in the original odes: "Man is great in adversities. Heated in the furnace, gold is purified from dust";[26] "Tempered like copper in a furnace."[27]

Derzhavin's sense of himself as a new kind of poet for Catherine and Russia—who fosters and even enhances their greatness—bears the clear mark of Friedrich's influence. What begins in the Chitagalai collection comes full circle about ten years later in "The Vision of the Mirza"[28] and the "Preliminary Sketch to the Vision of the Mirza."[29] In these, Derzhavin uses the fact that Catherine was not a poet to emphasize her need for the kind of poet he proposed to be. In the earlier "Ode to Felitsa"(1782), Derzhavin did not offer himself to the tsaritsa in this capacity, but rather commended himself very strongly to her notice. Yet almost immediately thereafter, he began to write documents and poems that showed signs of Friedrich's direct influence.

A New Kind of Poet for Catherine

Khodasevich has indicated that there are echoes of ideas and images from the odes of the "philosophe de Sans Souci" throughout the poetry of Derzhavin, but the strongest locus of the direct influence of the king poet comes the year after the "Ode to Felitsa" in its sequel "The Vision of the Mirza." In "Felitsa" Derzhavin had humorously written that Catherine imbibed poetry "like tasty lemonade in summer." This assessment of Catherine's view of poetry is not far from the truth. Catherine did not share Friedrich's abhorrence of flattery and, as time showed, was not that anxious to have a poet, however honest and well-intentioned, who criticized her or her policies. For Derzhavin to take on the role of Friedrich's enlightened poet, he had to remake his monarch, Catherine II, to some extent. He proceeded to do this in the unpublished—and not intended for publication—preliminary stanzas and prose sketch for "The Vision of the Mirza."

Yakov Grot appears to be the only scholar to take serious interest in this curious document, which he found penned in Derzhavin's hand in the archive and most of which he published as an appendix to Derzhavin's "Explanations" of his verse. The published excerpt contains some thirty poetic lines and a prolix prose justification for the piece, written in the form of an open letter to the tsaritsa herself. This document is cardinally important for the subject at hand, since its ideas about abject flatterers and the harm they do and about enlightened absolute monarchy and the proper function of the poet within it add up to a passionate outpouring of Friedrich-inspired ideas to the tsaritsa. Derzhavin expresses sincere admiration for her, crediting her with understanding the need for an honest poet, his value to the state, and, by implication, the need to protect and cherish him (Derzhavin) as he would inevitably have powerful enemies (calumniators!). More wide-ranging and unfocused than the poem that it spawned, the "Sketch" is an attempt to educate and persuade Catherine, who most probably never knew of its existence, to accept Friedrich's espoused positions. In it Derzhavin attempts to convince her to embrace the positions on poetry that he has her take in "The Vision of the Mirza."

The excerpt begins with a savage counter-attack on Derzhavin's enemies—both high courtiers and other poets who assailed him as a flatterer after "Felitsa." The "Sketch" is an anti-flattery discourse par excellence in which Friedrich's metaphor of "poison" is reprised:

> I consider those poets flatterers, who in their dreams, for example, praise some imaginary Tsar-Maiden, some Sublime-Tsaritsa, and having nothing glorious to say in her praise, wave their magic wands and drag out … the sun, the moon and the stars.… I consider those poets flatterers who from fear, wanting to please some evil old hag … whose heart craves only blood … [they] parade their verses before her and when their hearts are oppressed by terror and their heads all but on the chopping block, can produce nothing in her praise that is intricate, grandiose or sublime, they just bray in the voices of asses: there is such a woman, there is such an old hag.[30]

In this text—which, it should be emphasized, is a private document, written more than anything for Derzhavin himself and reflecting his internal dialogue—the harm flatterers cause to the sovereign and the Empire is such that Derzhavin adopts Plato's position that an ideal state would have no poets at all. The ideal monarchies of the future are described thus:

> They (the rulers) will build flowering cities, whose tops will reach the clouds. In steppes, which perhaps were never inhabited from the beginning of the world, they will sow meadows.... They will erect temples to a Divinity and the immortal marble of these temples will proclaim their virtue in the last times. They will institute schools for the people and bring all men to Enlightenment. They will erect the honorable edifices of love for man, of conscience and justice, whither widows, orphans, the sick and the lepers, fat spendthrifts and skinny *poets* will flock for protection, where they will be placed ... given to drink and eat, be shod and clothed, with the provision that *the poets never write any verse ... for the poets eeking out verse is harmful to their [the rulers'] well-being and health.* [Italics mine, A.L.C.][31]

The epistle bristles with disdain for ode writers of the old, existing type. Derzhavin claims that, had it not been for Catherine, a new kind of monarch, he would never have joined the "gang of verse writers":

> But, to tell the truth, if it were not for you [Catherine] I would not want even now to be counted among the gang of poets, whom I, and especially the writers of odes, consider to be like beggars sitting with outstretched hands and cups on the bridges, singing the praises of giants whom they hardly know or do not know at all. In order to avoid being a poet, what haven't I taken up?[32]

After listing his myriad activities, Derzhavin makes the case for the new type of poet. The truly enlightened monarch needs a "friend-poet," in Derzhavin's rather daring words. The great poet will be Catherine's fated partner, almost her twin:

> From all this you see that lowly adulation is not the cause of my praise for you, but your great virtues and my grateful and feeling heart for them, or to put it better, I could say it thus, that Fate threw two lots into a sack and began to shake the sack: one lot was drawn by you, godlike Tsaritsa, to rule and amaze the universe, and the other lot fell to me: to praise you and your deeds with my humorous Tatar songs.[33]

This notion of a foreordained fateful bond between Catherine and Derzhavin—indeed, the whole impetus of this letter-instruction to Catherine—is in total concert with the ideas of Friedrich. The entire letter to the "empress," i.e., to himself, is an attempt to re-create a

Catherine who completely shares Derzhavin's German-derived ideas on the role of poetry. The new kind of poet will praise what is truly beneficial and attack and correct all that is bad in Russia, guiding the tsaritsa when she errs. Derzhavin will have Catherine repeat Friedrich's standards "we are all human" and "even a king is human and therefore fallible," phrases that came more easily to Friedrich's lips than to Catherine's, to be sure.

Much of the anti-flattery rhetoric is toned down in the poem "The Vision of the Mirza," but the Felitsa who appears to the mirza (Derzhavin) and instructs him on how to be a virtuous poet is reformed in accordance with the odes of Friedrich, the king-poet:

> Felitsa's speech:
> Tremble in your footsteps, miserable mirza!
> And listen to the terrible truth,
> Which passionate poets
> Hardly seem to believe on this earth …
> When poetry is not sheer nonsense,
> But the highest gift of the Almighty,
> This gift must be turned to honor,
> And to the instruction of the path of kings.
> Not to flattery and the destructive adulation of people …
> The rulers of the world are human,
> There are passions within them though their heads be crowned.
> The poison of flattery harms them no less often.
> And where are poets who are not flatterers?
> A man who conserves honorable morals,
> Carrying out his duty, his actions
> Bring much more honor to his Tsar
> Than all the praises of poets.[34]

In this poem, moreover, the virtuous mirza (Derzhavin) is the victim of the calumny of lesser poets, striking another theme of the king-poet. In the "Ode to Felitsa" these calumniators are called "Zoily," after the Greek grammarian, Zoilus, who attacked Homer, as in Friedrich's ode, and Catherine was seen indulging them: "They say you tend to forgive those very crocodiles, the Zoiluses of all your kind mercies."[35] In "The Vision of the Mirza" they are called "satirists," and the reformed Catherine will not tolerate them. Attacked by Felitsa, who declares the error of his ways, the mirza defends himself, protesting that lowly flatterers have already attacked him too much:

> Enough without you too [Catherine],
> Has the poet had to answer for his every word,

For his every thought to evil society ...
And to defend himself from evil satires ...
Because this one wanted a watermelon,
And that one wanted salted pickles.[36]

The ploy used by Derzhavin in the "Sketch" and in this poem is brilliant. In the "Ode to Felitsa," he had implored the virtuous tsaritsa to give instruction on how to live righteously. After remaking Catherine largely in a Friedrich-inspired mold, Derzhavin has her instruct him on how to be a proper poet in "The Vision of the Mirza," putting his own views in Catherine's mouth and having her preach them back at him.

A Parting of the Ways: Independence from the Monarch of the "Poet-King"

Derzhavin, who sincerely admired Catherine throughout her reign, but who still idealized her somewhat in the early 1780s, was aware to some degree that their views of the role of the poet were not identical. Still he felt rather appreciated as a poet in that decade, until he was forced to resign from his position as governor of the Tambov province.[37] The protracted "Tambov affair" was the most difficult period in Derzhavin's professional life, resulting in more than a year of unemployment and twenty-two counts of insubordination and malfeasance as governor pending against him in the Senate. This experience permanently changed his attitude toward many things—toward his monarch and her valuation of the honest poet, toward governmental service, and toward his own role as poet in the real Russia. That Russia, with all its moral turpitude, is depicted most satirically in the unpublished (and unpublishable during Catherine's lifetime) "Ode to Fortune."[38] I have demonstrated in detail elsewhere[39] that this poem is an anti-ode in which Derzhavin effectively cancels out most of the laudatory predicates he had attributed to the tsaritsa in "Ode to Felitsa." "Ode to Fortune" was such an attack, albeit veiled, on the Russian monarch and monarchs in general that Derzhavin had to resort to lame and humorous excuses for its writing, saying that he wrote it at carnival time and in a drunken state. The fact remains that "Fortune," the ostensible hero of this "ode," is a consistently sinister and capricious god and that Catherine II is linked to him:

In these times when She, the incomparable one,
Is harnessed to you [Fortune].[40]

The satirical attack on Catherine is mild, however, compared to the attack on the terrible state of Russia, depicted as the opposite of the ideal monarchy portrayed in "Ode to Felitsa." The real hero and true patriot in "Ode to Forune" is the maligned and mistreated poet (Derzhavin), who, while appealing to Catherine in his troubles, implies that he knows she will not help him and proudly declares that he can go it alone without her:

> Alas, you still do not listen,
> Oh Fortune, to my pleas,
> You despise my vows,
> Which means that I am not pleasing to you.
> But whether you are on plush sofas,
> In the shade of myrtles or laurels,
> Or living in a land of gold,
> Listen, and whisper to your favorites
> Magnates, kings, and princes,
> My tranquility lies within me![41]

The mockery made of monarchs in this piece is extreme. Louis of France is shown spending most of his time at his hobby of metalwork, and Carlos of Spain is immersed in making fresh pasta. There is a fairly conspicuous reference to "Gore-Bogatyr," Catherine's fairy tale and comic opera savaging Gustavus of Sweden, a work so insulting to a reigning monarch that Potemkin opposed its being performed, even privately in the Hermitage theater.[42]

It is not surprising that even when Derzhavin's fate changed for the better and Catherine appointed him her State's Secretary in late 1791, the poet retained his skeptical attitude toward the tsaritsa's view of poets due to his previous "mistreatment" at her hands. Meanwhile, his own assessment of the importance of poetic service over and above all other administrative duties he had performed only increased. In "My Bust" (1794), which includes the image from Friedrich's poetry of the French king's bust being destroyed immediately after his death,[43] Derzhavin states unequivocally that the only reason for which he would deserve to have his bust in the Cameron Gallery with the great men of Russia is his poetic service. While he was State's Secretary, Catherine asked Derzhavin, personally and through another secretary, Khrapovitskii, to write praises of her in the old "Felitsa style." The poet refrained from doing so. He explains Catherine's growing coldness toward him in his autobiographical *Notes*:

> Perhaps it [her coldness] was due to the fact that he, seeing the insidiousness of the court and the constant sallies against him, could not pull

> himself together and write such refined praises as he had in the ode to Felitsa and similar works, which he had done when he was away from Court. This is so because from a distance those subjects that had seemed divine to him and enflamed his soul so much, when he found himself at Court, seemed all too human and even unworthy of the great Catherine; so his spirit so cooled that he was unable to write almost anything in her praise with a warm, pure heart.[44]

While Derzhavin was clearly influenced by the works of the king-poet Friedrich the Great, the author of the four original Chitagalai odes was *not* a king. Even in "Ode to Greatness" he wrote "If Fate did not give one a throne. . . ."[45] Oddly, Derzhavin repeatedly describes himself as "not a king" and as even better off as a free poet than if he were—so much that the suggestion that he was of a caliber to be king is clearly present. Recurrent statements such as "No one has handed me the scepter" and " I would never take the crown in exchange for the joys of poetic inspiration" can be dismissed as Derzhavin's brief fits of megalomania. However, although it cannot be proved, perhaps his early emulation of Friedrich provoked these numerous, and otherwise strange, disclaimers throughout his life.

Like an independent autocrat in his own exalted sphere of Poetry, Derzhavin put Catherine on notice. The clearest expression of this comes in "On Moderation,"[46] and we should not be fooled by its dedication to the empress's favorite, Platon Zubov. The poem is specifically about Derzhavin's dealings with Catherine as State's Secretary and went unpublished until Pavel's reign. (Works such as this and "Ode to Fortune" were still well known in manuscript, by Khrapovitskii, for example, and perhaps came to Catherine's attention.) The image of the poet in this Horatian-inspired poem is again in complete accord with Friedrich's portrayal. Here neither Catherine, Potemkin (Jason), nor Zubov comes off as a great hero; the poet Derzhavin is the heroic model of noble firmness. The poet will be unshakable in the presence of the king's anger or pleasure. His serenity and happiness are within him. "Moderation" here is Friedrich's "steadfastness," as seen in his ode on the subject, "La Fermete."

From the early 1790s onward Derzhavin desisted from praising Catherine and began a series of poems in praise of himself as a poet-hero and his own poetic gift. "My Bust," (1794), "The Nightingale" (1795), "Exegi monumentum"(1795), "The Lyricist" (1805), and the bantering "The Gatekeeper" (1808), which opens with the powerful line: "There is one God and one Derzhavin," are but a few of these.

With his independent and autocratic attitude, Derzhavin felt he could designate his successor or heir, something he first did in 1807, "passing his lyre" to Vasilii Zhukovskii, and in 1815, in an event now legendary, he corrected himself, dubbing the sixteen-year-old Alexander Pushkin the next leading poet of Russia. A true Catherinian magnate, Derzhavin was the Russian poet who held the most numerous and highest bureaucratic posts in the imperial state service, and, as we have seen, his conception of Russia's leading poet bore aspects of his "bureaucratic and monarchist mentality." His "very free treatment of language," remarked on by Bicilli and others, was the natural outcome of his imperious independence within his own "ministry," i.e., Poetry, as were his moral independence and his unwillingness to submit to the traditional patron-poet relationship, even with Catherine. Having made Friedrich's poetry his own, Derzhavin handed his "heavy lyre" to the young Pushkin. This almost regal concept of the poet's status was vastly more powerful and authoritative than the modest "lyre" he had taken up in the 1760s as a young soldier.

Notes

1. Friedrich the Great, *Vermischte Gedichte des Durchlauchtigen Verfassers Brandenburgischen Denkwurdigkeiten* and *Poésies diverses du Philosophe de San Souci*, both (Berlin, 1760), and *Die Oden der Friedrich dem Grossen* (Berlin, 1872).
2. Vladislav Khosadevich, *Derzhavin* (Paris, 1931), p. 85.
3. Efim Etkind, "Rozhdenie krupnogo sloga," in *Gavrila Derzhavin: A Symposium* (Norwich, Vt., 1995), pp. 158–84.
4. Ibid., pp. 164ff.
5. Ronald Vroon, "Chitagalaiskie Ody," in *Gavrila Derzhavin*, p. 199.
6. Etkind,"Rozhdenie krupnogo sloga," pp. 164–65.
7. Vroon, "Chitagalaiskie Ody," pp. 187ff.
8. Gavrila Romanovich Derzhavin, *Sochineniya Derzhavina*, ed. Yakov Grot (St. Petersburg, 1864), III, pp. 269–310.
9. Vroon,"Chitagalaiskie Ody," pp. 186–87.
10. The second original half of the Chitagalai collection does, as Vroon contends, answer the first four of the German monarch's poems. However, as indicated, Vroon's four thematic pairings of the odes are not equally convincing, nor are they all done on the basis of comparable criteria. As Vroon himself admits, "Ode to Greatness" is thematically a better mirror reflection of "On Steadfastness" than "Ode to Calumny," paired with Derzhavin's "On Nobility"; in fact, calumny forms a small part of Derzhavin's ode, which resonates more with "On Steadfastness," emphasizing as it does once more the theme of stalwartness in adversity. What Vroon's pairing here has in common is circumstantial and not explicit, i.e., the fact that the courts are full of ignoble noblemen who often

badmouth the virtuous ones. Still, a much wider catalogue of non-noble qualities is offered in Derzhavin's ode.

11. Etkind, "Rozhdenie krupnogo sloga," pp. 174–82
12. Derzhavin, *Sochineniya*, I, p. 142.
13. Friedrich, *Die Oden*, p. 18.
14. Jane Gary Harris, "The Creative Imagination in Evolution: A Stylistic Analysis of Derzhavin's Panegyric and Meditative odes" (Ph.D. diss., Columbia University, 1969), passim.
15. Derzhavin, *Sochineniya*, III, pp. 241–45.
16. Friedrich, *Die Oden*, p. 10.
17. Ibid., p. 12.
18. Derzhavin, *Sochineniya*, III, pp. 235–35.
19. This view, at least as far as living on for posterity is concerned, is extremely widespread in the late Derzhavin.
20. Derzhavin, *Sochineniya*, I, pp. 457–88. The pessimistic position is seen in stanzas 55–57 of "The Waterfall" and is emphatically countered in stanza 58.
21. Ibid., III, p. 293.
22. Ibid., p. 299.
23. Ibid., p. 304.
24. Vroon, "Chitagalaiskie Ody," p. 191.
25. Friedrich, *Die Oden*, p. 40.
26. Derzhavin, *Sochineniya*, III, p. 293.
27. Ibid., p. 303.
28. Ibid., I, pp. 157–69.
29. Ibid., III, pp. 605–9.
30. Ibid., pp. 605–6.
31. Ibid., pp. 606–7.
32. Ibid., p. 608.
33. Ibid.
34. Ibid., I, p. 166.
35. Ibid., p. 143.
36. Ibid., p. 166.
37. See detailed discussion in Anna Lisa Crone, "The Daring of Derzhavin: The Moral and Aesthetic Independence of the Poet in Russia," 370 pp. (in manuscript) forthcoming from Slavica Publishers, Bloomington, Indiana, 2000.
38. Derzhavin, *Sochineniya*, I, pp. 243–59.
39. Anna Lisa Crone, "'Na shchast'e' as the Undoing of 'Felitsa,'" *Russian Literature* (Amsterdam, 1998), pp. 17–40.
40. Derzhavin, *Sochineniya*, I, p. 250.
41. Ibid., pp. 258–59.
42. Ibid., pp. 253–54.
43. Friedrich, *Die Oden*, p. 6.
44. Derzhavin, *Sochineniya*, III, p. 497.
45. Ibid., p. 292.
46. Ibid., pp. 489–99.

Chapter 3

Dreams of a German(n) in Russia

German Heroes and Subtexts in *The Queen of Spades* and Other Works by Alexander Pushkin

Wolf Schmid

"Hermann is a German: he's calculating, that's all" (*Germann nemets: on raschetliv, vot i vse!*).[1] Not only do these words from *The Queen of Spades* (1834) reveal the Russian stereotype of the German, but Pushkin's own cliché seems to take shape in them. In any case, Hermann, the cold and calculating person for whom happiness consists in the accumulation of riches and who, in order to achieve that goal, is willing to deceive a young woman and to kill an old one, became the epitome of a German. He was predestined for this role by the semantics of his name and by an intertextual allusion. In Balzac's novel *L'Auberge Rouge* (1831) we read: "Il se nommait Hermann, comme presque tous les Allemands mis en scène par les auteurs."[2] In addition, prosaic, greedy *Hermann* is contrasted with mysterious *Saint-Germain*, whose merits are amiability, generosity, and unselfishness, as even his name suggests.

Pushkin's Hermann has obviously inspired Gogol to shape his Schiller. In "Nevsky Prospect," this craftsman is characterized as follows:

> Schiller was a perfect German—in the full sense of the word. As early as twenty years of age—that happy time when a Russian would be going wild (*zhivet na fufu*)—Schiller had already planned out his entire life and under no circumstances permitted himself any exceptions.... He set as his goal to accumulate a capital of fifty thousand

> in ten years, and this was as firm and unshakeable as destiny.... His sense of discipline went so far that he made it his rule to kiss his wife no more than twice in twenty-four hours.[3]

In more recent criticism, Hermann has become one of the stereotype characters of the calculating German. In 1945, Julia Sazonova still argued that Pushkin was in principle opposed to the condemnation of men who "had the misfortune to be Germans" (the last words are quoted from one of Pushkin's letters[4]). "But when [Pushkin] came to create an artistic image of [a German], he unconsciously gave it those same characteristics that he theoretically denied: dislike for Russia and the Russians, indifference to everyone, self-love, and avarice."[5] And all these traits Sazonova finds in Hermann:

> Amid the carefree, simplehearted company of Russian gamblers, Herman stands apart, neither mixing with them nor yet leaving them alone. Like all Germans in Russian literature, he is thrifty and unwilling to take risks: he is attracted not by the powerful emotion, not by the play of fortune, but by the possibility of getting rich quickly and without effort. He wishes to bet only on a sure thing, and cards attract him only by the hope they give of riches.[6]

Does Hermann not explicitly confirm these qualities himself? After a short consideration as to whether one could put any faith in Tomskii's anecdote, he makes a typically German decision: "No! Calculation, moderation, and industry [*raschet, umerennost' i trudoliubie*]; these are my three reliable cards. They will treble my capital, increase it sevenfold, and bring me ease and independence!"[7] But how reliable are Hermann's words? His "German decision" does raise some questions. Hermann proves to be "calculating" only in the figurative sense of the word. His impulse toward moderation is thwarted by his "unbridled imagination" (*neobuzdannoe voobrazhenie*).[8] And how serious is his "industry," if he sits night after night at the card tables watching the gamblers?

The motif Hermann uses to justify his abstention from playing, "reckoning [*raschital* (!)] that his circumstances did not allow him (as he was fond of saying) to sacrifice the necessary [*neobkhodimoe*] in the hope of gaining the superfluous,"[9] is disproved by the reality of his life. He does manage with his income only; he need not even touch the interest, let alone his father's inheritance. And when he waits for the countess at midnight in order to obtain her secret, "his heart beat[s] evenly, like that of a man determined to do something dangerous but necessary [*neobkhodimoe*]."[10] Hence, "necessary" to Hermann is not his father's capital (which he stakes on one card in the first game), but something different. This probably is why he who is hailed as "calculating" and who himself relies on "calculation" does not even think of gaining the gambler's riches by mathematical calculation—something many players dreamt of at the time and, if you followed the stereotype, something that would suit a German engineer well. Hermann's abstinence from calculation is all the more surprising, since in one of the German precursors of the novella, the novel *The Dutch Jew* (Der holländische Jude) by Karl Gottlieb Samuel Heun,[11] mathematical calculation results in trey and seven as winning cards of the banking game.

Thus, Hermann does not bear out the simple cliché that readers so often imputed to the novella, and appears as a more complex personality than many would like to admit.

* * * *

Before we investigate the problem of Hermann's Germanness further, let us take a look at earlier works by Pushkin. Of course, Pushkin has an ironic attitude toward Germans and ridicules their remoteness from worldly concerns and their enthusiasm for theory, for example, when the narrator in *Eugene Onegin* mocks at the German style acquired by Lenskii as a student in Göttingen. One can also sense irony when Pushkin alludes to German subtexts in his prose works. An example of this is "The Stationmaster" (Stantsionnyi smotritel'; 1831). This work is an ironical "re-make" of a tale by the same title (1827) by the German-Russian sentimentalist Vil'gel'm Ivanovich Karlgof; it is, in fact, a "re-make" that turns Karlgof's idyllic and moral story upside down.[12]

Karlgof's stationmaster, a merchant's impoverished son who has studied in Göttingen, marries his beloved Liza against the fierce resistance of her father, and together with her withdraws from St. Petersburg to a post-station somewhere in the country, where they lead a poor though happy life with their swarm of children. Pushkin's Dunia and Minskii must have encountered Karlgof's couple on their way from Vyrin's post-station to St. Petersburg. The author's ironic attitude toward the German-Russian subtext is indicated by a metapoetical detail: the extensive library in the house of Karlgof's stationmaster (which contains collected works by Karamzin, Zhukovskii, Genlis, Schiller, Goethe, Mendelssohn, and Herder) dwindles in Pushkin's tale to the German verses accompanying the four pictures of the life of the prodigal son.

As early as in the unfinished "Notes of a Young Man" (Zapiski molodogo cheloveka; written 1829 or 1830), Pushkin gives an ironic reply to

Karlgof's idyllic tale. The "humble dwelling" of the stationmaster is adorned with three pictures depicting the parable of the prodigal son. The narrator, a freshly commissioned officer, reads the German verses that are printed under each picture with such enjoyment that he copies them "in order to translate them in some future hour of leisure,"[13] entirely forgetting his original intention "never to utter one single word of German again"[14] after leaving the cadet corps. When the young man, suffering from the boredom of a Russian post-station, asks the serving woman whether she could find him something to read, she brings out "a worn primer and an arithmetic book published for elementary schools."[15]

The ironical portrait of Germans occasionally delivered by Pushkin and his ironical attitude toward German subtexts are, however, less interesting in our context than the part Germans play in the plot of some of his tales. Here it becomes apparent that Pushkin, an enemy of all fixed patterns, also undermines the national stereotypes, as we shall see by examining three of his works.

In "The Stationmaster" the German physician plays a fatal role in Samson Vyrin's misery. Called to the would-be ill Minskii, "he felt the patient's pulse, spoke with him in German; in Russian he declared that all the sick man needed was rest."[16] We know that this diagnosis, for which Minskii pays him twenty-five rubles and invites him to share a meal with wine, is false. In addition, the German physician will assure the deceived father "that the young man had been perfectly healthy, and that he, the doctor, had guessed his evil intentions even then but had kept his silence, fearing the young man's whip."[17] What narrative role is played by the bribable or cowardly German? If one looks at Dunia's story as a story of a lamb that has gone astray, as Vyrin does, or as one of a prodigal daughter who does not return, then the doctor objectively appears as a mediator of misfortune. If one concedes, however, that it is not the child who is prodigal but the father himself, that Dunia had to leave her father for the sake of her happiness,[18] and that she—as the coachman reports—went with Minskii "of her own free will," then that which is behind the narrative function of the German physician emerges to be something rather different and far more complicated.

A narrative role that should not be underestimated is played by the "land surveyor named Schmidt, wearing mustachios and spurs"[19] in "The Blizzard" (Metel'; 1831). Schmidt is one of the three men who not only accede to Vladimir's request to be witnesses to the secret marriage, but even swear they will sacrifice their lives for him. The three heroes' lives are, of course, not to be sacrificed, but how do they carry out their duty as witnesses? One has to consider: it is the three witnesses, shouting "Here! Here!," who direct the false bridegroom Burmin through the blizzard to the church and make him alight from the sledge and hurry to the desperate bride. During the ceremony, the girl is supported not only by her maid but also by the three heroes, as if the seventeen-year-old needed four people to keep her upright.[20] The three men were, says the text, "busy only with her."[21] This seems to explain the witnesses' error. When Burmin's recklessness becomes obvious, they don't threaten him physically but fix on him their eyes only, those eyes that were earlier much too exclusively turned on the beautiful bride. Hence, also in "The Blizzard," a German is partly responsible for the misfortune, but here, too, an apparent misfortune proves to be a good turn—at least for Maria Gavrilovna and Burmin. Because of their error and the dereliction of their duties, the witnesses have contributed to the wedding of those who will fall in love with each other after years of separation.[22]

In "The Undertaker" (Grobovshchik; 1831), the Germanness of the industrious craftsmen around the Nikita Gate plays a central role. The shoemaker Gottlieb Schulz thus answers the undertaker Adriian Prokhorov's question about his business: "I can't complain. Though, of course, my wares are not the same as yours: a man alive can go without shoes, but a dead man can't live without a coffin."[23] The German, who speaks "with the kind of Russian accent that we"—the narrator remarks—"to this day cannot hear without laughter,"[24] uses a Russian proverb. But he corrupts it a little and makes it, without intending to do so, appear absurd. The Russian undertaker, however, does not notice this absurdity but enhances it in his reply: "That's true enough. On the other hand, if the live man cannot afford to buy shoes, he can, forgive my saying so, just as well go barefooted; but a beggar's corpse will take a coffin free of charge."[25] One can see in the manuscripts how Pushkin carefully elaborated the absurdity of the stock phrases. And it was entirely important for the plot, because the absurd talk of the German shoemaker, taken up and continued by the Russian undertaker, lays bare the inherent absurdity into which Prokhorov turns the paradox of his trade: making a living out of the death of his customers. He speaks and thinks of the dead as of living people, inhabiting their coffins like houses. This thinking explains, too, the absurd sign above the gate of Prokhorov's new house: "Plain and Colored Coffins Sold and Upholstered; Available for Rent; Used Ones Repaired."[26] Prokhorov offers his coffins just as he does his old house, nailing "a notice on the gate advising that the house was available for sale or rent."[27] In the new house, however, which is yellow like the faces of the dead, he lives as if in a coffin.[28]

While the German shoemaker makes the Russian undertaker's absurd way of thinking obvious, the novella's central event, Prokhorov's invitation to the Russian Orthodox dead for his inaugural ceremony, is evoked by the laughter of all the craftsmen who have gathered at Schulz's silver wedding anniversary. Why do the Germans laugh? Lately, significance has been attached to this laughter that reaches deep into the differences between cultures and denominations.[29] Thus, some Russian Christian interpreters read the laughter as a sign of Western rationalistic denial of life after death.[30] But one must not see the simpleminded craftsmen around the Nikita Gate as representatives of the Enlightenment, of free-thinking, or even of a contemptuous notion of life after death. And by no means does their laughter make fun of their Russian colleague's trade, as Prokhorov believes or wants to believe. The Germans laugh because Iurko, the tipsy police guard, has demanded an absurd toast of the undertaker: "And how about you? Drink, brother, to the health of your corpses."[31] Iurko means no insult, and the merry Germans as well are only amused at the paradox made obvious. The undertaker, however, attaches a deeper importance to Iurko's words. Prokhorov, never having become aware of the paradoxical nature of his profession, sees his Orthodoxy questioned by the laughing Germans. What does he believe in? Not in a Christian dichotomy of body and soul, but rather that the dead are living physically in the coffins he makes for them, his customers. And why then should he not drink to the health of his customers? The harmless laughter of the Germans, thoroughly misunderstood by the Russian undertaker, finally causes his journey into the underworld, but it is also a prerequisite for his salvation, namely his ultimate return to life and the living.[32]

* * * *

Let us return to *The Queen of Spades*. We have found out that it is not his father's capital that is "necessary" to Hermann. What is it then that is so dangerous and necessary? The German engineer does not trust in calculation but in the miraculous. He dismisses Tomskii's anecdote as a "fairy tale" but still follows its magical logic. It is, as it seems, the secret of the cards that attracts him.

The following question now arises: is Hermann a representative of the Romantic Germany, a second Lenskii, a bourgeois with a soft spot for the fairy tale, a philistine who compensates for his narrow existence by dreaming of the miraculous? Far from being a prosaic engineer, Hermann becomes a writer by dint of his "fiery imagination."[33] As a tender and passionate letter-writer, he requires another's imagination only at the beginning. He even invents a whole novel, becoming "very much absorbed in his intrigue" (*ochen' zaniatyi svoei intrigoiu*).[34] In the countess's bedroom, he appears as a highly eloquent seducer, first addressing the old woman in the language of sentimental and romantic discourses as a wife, mistress, and mother of a newborn son, and then promising her, in religious diction, that his children, grandchildren, and great-grandchildren will "bless [her] memory and hold it sacred."[35] The dream, in which the countess reveals her secret, again proves Hermann's power of imagination, and only a small prosaic detail, the shuffling slippers of the apparition, betray a slight shakiness in the dreamer's style. Hermann the engineer is thus increasingly becoming caught up in the conventions of fantastic literature; when he eventually sets down an account of his vision, he here, too, acts as a writer. As a writing visionary, Hermann repeats the career of Swedenborg, to whom the half-ironic, half-prosaic epigraph to Chapter 5 is ascribed. Swedenborg started as a natural scientist, was—like Hermann—an army engineer, and ended as a visionary theosophist, dismissively called by Kant a "ghost-seer."

Hermann shows some more anti-prosaic traits, concerning love, gambling, and the secret, which lead to contradictions in his character. How open is the hero to the temptations of the goddesses Venus and Fortuna? Is he a lover or a gambler? Just as the text causes realistic and fantastic explanations to overlap, so, too, does it provide contradictory indications regarding Hermann's relationship to Venus and Fortuna. On the one hand, the engineer seems to use amatory discourse solely toward his own ends; on the other, he evinces unrequited erotic desires.[36] Likewise ambiguous is his relation to gambling. He enters the story asserting that "the game interests [him] very much but [that he is] not in a position to sacrifice the necessary in the hope of gaining the superfluous,"[37] and the reliable omniscient narrator affirms that Hermann is "a gambler at heart" (*v dushe igrok*).[38] On the other hand, Hermann's dream suggests that he is attracted by "heaps of bank notes and piles of gold coins"[39] rather than by gamesmanship, the thrill, destiny's challenge, the victory over chance. A third motive, which is left undecided by the text, is Hermann's interest in secrets. On the one hand, he seems to have a soft spot for the miraculous, but, on the other, there are signs that he does not want to fathom the fantastic but is instead interested only in "fantastic riches" (*fantasticheskoe bogatstvo*).[40]

Hermann thus appears to be a character with conflicting traits. But this split character, too, corresponds to a literary stereotype. Hermann, the "son of a Russified German,"[41] gives the impression that there are, in a Goethean manner, two souls dwelling in his breast. Tomskii draws a "portrait" of Hermann by referring to the fashionable Gothic literature of the period: "This Hermann is a truly romantic character: he has the profile of Napoleon and the soul of Mephistopheles. I think he has at least three crimes on his conscience."[42] In some passages, the text suggests a conflict between rationality and irrationality, coldness and passion, "resoluteness" (*tverdost'*) and "delusion" (*zabluzhdenie*) that is characteristic of E. T. A. Hoffmann. This Hoffmannesque split of personality is emphasized by allusions to German works. Hermann has a certain knowledge of German literature,[43] but he does not seem to have thoroughly read some of the subtexts to which the novella alludes. From Clas Johan Livijn's Swedish short novel *Spader Dame, en Berättelse i Bref, Funne på Danviken* (1824) or from Friedrich H. K. de la Motte-Fouqué's German translation of it, *Pique Dame. Berichte aus dem Irrenhaus in Briefen* (1826), he could have learned the lesson that an apparently safe stake on the queen can have fatal consequences: like Livijn's and Fouqué's hero, Hermann goes mad and is confined to a lunatic asylum.[44] And there is Hoffmann's tale *Gambler's Luck* (Spielerglück; 1819; Russian translation 1823), which has obviously inspired Hermann's dream of endless winnings. This story of the reversal of fortune, however, contains the ultimate message that gambler's luck brings ruin even on the most sensible of men.

On the other hand, there are numerous indications that modify Hermann's romantic split personality, four of which are pointed out here. Firstly, Tomskii's Gothic portrait of Hermann as a split romantic character is immediately undermined by the narrator: "Tomskii's words had been no more than a mazurka partner's chit-chat (*mazurochnaia boltovnia*)."[45] Secondly, Hermann's nature seems to be comparable to that of Julien Sorel in Stendhal's *Le Rouge et le Noir*. In this intertextual understanding, Hermann is guided by his greed for money and by the habit of using love, play, and secretiveness as a means to gain other ends. Thirdly, in *Gambler's Luck*, Hoffmann distinguishes between two types of gamblers. To the first type "gambling as game, regardless of winnings, grants an indescribable mysterious pleasure," while the second type "has an eye to the winnings only and regards the game as a means to enrich himself quickly."[46] Hermann rather seems to belong to the second type. Fourthly, in metapoetic terms, it can be argued that Pushkin, who had already ironically exposed romantic character contradictions in "The Shot" (1831) and "The Undertaker" (1831), might in 1833 not have taken such enigmatic contrasts of character all that seriously. On the other hand, the middle-class social climber striving to become a member of the aristocracy was a character type already so unoriginal that Hermann can hardly be seen to coincide with Julien Sorel.

Thus, a definitive assessment of Hermann's character becomes impossible. His character oscillates between two opposite poles—a Romantic and Gothic conflicting nature on the one hand, and a one-track mind ruled by greed on the other. This oscillation corresponds to the novella's wavering between magical fantasy and psychological realism and is eventually connected with the aleatory binarism of the game of *pharaon*. Thus, we can observe in Hermann's character how the game of cards is reified in the narrative. The vacillation between one set of traits and another in Pushkin's characterization of the hero corresponds to the movement in *pharaon* from left to right and from right to left or to Hermann's alternating muttering "Trey, seven, ace! Trey, seven, queen!"[47]

Hence, the question regarding Hermann's Germanness becomes unanswerable. The view that "Hermann is a German: he's calculating, that's all" is not the author's answer but only impatient Tomskii's objection: Tomskii simply wants to shrug off the general astonishment about Hermann's paradoxical abstention from playing *pharaon* since he wants to tell as quickly as possible the still more paradoxical anecdote about his grandmother, who does not play cards. The national stereotype characterized by Tomskii is a trap for the reader. The stereotype is not disproved by the narration but undermined by the alternation between two opposite possibilities. This is a typically Pushkinian way of treating the problem of national stereotypes.

Notes

I would like to thank Miss Nina Owen for her help in preparing the English text.

1. A. S. Pushkin, *Polnoe sobranie sochinenii*, 17 vols. in 21 (Moscow, 1937–59), vol. VIII, p. 227. Unless otherwise indicated the translation is mine.
2. See V. V. Vinogradov, *Stil' Pushkina* (Moscow, 1941), p. 609.
3. N. V. Gogol', *Polnoe sobranie sochinenii*, 14 vols. (Moscow, 1937–52), vol. III, pp. 41–42. For the translation of this passage, I am indebted to J. Douglas Clayton (Ottawa).
4. To I. I. Lazhechnikov of 3 November 1835 (Pushkin, *Polnoe sobranie sochinenii*, vol. XVI, p. 62).
5. J. Sazonova, "The German in Russian literature," *American Slavic and East European Review* 4 (1945), p. 59.
6. Ibid.

7. Pushkin, *Polnoe sobranie sochinenii*, vol. VIII, p. 235. English translation: A. S. Pushkin, *Complete Prose Fiction*, translated by Paul Debreczeny (Stanford, 1990), p. 219. In Pushkin's letters and works, there are two variants of this German triad. In the letter to P. A. Pletnev of 26 March 1831, Pushkin writes: "Pogodin is ... a true German (*istinnyi nemets*) in his pure love for science (*nauka*), his industriousness (*trudoliubie*), and his moderation (*umerennost'*)" (*Polnoe sobranie sochinenii*, vol. XIV, p. 158; translation: *The Letters of Alexander Pushkin*, translated by J. Thomas Shaw, 3rd ed., revised [Los Angeles, 1997], p. 483). In the *Scenes from Knightly Times* (Stseny iz rytsarskikh vremen; 1835), the German Martyn praises his three virtues: "thrift" (*berezhlivost'*), "patience" (*terpenie*), and "industry" (*trudoliubie*) (*Polnoe sobranie sochinenii*, vol. VII, p. 215). See Iu. Mann, "Aleksandr Pushkin über die 'göttingensche Seele,'" *West-östliche Spiegelungen*, Band 3. B: *Deutsche und Deutschland aus russischer Sicht. 19 Jahrhundert: Von der Jahrhundertwende zu den Reformen Alexanders II*, ed. D. Herrmann and A. Ospovat (Munich, 1997), p. 407.
8. *Polnoe sobranie sochinenii*, vol. VIII, p. 238; translation: Debreczeny, p. 221. This characterization is an allusion to A. Pogorel'skii's *The Baleful Consequences of an Unbridled Imagination* (Pagubnye posledstviia neobuzdannogo voobrazheniia; 1828), a weak imitation of E. T. A. Hoffmann's *The Sandman* (Der Sandmann; 1817).
9. Pushkin, *Polnoe sobranie sochinenii*, vol. VIII, p. 235; translation: Debreczeny, p. 219.
10. Pushkin, *Polnoe sobranie sochinenii*, vol. VIII, p. 240.
11. Pseudonym: Heinrich Clauren. A Russian translation of the novel was printed 1825 in *Syn otechestva* under the title *Gollandskii kupets*.
12. See V. N. Turbin, *Pushkin. Gogol'. Lermontov. Ob izuchenii literaturnykh zhanrov* (Moscow 1978), pp. 70–77; W. Schmid, *Puschkins Prosa in poetischer Lektüre. Die Erzählungen Belkins* (Munich, 1991), pp. 130–33, in Russian: V. Shmid, *Proza Pushkina v poeticheskom prochtenii. Povesti Belkina* (St. Petersburg, 1996), pp. 118–21.
13. Pushkin, *Polnoe sobranie sochinenii*, vol. VIII, p. 404; translation: Debreczeny, p. 453.
14. Pushkin, *Polnoe sobranie sochinenii*, vol. VIII, p. 403; translation: Debreczeny, p. 452.
15. Pushkin, *Polnoe sobranie sochinenii*, vol. VIII, p. 405; translation: Debreczeny, p. 454.
16. Pushkin, *Polnoe sobranie sochinenii*, vol. VIII, p. 101; translation: Debreczeny, pp. 98–99.
17. Pushkin, *Polnoe sobranie sochinenii*, vol. VIII, p. 102; translation: Debreczeny, p. 99.
18. For a detailed argumentation, see Schmid, *Puschkins Prosa*, pp. 103–70; Shmid, *Proza Pushkina*, pp. 94–152.
19. Pushkin, *Polnoe sobranie sochinenii*, vol. VIII, p. 79; translation: Debreczeny, p. 78.
20. See V. E. Khalizev and S. V. Sheshunova, *Tsikl A. S. Pushkina 'Povesti Belkina'* (Moscow, 1989), p. 26.
21. Pushkin, *Polnoe sobranie sochinenii*, vol. VIII, p. 86; translation: Debreczeny, p. 86.

22. For a more detailed analysis, see Schmid, *Puschkins Prosa*, pp. 221–59; Shmid, *Proza Pushkina*, pp. 200–225.
23. Pushkin, *Polnoe sobranie sochinenii*, vol. VIII, p. 90; translation: Debreczeny, p. 88.
24. Ibid.
25. Ibid.
26. Pushkin, *Polnoe sobranie sochinenii*, vol. VIII, p. 89; translation: Debreczeny, p. 87.
27. Ibid.
28. The equivalence of house and coffin (referring to the Russian *domovina*) is activated by the contrast between Pushkin's "morose and pensive" undertaker and Shakespeare's (and Walter Scott's) "merry and jocular" gravediggers in the beginning of the tale. In the churchyard scene of *Hamlet* (act V, sc. 1) we find the conundrum "Who builds stronger than a mason, a shipwright, or a carpenter? … a grave-maker; the houses he makes last till doomsday."
29. Cf. A. Ebbinghaus, "Handlung und Ereignis in A. S. Pushkins Erzählung 'Der Sargmacher' (Grobovshchik)," *Arion. Jahrbuch der Deutschen Puschkin-Gesellschaft* 1 (1989), pp. 79–89.
30. O. Povolotskaia, "'Grobovshchik': kolliziia i smysl," *Voprosy literatury*, No. 12 (1989), pp. 210–24.
31. Pushkin, *Polnoe sobranie sochinenii*, vol. VIII, p. 92; translation: Debreczeny, p. 90.
32. See Schmid, *Puschkins Prosa*, pp. 259–338; Shmid, *Proza Pushkina*, pp. 259–93.
33. Pushkin, *Polnoe sobranie sochinenii*, vol. VIII, p. 235; translation: Debreczeny, p. 218.
34. Pushkin, *Polnoe sobranie sochinenii*, vol. VIII, p. 238; translation: Debreczeny, p. 221.
35. Pushkin, *Polnoe sobranie sochinenii*, vol. VIII, p. 242; translation: Debreczeny, p. 225.
36. See G. Williams, "The Obsessions and Madness of Germann in 'Pikovaia dama,'" *Russian Literature* 14 (1983), pp. 383–96.
37. Pushkin, *Polnoe sobranie sochinenii*, vol. VIII, p. 227; translation: Debreczeny, p. 211.
38. Pushkin, *Polnoe sobranie sochinenii*, vol. VIII, p. 235; translation: Debreczeny, p. 219.
39. Pushkin, *Polnoe sobranie sochinenii*, vol. VIII, p. 236; translation: Debreczeny, p. 219.
40. Pushkin, *Polnoe sobranie sochinenii*, vol. VIII, p. 236.
41. Pushkin, *Polnoe sobranie sochinenii*, vol. VIII, p. 235; translation: Debreczeny, p. 218.
42. Pushkin, *Polnoe sobranie sochinenii*, vol. VIII, p. 244; translation: Debreczeny, p. 226.
43. The first letter he writes to Lizaveta Ivanovna is "translated word for word from a German novel" (Pushkin, *Polnoe sobranie sochinenii*, vol. VIII, p. 237; translation: Debreczeny, p. 220). This novel could have been *The Sufferings of Young Werther* by Goethe (cf. M. D. Shrayer, "Rethinking Romantic Irony: Pushkin, Byron, Schlegel and the 'Queen of Spades,'" *The Slavic and East European*

Journal 36 [1992], p. 402). In one of the outlines of Pushkin's novella, the German hero named Hermann and the German neighbour girl Charlotte (!) "fall in love so deeply as only Germans are capable of in our day and age" (Pushkin, *Polnoe sobranie sochinenii,* vol. VIII, p. 835; translation: Debreczeny, p. 516).

44. See Véra Nabokov and G. Barabtarlo, "A Possible Source for Pushkin's 'Queen of Spades,'" *Russian Literature Triquarterly* 24 (1991), pp. 43–62.
45. Pushkin, *Polnoe sobranie sochinenii,* vol. VIII, p. 244; translation: Debreczeny, p. 226.
46. E. T. A. Hoffmann, *Gesammelte Werke in Einzelausgaben,* vol. 5 (Berlin and Weimar, 1994), pp. 259–60.
47. Pushkin, *Polnoe sobranie sochinenii,* vol. VIII, p. 252; translation: Debreczeny, p. 233. For *The Queen of Spades* as a metatextual novella and a multilayered narrative realization and unfolding of the game of *pharaon,* see W. Schmid, "'Pique Dame' als poetologische Novelle," *Die Welt der Slaven* 42 (1997), pp. 1–33; V. Shmid, "'Pikovaia dama' A. S. Pushkina (problemy poetiki)," *Russkaia literatura,* No. 3 (1997), pp. 6–28; V. Shmid, "'Pikovaia dama' kak metatekstual'naia novella," *Proza kak poeziia. Pushkin – Dostoevskij – Chekhov – avangard,* 2nd ed., revised and enlarged (St. Petersburg, 1998), pp. 103–44.

Chapter 4

THE FANTASTIC IN THE EVERYDAY

Gogol's "Nevsky Prospect" and Hoffmann's "A New Year's Eve Adventure"

Priscilla Meyer

Gogol's laughter is man's struggle with the Devil

—Dmitri Merezhkovsky[1]

In his article about "Nevsky Prospect" (1835), V. V. Vinogradov suggests that Gogol uses Thomas de Quincey's *Confessions of an English Opium-Eater* to move from the German romantic tradition toward the realism of the French *roman furieux*.[2] The commentators of the Academy edition of Gogol's works[3] and Mikhail Gorlin[4] follow Vinogradov's lead, seeing in "Nevsky Prospect" a turning away from the fantastic of Hoffmann toward the realism of Jules Janin, who finds "the fantastic in the everyday."[5] But with Yury Mann's insight into Gogol's method of disguising the demonic in the everyday,[6] we may wonder if "Nevsky Prospect" is as abrupt a turning point as scholars have suggested. Reading it together with "A New Year's Eve Adventure" suggests that in "Nevsky Prospect" Gogol reaffirms rather than abandons Hoffmann's supernatural world.

In all that has been written about Gogol's use of the works of E. T. A. Hoffmann, "Nevsky Prospect" has nowhere been paired with "A New Year's Eve Adventure" (1815),[7] although the publication of the Russian translation of Hoffmann's story in 1830 in *Son of the Fatherland*[8] makes it almost certain that Gogol would have read it shortly before beginning to write his own story in 1831.[9] Rather, Gorlin and others have seen "A

New Year's Eve Adventure" as a source for "The Nose" because of the thematics of Schlemihl's lost shadow and Spikher's lost reflection.[10]

There is, however, a no less important theme shared by "A New Year's Eve Adventure" and "Nevsky Prospect": the discordancy between dream and reality, especially as experienced by the artist,[11] that is central to the work of both Hoffmann and Gogol. The close parallels between theme, structures, and motifs leave little doubt that Gogol had Hoffmann's tale clearly in mind when he wrote "Nevsky Prospect." Reading Hoffmann's story as the pentimento of Gogol's heightens the supernatural coloring of "Nevsky Prospect," and suggests that Gogol's dialogue with Hoffmann involves more than parody of an outmoded school of literature.

Hoffmann's tale begins with a "Foreword from the Editor," who introduces the Travelling Enthusiast (TE), and apostrophizes the reader on the blurred boundary between the inner life and the outside world. Taken together with the TE's "Postscript," these first and last sections parallel Gogol's introductory and concluding descriptions of Nevsky Prospect both structurally and thematically. They frame the four sections narrated by the TE. In the first, "The Beloved," the TE tells how after many years he encounters his former love at a New Year's Eve party at the Justizrat's in Berlin. Julia treats him both warmly and coldly. When her "spindle-shanked cretin"[12] of a husband appears, the TE flees in despair. In the second section, "In the Beer Cellar," the TE enters a beer cellar where he drinks with the legendary Peter Schlemihl and Erasmus Spikher, a little man with two contrasting faces—one young, the other old. Spikher's horror when the TE offers him his "small steel tobacco box, polished like a mirror" suggests that Spikher has lost his reflection, as Schlemihl has lost his shadow. In the third section, "Manifestations," the TE has to spend the night at an inn because he has left his keys at the Justizrat's. He finds Spikher asleep in his room. Spikher awakens and confesses that he has given his reflection to Giuletta. They go to sleep, but the TE wakes to find Spikher writing at a table. Feeling "the chill of the supernatural" (114), the TE falls into a waking dream in which Julia appears, looking like "one of the alluring young women, surrounded by demonic monsters" from the work of Brueghel, Callot, or Rembrandt (115).

When the TE finally awakens, he thinks the adventure with the little man may have been only a dream, but he finds Spikher's manuscript, "The Tale of the Lost Reflection," on the table. This, the fourth section of the TE's narrative, is Spikher's account of meeting the irresistible Giuletta in Italy. She is associated with a "tall, thin, dried-out looking man" (118) named Dapertutto. When Spikher murders one of Giuletta's suitors and is forced to flee Italy, he leaves her his reflection. Returning to Germany, he is renounced by his wife, and now wanders the wide world with Schlemihl.

At the end of the tale, the TE appends a postscript addressed to Hoffmann. He bemoans the "strange dark power" that manifests itself in his life, and that made him believe that "the good Julia was a picture of a siren by Rembrandt or Callot who betrayed the unfortunate Spikher to get his alter ego" (129). In his conclusion, the TE reveals that Spikher's adventure is a hallucinatory variation-interpretation of the TE's meeting with Julia, an illusion inspired by the "dark power."

Both "A New Year's Eve Adventure" and "Nevsky Prospect" are tales of doubles. In Hoffmann's story each character has a counterpart: the TE is doubled by Spikher. Both grieve for their lost loves, Julia and Giuletta, the first the German and the second the Italian version of the same name. The women's protectors are also doubles: Julia's old husband resembles the diabolical Dapertutto. The narrator TE includes Spikher's manuscript in his account, so that the very narrative is doubled. The TE's tale takes place in Berlin's polite society and, although seen through the eyes of the fanciful TE, is thoroughly realistic. Spikher's tale takes place in Italy and is overtly supernatural: the beautiful Giuletta is the devil's snare who captures Spikher's reflection.

In "Nevsky Prospect," Piskarev's and Pirogov's paired romances are equally symmetrical. Piskarev pursues a brunette, Pirogov a blonde. Piskarev, the idealist, is an artist who sees the brunette as his muse; Pirogov, the realist, is an officer who only wants to seduce the blonde. The first woman turns out to be a prostitute, the second to be the virtuous wife of an artisan. As in Hoffmann's tale, both men lose the women they desire. Piskarev's tale follows the romantic pattern of the failure of the artist to embody his ideal in everyday reality, while Pirogov's appears to be a comic burlesque of that topos, well-worn by 1834 when Gogol was writing "Nevsky Prospect."

The TE and Piskarev are typical romantic idealists and dreamer-artists; the first, an appreciator of painting and music, writes, and the second paints. The TE records his experiences in journals that he sends to Hoffmann, addressing him as "dear Theodor Amadeus Hoffmann" in his postscript. The TE is given to absurd mishaps typical of the dreamer who lives not in this world but in his vivid imagination, spilling tea on the Justizrat and rushing off into the Berlin winter night without his coat and hat, on the one hand, but, on the other, writing the supernatural, dream-induced tale he sends to his alter ego, Hoffmann.

The narrator of "Nevsky Prospect" emphasizes the contradiction between Piskarev's inner world and outer reality: "He was an artist. A strange phenomenon, is it not? A Petersburg artist. An artist in the land of snows. An artist in the land of Finns where everything is wet, flat, pale, grey, foggy. These artists are utterly unlike Italian artists, proud and ardent as Italy and her skies."[13] Gogol opposes the reality of Piskarev's physical surroundings (the cold north) to his art (the ardent south), just as Hoffmann opposes Germany to Italy. Berlin, too, is cold and gray; the TE is entranced by the Christmas holidays, "but after the holidays everything becomes colorless again, and the glow dies away and disappears into drab darkness."[14]

Each artist encounters his beloved unexpectedly and, in a state of ecstasy, describes her as a divine being. The TE sees Julia at the Justizrat's: "'Yes, it was Julia, Julia herself, as pretty as an angel and as demure.... I will never let you go.... Your love is the spark that glows in me, kindling a higher life in art and poetry. Without you, without your love, everything is dead and lifeless.'"[15] Piskarev sees his lady as "an exquisite creature who seemed to have flown straight down from heaven onto Nevsky Prospect,"[16] a "divine being" (215), "the original of his dream pictures" who is to "inspire [his] work" (226). The TE compares his beloved with paintings by Mieris, Callot, Brueghel and Rembrandt; Piskarev sees his as "a Bianca of Perugino" (212).

But both divine beloveds prove to be all too much of this world. For the TE, "there was something strange about [Julia's] whole figure, I thought. Somehow she seemed larger, more developed, almost lush. Her blouse was cut low, only half covering her breasts, shoulders and neck...."[17] Piskarev follows his beauty to a place where woman is "transformed into a strange equivocal creature, where she loses with the purity of her heart all that is womanly ... and ceases to be the delicate, the lovely creature so different from us" (217). Julia's fall from divine to degraded status is heralded when the TE first recognizes her at the Justizrat's: "You see a miraculously beautiful flower, glowing with beauty, filling the air with scent, hinting at even more hidden beauty. You hurry over to it, but the moment you bend down to look into its chalice ... out pops a smooth, cold, slimy little lizard...."[18] Piskarev thinks of his beauty: "[B]y some terrible machination of the fiendish spirit, eager to destroy the harmony of life, she had been flung with satanic laughter into this horrible swamp."[19]

Piskarev dreams of his brunette, transposing her from her house of ill repute to an aristocratic setting. The only way he can reconcile his idea of her with her degraded existence is in opium-induced dreams. Similarly, the TE falls into a "waking dream" in which Julia tempts him to drink from a flaming goblet.[20] His dream is less realistic than Piskarev's—in it Julia says "I have you and your reflection, once and for all."[21] By her theft of his reflection, Julia merges in the TE's mind with Giuletta in Erasmus Spikher's tale that he has yet to read. Thus Hoffmann indicates the unity of the idealized woman with the diabolical one, a unity that underlies the doubling of both Hoffmann's and Gogol's stories.

The first parts of "A New Year's Eve Adventure" and "Nevsky Prospect" depict the romantic artist's horrified encounter with a divine muse-beloved who turns out to be an exaggeratedly sensual woman. The TE runs out of the Justizrat's party into the stormy night: "Bareheaded and without a coat I finally felt the cold when icy shivers began to interrupt my feverishness."[22] He goes off to a beer cellar and then spends the night at an inn, dreaming of Julia and murmuring her name. Piskarev rushes away from his final interview with the brunette "with every thought and feeling in turmoil. His mind was clouded: stupidly, aimlessly, he wandered about all day, seeing nothing, hearing nothing, feeling nothing. No one could say whether he slept anywhere or not...."[23] He cuts his throat in despair at the disparity between his dreams and reality.

The resolutions of the TE's and Piskarev's tales are followed by the stories of their doubles. Hoffmann pays homage to Chamisso's *Peter Schlemihl* by having its hero appear in the beer cellar. Erasmus Spikher is the double of Schlemihl, who traded his shadow for gold. Spikher's "Story of the Lost Reflection" is Hoffmann's variation on *Peter Schlemihl.* Gogol, in turn, pays homage to Spikher's author by having him turn up in Piskarev's tale.

Lieutenant Pirogov and Erasmus Spikher are quite unaware of their authors' subtexts; they are from simpler strata of society than their idealist counterparts Piskarev and the TE. Both are captivated by the sensual aspect of the women they pursue. Pirogov follows his blonde into the German part of St. Petersburg; Spikher meets Giuletta in Florence. Cultural opposition is an element in both tales. The blonde's husband, the tinsmith Schiller, is "a real German in the full sense of the word" who has planned his life precisely from the age of twenty and kept to his plan, as opposed to "the Russian [who] lives at that age without a thought of the next day."[24] Schiller also keeps his word, a quality that explains his acceptance of Pirogov's commissions despite the latter's flirtation with Schiller's wife.

Spikher initially shares this steadfastness, resisting his German companions' urging to take a mistress while in Italy. He objects, "If I took up with a girl for even one night it would be betraying my wife," for which one of the Italian women calls him a "cold-blooded, heartless German."[25]

But Giuletta inflames Spikher. He spends the evening with her among his friends, but doesn't dare follow her as she leaves. When he calls on her some time later, "to Erasmus's fiery passion she opposed a mild indifference." His friends observe that he is "totally under her spell" (119), but he seldom sees them anymore.

Pirogov, "smoking a pipe with his fellow officers … alluded significantly … to his little intrigue with the pretty German, with whom he was, according to his account, already on the best of terms, though as a matter of fact he had almost lost all hope of winning her favor"[26]—the blonde does more than "oppose a mild indifference," but here too the suitor's infatuation is revealed against the background of his relationship to his friends.

In Hoffmann's tale, the TE and Spikher have parallel experiences: both are deceived by the women they adore. Julia no longer reciprocates the TE's passion and has married a "spindle-shanked cretin … who said, in a mixture of croak and cackle, 'Where the Devil is my wife?'"[27] The TE's "former life of love and poetry" with Julia (106) has been corrupted by the devil of mundane reality so offensive to the artist—a husband. In the TE's dream-tale of Spikher, he transforms the diabolical mundane into the diabolical supernatural: the husband becomes Dapertutto, and Julia's charms become the snare of Giuletta's sensual attraction. In recasting his lost love for Julia in his tale of Spikher's for Giuletta, the TE reveals the supernatural potential of our passions. As the TE's naive double and mirror image, Spikher loses both his soul and his virtuous, mundane existence when he falls from grace by succumbing to sensual passion. The idealist TE gives himself a lesson in the folly of rejecting the mundaneness of reality by creating a realist hero who loses his happy everyday existence to the supernatural.

Gogol teaches the opposite lesson, but also uses inversion to do so: Piskarev's tragedy is paired with Pirogov's burlesque. Piskarev cannot bear the dissonance between his idealism and reality; Pirogov, unaware of any supernatural presence, is soundly thrashed by indignant Germans and simply forgets about the whole thing, distracted from plans for revenge by two *pirozhki* at a café. "A New Year's Eve Adventure" shows us that Pirogov's seduction has been thwarted by the same virtuous everyday reality that Spikher forfeited to the Devil—an upright German woman. And she is married to a realist version of the romantic writer Schiller. Gogol's Pirogov-blonde-Schiller triangle corresponds to Hoffmann's Spikher-Giuletta-Dapertutto triangle. The blonde is an unwitting but no less diabolical snare than Giuletta, and Pirogov's adventure appears to be merely comic; but in Gogol's tale the tragedy is that Pirogov pays no penalty for his sins.

The comic zero ending of "Nevsky Prospect" is seen by Ingham as a parody of Hoffmann's supernatural tales: "Gogol is telling us that the romantic vision is no match for cold reality; and he is parodying a convention of fantastic literature by having the ideal yield place to that reality instead of vice-versa."[28] But as Mann says, although there is no literal presence of the diabolical in "Nevsky Prospect," it is nonetheless present in veiled verbal traces.[29]

These traces take on coherent meaning when seen through the subtext of "A New Year's Eve Adventure." One such trace is the dagger. Spikher, inflamed by his adulterous passion for the courtesan Giuletta, kills a rival in a fit of jealousy: "[A] dagger gleamed in the Italian's hand, but Erasmus dodged, seized him by the throat, threw him to the ground, and shattered his neck with a kick." Spikher is forced to flee Italy, which provides Giuletta the occasion to ask him to leave her his reflection: "Won't you even give me this trivial thing?"[30]

Dapertutto tells Spikher, "'That was a real nice German kick in the neck' ... his words felt like dagger blows in Erasmus's chest" (122). Dapertutto proposes a way for Spikher to stay in Italy, promising to "keep your enemies' daggers away from you" (122). The dagger, then, is associated with Spikher's sins, murder and adultery, and the Devil's temptations that lure Erasmus away from his family and into an adulterous affair.

For Pirogov, the dagger serves a similar purpose. Having kept his foot in the door of Schiller's house by ordering spurs from the tinsmith, he comes up with another commission when the spurs are ready: "Make me a sheath for a dagger." As the Italian weapon is foreign to the German Spikher, so Pirogov's dagger is a "fine Turkish dagger";[31] it belongs to a non-Christian culture, as befits a diabolic weapon for Gogol. It is to be Pirogov's means to lead Frau Schiller into adultery, a mirror image of Spikher's infidelity to Frau Spikher.

Another revealing link between the two tales is Gogol's carefully placed use of red clothing. When Spikher leaves the party where he has just met Giuletta, he passes "a strange figure ... with ... glowing eyes,... wrapped in a flame-red cloak with brightly-polished steel buttons"[32] who makes fun of his cloak. Spikher retorts, "What the Devil is my clothing to you?" and shoves the "red-clad stranger" aside (119). This is Dr. Dapertutto, the Devil himself, who uses Giuletta to snare men's souls. The everyday use of "the devil" when it *is* the Devil is a device typical of both Hoffmann and Gogol; together with his red cloak, it identifies him to the reader instantly.

When Pirogov finally manages to see the blonde alone at her home, he is caught kissing her by her husband Schiller and his friend Hoffmann.

Schiller cries, "'You are a scoundrel, and not a Russian officer. The devil take it, my friend Hoffmann, I am a German, and not a Russian pig!... Take him, my friend Hoffmann, by the collar ...,' he continued ... at which his face resembled the red cloth of his waistcoat..."[33]

Gogol uses the names Hoffmann and Schiller for the German bootmaker and tinsmith for more than "making fun of Germans in general."[34] The duality of German culture as seen through Russian eyes had already been captured by Pushkin in *The Queen of Spades*—practical and methodical on the one hand, idealist, fantastical and supernatural on the other. Gogol too plays on this duality, but in order to comment on Pirogov's world, which interests him more than a discussion of Germany.

Just as Hoffmann described the circle of young Germans with their Italian girlfriends at the beginning of Spikher's tale, Gogol gives a long description of the low tastes of Pirogov's circle as an introduction to the lieutenant's adventure with the German blonde. Members of the circle of civil councillors he frequents equate Bulgarin and Pushkin, pick husbands for their phlegmatic daughters by their rank, smell of cabbage, and "pass for well-bred, highly educated men."[35]

The Devil, as we have seen, is responsible for Spikher's downfall. His name conceals his supernatural nature from Spikher, who speaks no Italian; had he known that Dapertutto means "here, there, and everywhere," he might have taken the cue from Dapertutto's red cloak and resisted temptation. The Devil is also involved in Pirogov's "ruin"—his failure to seduce the blonde and the thrashing he receives from the Germans. Gogol's Devil is masked from the Russian Pirogov by the German names Schiller and Hoffmann. Pirogov as a reader of the "Northern Bee" may know the names, but is unlikely to have understood their idealism. In keeping with the tastes of Pirogov's circle, the narrator straight-facedly insists on the superior fame of the German craftsmen over the romantic writers: "Not the Schiller who wrote *William Tell* and the *History of the Thirty Years' War*, but the famous Schiller, the ironmonger and tinsmith of Meshchansky Street ... not the writer Hoffmann, but a rather high-class bootmaker who lived in Ofitsersky Street."[36] The street names emphasize the contrast between the German idealists and Pirogov, the *officer* who frequents a *meshchansky* (petit bourgeois) circle.

Pirogov's "ruin" is effected by Schiller and Hoffmann. As we have seen, they are associated with the Devil by Schiller's red waistcoat and the phrase "the devil take it!" The masking of this association is analogous to the masking of German romantic literature by the craftsmen's names. In St. Petersburg, these masters of idealist poetry and prose are valued, in

Gogol's tale, for their materialist artistry, making boots, coffee pots and samovars. The Piskarevs perish, but the Pirogovs flourish. Just as in "The Nose," an experience that should lead to a spiritual epiphany ends in nothing and the heroes go about their godless ways unchastened.

Vinogradov has shown that Piskarev's dream of seeing his beloved at the ball derives from Thomas de Quincey's *Confessions of an English Opium-Eater.*[37] But the figure of the brunette prostitute need not derive from de Quincey or from Jules Janin's "L'Âne mort et la femme guillotinée"; she has little in common with either, besides her profession, which is in any case represented in the works of several French writers of the period.[38] The Italian courtesan Giuletta provides Gogol with both the ambiguous social status of Piskarev's brunette and the cause of her lover's ruin. Henriette and Ann do no harm to their admirers, who are not even in love with them as Hoffmann's heroes are with Julia/Giuletta.

The subtext of "A New Year's Eve Adventure" suggests a quite different interpretation of "Nevsky Prospect" from the one given by students of Hoffmann's effect on Gogol's work. Pirogov's tale is not a refutation of idealist poetics, but a confirmation of them. The real/ideal opposition is not undermined by Pirogov's trivial tale. His narrow world of petty officialdom is no more comically mundane than that of Conrektor Paulmann and his earring-loving daughter Veronika in Hoffmann's romantic "Golden Pot." Anselmus rejects her world in favor of the life of the spirit with his beloved snake-bride Serpentina, affirming the idealism of Hoffmann (the writer). Pirogov is totally oblivious to the existence of an ideal world, although he is beaten over the head by Hoffmann (the bootmaker). Gogol insists on the value of the writer's idealism by the implied juxtaposition of the two Hoffmanns, representative of the ideal and the mundane, and the absence of the writer's vision from the characters who people Pirogov's world. Hence, "the devil take it, my friend Hoffmann": Pirogov is as unsuspecting of the presence of the Devil as Erasmus Spikher, and the Devil will indeed take him, too. As Dmitri Merezhkovsky put it, "Gogol was the first to detect invisible evil … not in tragedy, but in the absence of everything tragic."[39]

"Nevsky Prospect" affirms this reading in its concluding paragraph, which forms an epilogue to the adventures of Piskarev and Pirogov. It follows the structure of "A New Year's Eve Adventure" in which Hoffmann provides the epilogue of the "Postscript by the Travelling Enthusiast." "Julia, Giuletta—divine image, demon from Hell,"[40] he meditates, thereby summarizing the paired tales and the battle between the divine and the diabolical that takes place in them. "A strange, dark power manifests itself in my life all too often ... I more than half believe that ... the good Julia ... betrayed the unfortunate Spikher to get his reflection in the mirror" (129). The dark power deceives the TE into finding the diabolical Giuletta in the good Julia. The entire story is the TE's record of the conflict between his imagination, which is prey to the "dark forces," and his rational understanding of reality.

Gogol's narrator too stands back and draws a larger moral to his tale: everything on Nevsky Prospect breathes deception. All of one man's wealth lies in his coat; two others are talking not about the architecture of a church but about two crows; the trifles in the shop windows "smell of a terrifying quantity of money."[41] Instead of inner worth or the church, people are preoccupied with trifles, with goods, and the terror that should be for their souls is misplaced. "But God preserve you from looking under the ladies' hats!... Keep your distance, for God's sake, keep your distance from the street lamp!" Since the "Devil himself lights the street lamps to show everything in false colors," God is twice invoked as protection. Piskarev's and Pirogov's adventures illustrate that it is precisely the ladies that the Devil shows in false colors and are his snares, as Dapertutto uses the deceptive Giuletta to snare Spikher's soul.

This is far from the move to realism that critics have found in "Nevsky Prospect." The eerie ending is quite serious; rather than abandoning Hoffmann's fantastic tales to depict the everyday, Gogol insists on the uncanny and the continuing battle between God and the Devil. Reading Gogol's tale through Hoffmann's subtext reveals the connection between Pirogov's comic burlesque and the Devil's lighting of the lamplights. But while Hoffmann ends his tale with the Travelling Enthusiast's confession that Giuletta's tale is merely a product of his own fantasy inspired by a "strange dark power," Gogol reverses this ending and concludes on a sudden note of cosmic terror. His Devil does not disappear with Pirogov's comic denouement but rather continues to be regularly present on Nevsky Prospect. This ending is not "an unmotivated sad chord at the finale."[42] Seen in the supernatural light of "A New Year's Eve Adventure," "Nevsky Prospect" does indeed show "the fantastic in the everyday," but in Hoffmann's sense, not Janin's—with the emphasis on the fantastic.

Notes

I am grateful to my colleague Susanne Fusso for her help with this essay.

1. Dmitri Merezhkovsky, "Gogol and the Devil" [1911], *Gogol from the Twentieth Century*, ed. Robert A. Maguire (Princeton, 1974), pp. 57–102, here p. 57.
2. V. V. Vinogradov, *Izbrannye trudy: poetika russkoi literatury* (Moscow 1976), *Evoliutsiia russkogo naturalizma,* "O literaturnoi tsiklizatsii," pp. 45–62; "Romanticheskii naturalizm," pp. 76–100.
3. N. V. Gogol', *Polnoe sobranie sochinenii,* 14 vols. (Moscow, 1937–52), vol. 3, pp. 646–48.
4. Michael Gorlin, *N. V. Gogol und E. T. A. Hoffmann* (Leipzig, 1933), pp. 54–56.
5. Vinogradov, "Romanticheskii naturalizm," p. 77; Gorlin, *Gogol und Hoffmann,* pp. 54–56. Adolf Stender-Petersen ("Gogol und die Deutsche Romantik," *Euphorion* XXIV, [1922] iii, pp. 628–53, p. 651) and Alain Montandon ("Une source peu connue de la *Perspective Nevsky* de Gogol," *Revue de littérature comparée* 50, no. 3 [1976], pp. 291–95) also present this view: "Gogol condamne définitivement et sans recours l'idéalisme romantique, dans lequel il voit une imposture" (p. 295).
6. Yury Mann, *Poetika Gogolia* (Moscow, 1988).
7. "Die Abenteuer der Silvester-Nacht" appeared in volume four of the *Fantasiestücke.* S. I. Rodzevich ("K istorii russkago romantizma [E. T. A. Gofman i 30-40 gg. v nashei literature]," *Russkii filologicheskii vestnik* LXXVII [Petrograd, 1917], section V) finds that only "Portret" contains enough material to justify speaking of Hoffmann's influence on Gogol. Charles E. Passage (*The Russian Hoffmannists* [The Hague, 1963]) relates "Nevsky Prospect" to "A Fragment from the Lives of Three Friends" (pp. 157–59) and "A New Year's Eve Adventure" to "The Nose" (pp. 168–70). Norman W. Ingham (*E. T. A. Hoffmann's Reception in Russia* [Wurzburg, 1974]) sees "Nevskii Prospekt" as a general response to the whole German school (pp. 171–73). Montandon, "Une source," considers "Das Öde Haus" a source for "Nevsky Prospect."
8. [A Tale], *Syn otechestva,* 1830, ch. 131, N. 4, pp. 219–28, in the text of the article "Poslednie dni zhizni i smert' Gofmana." Translated from French. See Ingham, *Hoffmann in Russia,* p. 272.
9. See Gogol', *PSS,* vol. 3, pp. 645–48, for the three-stage history of the tale's composition (1831–33; 1832–33; 1833–34).
10. Rodzevich, "K istorii russkago romantizma," p. 221; Gorlin, *Gogol und Hoffmann,* pp. 79–80; Mann, *Poetika Gogolia,* pp. 80–81).
11. Vinogradov, "Romanticheskii naturalizm," p. 47; Gorlin, *Gogol und Hoffmann,* p. 50.
12. E. T. A. Hoffmann, *The Best Tales of Hoffmann,* ed. E. F. Bleiler (New York, 1967), pp. 104–29, here p. 108. Citations from the two tales are taken from the English translations.
13. N. V. Gogol, *The Complete Tales of Nikolai Gogol,* 2 vols., ed. Leonard J. Kent (Chicago, 1985), vol. 1, pp. 207–38, p. 106. I have slightly amended Leonard J. Kent's translation of "Nevsky Prospect."

14. Hoffmann, "New Year's Eve," p. 105.
15. Ibid., p. 108.
16. Gogol, "Nevsky Prospect," p. 213.
17. Hoffmann, "New Year's Eve," pp. 106–7.
18. Ibid., p. 106.
19. Gogol, "Nevsky Prospect," p. 218.
20. Montandon briefly notes the parallel between Piskarev's dream and section one ("The Beloved") of "A New Year's Eve Adventure" ("Une source," p. 294).
21. Hoffmann, "New Year's Eve," p. 115.
22. Ibid., p. 108.
23. Gogol, "Nevsky Prospect," p. 227.
24. Ibid., p. 234.
25. Hoffmann, "New Year's Eve," p. 117.
26. Gogol, "Nevsky Prospect," p. 234.
27. Hoffmann, "New Year's Eve," p. 108.
28. Ingham, *Hoffmann in Russia*, p. 173.
29. Mann, *Poetika Gogolia*, p. 76.
30. Hoffmann, "New Year's Eve," p. 121.
31. Gogol, "Nevsky Prospect," p. 234.
32. Hoffmann, "New Year's Eve," p. 118.
33. Gogol, "Nevsky Prospect," p. 236.
34. Ingham, *Hoffmann in Russia*, p. 172.
35. Gogol, "Nevsky Prospect," pp. 228–29.
36. Ibid., 230.
37. Vinogradov, "Romanticheskii naturalizm," pp. 48–51; Gorlin, *Gogol und Hoffmann*, p. 56.
38. For example, Abbé Tiberge (pseud. Regnier-Destourbet), *Louisa, ou les malheurs d'une fille de joie* [1830] (Paris, 1865).
39. Merezhkovsky, "Gogol and the Devil," p. 57. This understanding of Gogol's diabolical in the everyday has become an accepted reading of Gogol's Petersburg tales. Cf. Michael J. Holquist, "The Devil in Mufti: The *Märchenwelt* in Gogol's Short Stories," *PMLA* 82 (October 1967): pp. 352–62, especially pp. 58–59.
40. Hoffmann, "New Year's Eve," p. 128.
41. Gogol, "Nevsky Prospect," p. 218.
42. Mann, *Poetika Gogolia*, p. 23.

Chapter 5

THE TYRANNY OF DIFFERENCE

Gogol and the Sacred

Michael Holquist

> I remember a conversation with Kafka which began with present-day Europe and the decline of the human race. "We are nihilistic thoughts, suicidal thoughts that come into God's head," Kafka said. This reminded me at first of the Gnostic view of life: God as the evil demiurge, the world as his Fall. "Oh, no," said Kafka, "our world is only a bad mood of God, a bad day of his." "Then there is hope outside this manifestation of the world that we know." He smiled. "Oh, plenty of hope, an infinite amount of hope—but not for us."
>
> —Max Brod, as quoted by Walter Benjamin

Gogol, perceived by many to be one of the most quintessentially Russian of Russian writers, is deeply connected to Germany. He came to maturity at a time when German Romantic philosophy and literature were everywhere in his homeland.[1] Nicholas I, as is well known, was an admirer of German order and sought to create a new sense of national identity in Russia that might match the resurgent patriotism of the Prussians after Napoleon's defeat. Under the ambiguous impact of Hegel,[2] history takes on an overwhelming importance during the years when Gogol's literary personality evolved. Nicholas Riasanovsky has argued that history "in one form or another became the center of attention and controversy. 'The historian represented,' in the words of Pogodin '… the crowning achievement of a people, for through him the people came to an understanding of itself.'"[3] German historians set the pattern for recovering a national identity through creative discovery of the past: even the great Niebuhr admitted that the "evil time of Prussia's humiliation [under Napoleon] had some share in the production of my history."[4] It was under the influence of these German-inspired ideas about the need to forge a national consciousness that the completely untrained Gogol tried to teach history during his first years in St. Petersburg.

The year 1829 may serve to exemplify just how important Germany was in the imagination of the young Gogol. In that year he published his first volume (under the pseudonym V. Alov), a longish poem, *Hans Kuechelgarten*, which he had written two years earlier while still in high school in the Ukraine. The German influence is obvious not only in the hero's nationality and library (where Tieck figures prominently between Petrarch and Aristophanes, a kind of unintended bracketing of the sublime and the ridiculous), but also in the source of the work, J. H. Voss's *Luise*. Although the unanimously dismissive reviews of *Hans* caused Gogol to buy up remaining copies and burn them, the attraction of Germany was undiminished: he immediately took a steamer to Lübeck and embarked on what Nabokov described as "a freak trip through Northern Germany."[5] Quickly returning to St. Petersburg, Gogol immediately started publishing the Dikanka stories that would secure his first fame. For the next decade, Gogol would publish tales that may have been set in the Ukraine or in St. Petersburg, but were all heavily influenced in their formal characteristics by the distinctively German genre of the *Kunstmärchen*.[6]

German Romanticism is an important context for analyzing Gogol not only because of the particular and obvious connections he had with that country, but because of a model of the author's task that arguably originated in nineteenth-century Germany. Gogol abstracted from the Germans a model of the author as national prophet, the keeper and producer of national identity. Much as ancient Greece exercised a tyranny over Germans from at least Winkelmann on, so German Romanticism exercised a tyranny over Gogol from the beginning to the end of his life. By the tyranny of German Romanticism I do not mean simply that Gogol chose to write many of his most famous works in genres influenced from abroad, such as messianic national history or the *Kunstmärchen*. I mean to suggest rather that there is a quality of mind, a kind of existential set toward things, a basic pattern that structures both the life and work that Gogol derived from German Romanticism, but which he made incomparably his own.

The characteristic Gogol shared most deeply with German Romantics was his homesickness and sense of exile. Think of the central role that *Wanderlust* plays in the German literature that most influenced Gogol; consider the ritual role that travel plays in the Romantic imagination, as captured in Goethe's Roman sojourn, or the symbolic features of Winkelmann's pathetic death in Trieste, on the road finally to the actuality of Greece, the holy place which he, like Moses, would prophesy, but never see.

In the case of the Germans, as—with different nuances—the Russians, an enormous sense of lack is betrayed by all this travel and searching. In the hundreds of small duchies and principalities into which the German folk were splintered, it was impossible to erect a whole sense of self. The well-known "tyranny of Greece over Germany"[7] arose from a need felt by the most gifted Germans on the cusp of the nineteenth century to find a home where they could achieve a full identity of the kind they felt the ancient Athenians had enjoyed. The celebration of "the South," or, more specifically, of ancient Greece and Rome, was a characteristic way for German Romantics to express their longing for a nobler fate for themselves and the German people.

Gogol shared this overwhelming sense of longing. His last words, as he lay dying from the effects of a self-inflicted fast (and the brutality of his doctor), were: "A ladder, faster, give me a ladder!" It could be argued that he was once again seeking a means to climb out of this world of flux and imperfection to the higher realm of static beauty he had devoted his life to seeking. It is this sense of an ontological crack in the surface of existence that accounts for so much of the grotesque and pathos in Gogol's work. In the years from 1831 to 1834, when Gogol was still teaching history (after his return from the abortive German trip), he sought in the past the same kind of wholeness that served as the dream object for the Germans. Beginning with a plan to write a history of his native Ukraine, he soon moved on to a grander design—a universal history that would overcome his sense of schism in the world, bringing together all the scattered pieces. As he said, such a work "should gather into one all the nations of the world, divided by time, accident, mountains, seas, and unite them into one harmonious whole, forming out of them a majestically perfect poem."[8] In what follows, I will show how this sense of exile shaped Gogol's oeuvre in fundamental ways.

Of all the authors whose works are considered nineteenth-century classics, perhaps none has had a more mixed reputation than Gogol. His status as a pillar of Russian literature was secured early in his career, and has remained unshaken; it is not his eminence that is at issue. What is bewildering is the wide array of reasons that have been put forward over the years by admirers wishing to explain Gogol's place in the Russian pantheon. He was perceived to be master of a unique style by fellow writers such as Pushkin; thus his reputation as a serious master of technique was early established. The public at large, however, read him either as a homey Realist, a source of funny stories about eccentric country bumpkins, or as a chronicler of life among the expanding urban lower middle class. These two versions of Gogol found a sharper expression in two traditions that polemicized with each other after 1917: the Formalists took claims for Gogol's stylistic mastery to an extreme, while critics more sympathetic to the Bolsheviks pictured Gogol as a precursor of Socialist Realism, an observer of peasant life in the Ukraine, or a reporter of social injustice in St. Petersburg.

These developments took a further twist in the West, where, as was the case with Tomasso Landolfi, Gogol was perceived as a precursor of the Dadaists, a master of the grotesque. In the wake of Structuralist and Post-Structuralist emphasis on the linguistic substrate, Gogol's work has been seen as curiously post-modern, self-conscious parables allegorizing the pitfalls and absurdities of the arbitrariness of language.

One of the more curious aspects of this reception history is that for all its roiling diversity it has little scope for at least one significant version of Gogol, and that is Gogol's own. By invoking Gogol's 'own' view of his achievement, I mean writings such as his "Author's Confession," the added-on final "allegorical" scene to *The Inspector General*, or, most notoriously, *Selected Correspondence with Friends*, in which claims are made for political, stylistic, and, above all, *religious* significance in the Gogolian canon in Gogol's name. In this brief chapter, I shall argue that the time has come to reconsider the importance of such texts as they offer a key to understanding the distinctiveness of Gogol's oeuvre.

Since we enter murky depths by invoking Gogol's religiosity as a key to understanding his work—not just a better comprehension of the thematics of that work, but, more precisely, a better understanding of the *formal* elements in Gogol—I should begin by marking off my own understanding of 'religious' from those views that are more traditionally invoked in connection with Gogol. Scholars as disparate as Konstantin Mochulsky,[9] Lorenzo Amberg,[10] and Robert A. Maguire,[11] to name but a few, have addressed the presence of religious motifs in Gogol's works, usually relating back to the author's eccentric, but intense, identification with Russian Orthodoxy in one form or another. Maguire, for instance, writes that "Central to Gogol's view of the human condition is the idea of 'place' (*mesto*) ... Implicit here is the idea that each person has his proper place in the world, which is assigned to him immutably by God and defined according to geography, class, and vocation. This is a key idea of Orthodoxy …" (46).

Rather than looking for specific doctrinal associations in Gogol, I propose to go back to Mochulsky's doubts about Gogol's Christianity, concentrating on the aspect of transcendence in Gogol that rests "on a pagan basis," or more particularly, "the experience of cosmic horror and an elemental fear of death" (Mochulsky, 10 [a citation also used by Maguire]). That is, I believe the core of Gogol's religiosity is to be located beyond the level of doxology in any particular religion; the anxiety and distance that is everywhere in Gogol, even in his most strikingly comic effects, has roots deeper than religion, which is always only one of the ways in which the more comprehensive category of the sacred masks itself. My argument, briefly stated, will be that the animating principle in Gogol's life as in his work was a compulsion to overcome the yawning gap he perceived between the banalities and beastliness of the profane, and the heights of the totally other realm of the sacred. Failing to find links between the two, he condemned himself to death, driven to do so by the very intensity with which he sensed the cut-off, the same obsession that provides the specific color of 'Gogol' to his work.

A first step might well be to think of Gogol not as a historically instanced man who existed from 1809 to 1852, nor even as author of the works ascribed to his name, although both these aspects of his life cannot, of course, be ignored. What makes the existence of the man Gogol worth remembering, and what is significant in his fiction and drama, is a recurring pattern that we might rather conceive as a trope, a figure of great complexity, but one that is everywhere recognizable in both his life and his works. An opening suggestion, then, is that when later readers and critics invoke 'Gogol,' it is a trope they are naming. When Belinsky made his famous remark to the effect that "We all climbed out from under Gogol's overcoat," he was merely stating the pervasiveness of the figure I am calling 'Gogol.'

This figure can be characterized in a number of different ways. But despite its protean appearance, at bottom it is always a name for the ineluctability of distance, the tyranny of separation from desired ends, a trope figuring alienation and rift, perhaps best summed up in its obsessive enactment of incommensurability mandated by differences too great to be comprehended. Gogol is the archetypal Russian writer in this view, since in him a pattern that is present in most other classic Russian authors is most apparent. The figure of Gogol, understood as naming the condition of longing for reconciliation with an impossible and distant goal, manifests itself on several levels. In Gogol's biography it is obvious in the constant sense of discomfort and unease the writer demonstrates at every point in his career, the phenomenon, paraphrasing V. S. Naipaul, we might call the anxiety of *non-arrival.* Another name for the figure of Gogol understood in this way might well be 'homesickness.'

Homesickness that is more than a banal sentiment of longing for familiar haunts is complicated by the possibilities inherent in both the term 'home' and the term 'sickness.' Such a complication might well begin with certain aspects of the topic that are so simple they are often not taken into account; therefore, a touch of kitchen phenomenology may well be in order. The malady is first and foremost an instance of longing. It is an illness that derives from consciousness, for whatever else this vexed term names, it requires awareness *of* something. It always has two parts: a gap or separation of some kind and the desire to overcome the distance occasioned by that lack. So if we were to do an etiology of homesickness, we might say its cause is, *au fond*, a lacuna, helpful here in its Latin origin as 'lacuna,' the diminutive form of '*lacus*,' the lake: it is the little pool that is separated from a larger body of water by a strip of sand. The greater entity of which it is a part is simultaneous with the lacuna, within view, but separated from it. So homesickness is a hopelessly binary term. But the two divisions comprising it are unequal: it is not home that is sick, but the one who is *not* at home. Homesickness is, then, closely related to the uncanny, *das Unheimliche,* that version of suppression of which Freud says the essence is recognition of the familiar in a new and disturbing light.

If something like these elementary considerations are indeed the case, then it follows that the everyday garden variety of homesickness—experienced when we are away from the place where we normally reside and feel a pang of longing for its familiar material and psychological comforts—is an instance in small of a malady that is potentially far more serious. Taken to its most radical extreme, homesickness might be understood as the ultimate desire to overcome all difference and to return to a presumed state of wholeness in which difference has no power to effect separation that can engender longing—something closer to the wish for non-being in Freud's drive to death in *Beyond the Pleasure Principle,* than the uncanniness of his "Unheimliche."

Gogol began as a provincial who set out to make his way in the capital. The distance between the muddy village outside Poltava in the Ukraine where he was born and the glitter of St. Petersburg was something he never really overcame. He always felt uncomfortable among the brilliant circles into which his early fame catapulted him, where hostile critics sometimes ascribed his outlandish style to his not being a native speaker of Russian (the fact that 'Hohol' is a Russian term of abuse for Ukrainians was never completely suppressed by the softening of the final 'l' in Gogol's name). His uneasiness in the capital was compounded after he was quietly released from teaching history at the aristocratic Patriotic Institute, a subject about which he knew no more than most provincial schoolboys. This lack of expertise led him to invent fantastic events and theories whenever he did show up for lectures, and to create even more bizarre excuses when he did not, as more often than not he failed to appear.

The distance between Gogol and his audience was a reflection of the distance between his cosmic aspirations and mundane reality: when the public roared with laughter at his comedy *The Inspector General*, he reviled them for missing the point of the great lesson he was teaching in his play. When he sought to make clear what the core of his message for Russia was in *Selected Correspondence with Friends*, he merely compounded the confusion by enraging his audience with his gratuitous and often grotesque advice. In 1848, when Europe was convulsed with revolution and a reaction against the excesses of the so-called Holy Alliance, Gogol chose just that moment to visit the Holy Land on a pilgrimage he vainly hoped would provide him final illumination. He sought to convert the success of his novel *Dead Souls* into a trilogy that could replicate the three divisions in Dante's *Divine Comedy*. Discovering that the distance between his intention and his ability to carry out his intention was insurmountable, he burned the manuscript and virtually killed himself through starvation and willing participation in the schemes of his physician, a killer-quack so comic, and at the same time demonic, that one is tempted to regard him as another of Gogol's inventions.

The tyranny of difference, the homesickness that covets a return to an all-embracing oneness, is the absolute hallmark not only of Gogol's life, but also of his work (suggesting the relevance of another of Freud's concepts, the will to death that is even more powerful than the 'pleasure principle.'). There is an alienation built into the very poetics in which Gogol chose his genres, since they all explore the very edges of the forms they purport to be: his greatest drama ends in living statues; he calls his novel a *poem*; and his short stories are for the most part exercises in the bastardized form of the *Kunstmärchen,* tales that fall somewhere in between folklore and literary conventions, but most importantly a genre in which the borders between the real and the supra-real are broken down, resulting in devils, ghosts, and witches appearing in great modern cities instead of the never-never land of fairy tale settings. The unbridgeable geographical and cultural distance between Poltava and St. Petersburg is replicated formally in the generic distance between *Märchen* and *Kunst*.

And of course in the works that have so programmatically come to be called *The* Petersburg stories, the distance that comes with difference is everywhere. In "Nevsky Prospect," the distance exists between a young artist's exalted conception of a young woman he longs to worship and the reality of her actual state—she is a particularly dissolute whore in one of the city's more degraded bordellos. Coats separate from their proper owners, only to be seen again worn by ghosts, and kings of Spain turn out to be madmen with ridiculous names such as Poprishchin. Most notoriously, even noses are alienated from the faces of their putative owners to the point that they may even outrank them. Gogol is only being the classic realist within his personal cosmology when he writes, "Oh, do not trust that Nevsky Prospect!... all is a deception, all is a dream, all is not what it seems ... it lies at all times, this Nevsky Prospect!"

But of all the distances that haunt the figure of Gogol, none occurs more frequently than that separating the reality of Russia from the dream of what it should be. We enter here a strange landscape, where geopolitical space is grafted onto maps drawn for the kingdom of ends, the sphere of ethics, the land of *ought*. The first (and only completed) part of *Dead Souls* ends with two kinds of flight superimposed on each other. The first is a rhetorical flight, as the speeding carriage of the hero, Chichikov, is transformed into a metaphor, moreover a metaphor for Russia herself: "And art not thou, my Russia, soaring along, even like a spirited ne'er-to-be-outdistanced troika?... Whither art thou soaring away, then, Russia? Give me thy answer! But Russia gives none."[12] The second flight in this passage is the whooshing passage of 'Russia' as it disappears beyond the horizon. It fails to respond to Gogol's anxious questions because it is not only the figure of a metaphor, but the figure of Gogol, the trope of distance, in this case the incommensurable gap between Gogol's exalted vision of Russia and Russia as it was outside that dream. The distance between the two is so great that no communication between them is possible: there is no medium of exchange because there is nothing in the middle between the two. The pain of incommensurability can be glimpsed in yet another of Gogol's troika images, this time the one that sails through the concluding passage of "Diary of a Madman":

> What do they want of me? What can I give them? I have nothing. My strength is gone. I cannot endure this torture. My head is aflame and everything spins before my eyes. Save me, someone! Take me away!

> Give me three steeds [a troika], steeds as fast as the whirling wind! Seat yourself driver, ring out little harness bells, wing your way up, steeds, and rush me out of this world![13]

What flight means in such a troika is the ability to get from an intolerable quotidian to a holistic paradise. Gogol, in other words, suffered from a common writer's malady, one that occurred especially in Russia: not just homesickness, but the ultimate version of longing that can be satisfied only by reunion with paradise. Gogol was allergic to the world as he found it, and longed for one that was higher, better—indeed, a world that was perfect in its oneness. In order to understand better the pathology of this particular homesickness, it will be helpful to take seriously the spiritual claims made by Gogol, at least to the extent of exploring their filiations with religious thought, or, more precisely, with the *decline* of religious sensibility as it is enacted in the phenomenon of secularization.

The longing that defines the figure of Gogol is of such ubiquity and intensity that it can be understood only within the context of goals that are unachievable: it is a homesickness that at a deep level knows you cannot go home again … *ever*! The distance of the Gogol figure is, then, of the ultimate sort that exists between the unbridgeable categories of the sacred and the profane. Mircea Eliade has spoken of the two spheres as "two modes of being in the World … a precipice separates the two modalities of experience, the sacred and the profane."[14] Rudolf Otto has described it as "the completely other."[15] But no one has done more to articulate the complexities of the sacred than Durkheim, who argues that: "All known religious beliefs, whether simple or complex, present one common characteristic: they presuppose a classification of all the things, real [or] ideal, of which men think, into two classes or opposed groups, generally designated by two distinct terms which are translated well enough by the words profane and sacred."[16] He elaborates on the impossibility of mixing the two categories: the heterogeneity of the sacred and the profane

> is sufficient to characterize this classification of things and to distinguish it from all others, because it is very particular; it is absolute. In all the history of human thought there exists no other example of two categories of things so profoundly differentiated or so radically opposed to one another. The traditional opposition of good and bad is nothing beside this; for the good and the bad are only two opposed species of the same class, namely morals, just as [he adds in a remark not without

> significance for our discussion of homesickness] sickness and health are two different aspects of the same order of facts, life, while the sacred and the profane have always and everywhere been conceived by the human mind as two distinct classes, as two worlds between which there is nothing in common.[17]

The argument I am making here is that the various particular instances of homesickness in Gogol are aspects of the larger nostalgia he experiences for the sacred. The problem he confronts is a very old one—indeed, it could be argued (and has been argued in a series of works by René Girard,[18] and in a new book by Caesário Bandera[19]), one that is as old as the origin of human language: how to negotiate the distance between the sacred and profane, which by definition are absolutely opposed. In Plato's works, the medium of the space that exists between the perfect world of ideas, home of eidos, and the fallen world of human existence, where only mere reflections, eidola, are to be found, is never rendered conceivable.

But of course the problem has a history; it confronts different societies differently at different times. The easiest way to read the figure of Gogol is, then, as a particularly paradigmatic example of the effects wrought by secularism. In order to see why this easy take on the problem is wrong, or at least insufficiently theorized, we should distinguish between the historical phenomenon of 'secularization' and the possibly more interesting thought that 'the secular' may be a category in its own right, just as fundamental as the status assigned by Durkheim or Otto to 'the sacred' or the 'profane.' Let us first distinguish between secularism and the secular.

Secularism may be read as a story, the tale Richard Rorty has called "the typical German story of the self-consciousness of the modern age (the one that runs from Hegel through Marx, Weber, and Nietzsche) [that] focuses on figures who were preoccupied with the world we lost when we lost the religion of our ancestors."[20] As such, it has been appropriated almost unthinkingly as the latest chapter in the larger narrative of the conventional history of the West, precisely because we, as opposed to other parts of the world, are at the end of the process of secularism, resulting in the agonistic politics Samuel P. Huntington, director of the Olin Institute at Harvard, calls "the West vs. the Rest."[21]

More often than not, secularism has insufficiently been distinguished from its use as a historical marker, either at the level of theory, or at the level of practice. This state of affairs has resulted in problems of theory that still haunt profession formations in the social sciences, some of which, such as sociology, are unthinkable without the generative role of the secularization theory, or anthropology, which has never overcome the opposition between cultural relativism on the one hand, and, on the other, the necessary conviction that science, the great product of European Enlightenment, has a transcultural value as a good thing. The confusion that results from conflating secularism as a process with secularism conceived as a normative chronology is also largely responsible for the breakdown of narrative authority across the boards that is now swept under the conceptual rug of a catch-all 'post-modernism.'

At the level of practice, the imprecision conduces to the sinister politics found in rhetoric of the kind Paul Johnson, author of *The Birth of the Modern*, permits himself when he writes: "Unlike Christianity and Judaism, Islam has not passed through a period of Reformation or Enlightenment, with a consequent loss of pristine theocratic beliefs. But such a change is inevitable and is probably on its way already … [Islamic Fundamentalism] may not be a portent of a perilous future but the last eruption of an ancient way of looking at the world."[22] Secularism is a historiography that equates religion with superstition and thus relegates belief to an atavistic past separated from the enlightened present and even more enlightened future in which reason shall reign. As Tom Stoppard has one of his characters in *Jumpers* say, there came a day in the calendar when "the onus of proof passed from the atheist to the believer, when, quite suddenly, the noes had it." Perhaps it was just such a date Virginia Woolf had in mind when she famously announced somewhere that "on or about October 1910 human nature changed."

Conceived in such stark terms, secularism becomes an inert force field of either/or across which there can be no negotiation. The historical break that the term "secularism" names becomes a barrier that cannot be breached, separating as it does two different ontologies and two different ways of knowing the world that each ontology underwrites. As such, secularism becomes a powerful force of repression, one of the more insidious, diverse, and ubiquitous masks for censorship. It is not only a way to cut off possible connections not only between church and state, reason and superstition, but also a means to sever as well possible filiations between the sacred and the profane, for secularism assumes that the sacred is a category humankind has outgrown with the passing of religion. What we are left with, if only people in the Third World had the education to see it, is the kingdom of the merely profane. This is an absolutist creed that logically is no more well grounded than the attempt from the other side to proclaim in its lingering Neo-Platonism that only the sacred is real, with the realm of the profane conceived as a barrier of illusion to be combated and overcome. These are some of the absurdities of secularism.

In order to grasp better the phenomenon of cosmic homesickness of the sort we encounter in the figure of Gogol, it will be necessary to discard secularism, for too often it is conceived as no more than the sacred turned on its head, nothing more than a dead inversion of definitions. If we were to do so, would we have to discard the historical effects that have been ascribed to it? I believe not, if we choose rather to concentrate on the aspect of secularism that relates it to the larger phenomenon of the secular.

In the compass of so short an essay there is no scope to adequately develop this concept, but it might well be regarded as a category in the technical Kantian sense, since it is a transhistorical phenomenon that all religions have entertained as the complex space of negotiation between the timeless home of the gods and the history-drenched world of mere human beings, who are thrust into and snatched out of the flux of time. In its most brutalized truncation, the secular might be defined as the necessity of thinking the boundaries of boundaries, the manifold specific ways the general problem of separating the sacred from the profane gets resolved at punctual times in particular places. The trepidation I feel in introducing the secular as a category in this sense is only partly overcome by the knowledge that Durkheim himself was heavily influenced by (if only to resist) Kant's epistemology, as the theoretical introduction to *The Elementary Forms of Religious Life*[23] attests. The sacred and the profane operate in Durkheim's system precisely as categories that are prior to any specific, historically instanced manifestation of the difference between them. Since some kind of negotiation of that difference is a necessary component of conceiving sacred and profane as an ultimate dyad, it follows logically that such a negotiation, too, has the character of a category. If we accept this working definition, then it will follow that the secular is a version of logic, as Durkheim suggests when he says,

> Since the idea of the sacred is always and everywhere separated from the idea of the profane in the thought of men, and since we picture a sort of logical chasm between the two, the mind irresistibly refuses to allow the two corresponding things to be confounded, or even to be merely put in contact with each other; for such a promiscuity, or even too direct a contiguity, would contradict too violently the dissociation of these ideas in the mind. The sacred thing is par excellence that which the profane should not touch, and cannot touch with impunity.[24]

As a version of logic, the separation of the sacred and profane has historically shifted between two different strategies of formalization, both well known in the sphere of logic as the syllogism and the enthymeme. A syllogism is a form of reasoning that consists of three parts: a major premise, a minor premise, and a conclusion. The classic example, deriving from Quintilian but quoted from Kiesewetter in Tolstoy's *Death of Ivan Ilych*, is: major premise, all mortals die; minor premise, Socrates is a man; conclusion, Socrates will die. The enthymeme, by contrast, is a form of reasoning that consists of only two parts; it is a truncated form of the syllogism that excludes one of the syllogism's premises. An example from literature is provided in Shakespeare's *Timon of Athens*, when Timon asks Apemantus, "Why dost thou call them knaves? Thou knowst them not." To which Apemantus replies: "Are they not Athenians?" And when Timon replies that they are, Apemantus says,"Then I repent not." The suppressed premise is that all Athenians are knaves.

The enthymeme is preeminently a form of the excluded middle. It is significant, I think, that 'thymos,' the Greek root of enthymeme, also grounds the psychiatric disorder cyclothymia, a bipolar disorder characterized by instability of mood, with swings from depression to euphoria. This illness, too, is organized around the absence of a middle. As the figure of excluded middles, the enthymeme is the most frequently encountered construction given to the sacred/profane polarity. But the middle that is excluded does not thereby cease to be; those who conceive the world in this way by suppressing thirdness, suppress themselves. By modeling the distinction between sacred and profane as an absolute cutoff, radical cutoffs become obligatory: the question is not only what term can encompass so radical a polarity (that is, what is the envelope that can hold both concepts, even when they are structured as mutually exclusive), but who can think it? Where is the subject—lost in the logical black hole that separates the two?

Enthymemic models of sacred/profane thinking are subject to a kind of willed amnesia; they suppress the ineluctable threeness of all comparisons (for that is what all distinctions come down to in the end). There is another relevant form of threeness we should consider here, one known in rhetoric—an ordering system that is closely related to logic—as metaphor, in which two things are compared not to end in the conclusion that they are finally incomprehensible, as in the enthymemic contrast of sacred and profane. The metaphor has the opposite tendency; its closeness to dialectic is made clear in its structural assumption that in its yoking of two things a third will emerge, a new meaning that was not present by itself in either of the things it brings together. The *tertium comparitionis* of the trope "Achilles is a lion" is an Achilles different from the hero before he was figured in metaphor—and so, of course, is the lion. The secular, then, may be conceived as an operation that is both syllogistic *and* metaphoric insofar as it posits the necessity of threeness in all comparisons, especially in thinking about the relation between the sacred and the profane. An important consequence of this distinction is that the secular so conceived is not antithetical to the sacred as such: it is still with us, if in different form, as the historically shifting name of the category of difference beyond which mediation becomes impossible, the outside envelope of thinking differences. We may no longer conceive the irreconcilability of the world to equate with our desire as a gap between the sacred and the profane, but the tyranny of difference is no less diminished.

One of the ways in which the transhistorical problem of negotiation we have been calling 'the secular' has been historicized in a particular period is in nineteenth-century secularism, and one of the effects of the secular so conceived in this period is the rise of nationalism. There is no cause and effect relation between the decline of institutionalized religions and the rise of nationalist consciousness, there are some linkages nevertheless that may help us understand the distinctive relation between Gogol's religious impulse and his investment in the high fatedness of Russia. As a leading student of the subject has pointed out,

> in Western Europe the eighteenth century marks not only the dawn of the age of nationalism but the dusk of religious modes of thought. The century of the Enlightenment, of rationalist secularism, brought with it its own modern darkness. With the ebbing of religious belief, the suffering which belief in part allayed did not disappear. Disintegration of paradise: nothing makes fatality more arbitrary. Absurdity of salvation: nothing makes another style of continuity more necessary. What then was required was a secular transformation of fatality into continuity, contingency into meaning ... It is the magic of nationalism to turn chance into destiny.[25]

Thus we may speculate that at the heart of Gogol's immense sense of division in the world, between this world and a higher order of things, and finally between his exalted vision of Russia and the venality of the Russia he saw every day around him is a structuring principle. It is this fundamental force of structuralization, not the structures themselves, we are seeking to identify. Gogol is the name of a figure who has come to represent a historicized version of enthymeme, a figure of longing for unity and wholeness whose unrealizability results in the obsessive emphasis on gaps, differences, and incommensurabilities in his works. It is in Gogol's constant staging of lost harmony that I would locate the relevance of Durkheim's concept of meditating the sacred and the importance of viewing the secular as a category devoted to joining the timeless world of ideals with the contingent world of everyday life.

It was in this context that Durkheim sought to overcome the alienation and loneliness of the individual by resuscitating a different sense of community. Recognizing that religion had been the traditional glue that had held communities together, he accepted the view that the decline of religion (whether believed to be a good or a bad thing) had left a great sense of emptiness that successor ideologies had been unable to mitigate, resulting in what a later sociologist would call 'the lonely crowd.'

It was in his attempts to rethink the nature of relations between individuals and groups that Durkheim turned to a concept that preceded religion as a cohesive force. His idea, shamelessly stripped of its subtleties, might be outlined as follows: although the binding force of religion has since the Enlightenment been lost to the community, the kernel of religious experience—the completely other of the sacred—is still with us, though screened in a number of different ways. If we understand that what gave religion the power to shape collectivity was not its doctrines but rather its rites and the joint performance of worship, we shall be in a better position to understand how the sacred works to erase individuality and create the sense of sharing that is necessary if communities are to have common values. The rituals of tragedy and religion were intended to negotiate the difference between the sacred and the profane, to create ways that the sacred, without ever being broached as such, could at least be appropriated to forms of representation that would provide the illusion of minimal control over the forces of the unknown, the unspeakable realm of the sacred. In his study of Australian aboriginals and through Marcel Mauss's studies of Eskimos, Durkheim became convinced that certain properties pertained to the sheer act of assembling for ritual events. The intensity of such meetings had the effect of convincing those attending them that they had emerged out of the dull repetition of quotidian life, erupting into a different, higher reality. He describes a fire ceremony of the Warramunga:

> Commencing at nightfall, all sorts of processions, dances and songs had taken place by torchlight; the general effervescence was constantly increasing. At a given moment, twelve assistants each took a great lighted torch in their hands, and one of them holding his like a bayonet, charged into a group of natives. Blows were warded off with clubs

> and spears. [A] general mêlée followed. The men leaped and pranced about, uttering savage yells all the time; the burning torch continually came crashing down on the heads and bodies of the men, scattering lighted sparks in every direction. "The smoke, the blazing torches, the showers of sparks falling in all directions and the masses of dancing, yelling men," say Spencer and Gillen, "formed altogether a genuinely wild and savage scene of which it is impossible to convey any adequate idea in words."[26]

This extraordinary scene of heightened feeling is used by Durkheim to explain the role of the sacred in creating communal bonds. "One can readily conceive how," he adds,

> when arrived at this state of exaltation, a man does not recognize himself any longer … It seems to him that he has become a new being … And at the same time all his companions feel themselves transformed in the same way and express this sentiment … everything is just as though he really were transported into a special world, entirely different from the one where he ordinarily lives … *How could such experiences as these … fail to leave in him the conviction that there really exist two heterogeneous and mutually incomparable worlds? One is where his daily life drags wearily along; but he cannot penetrate into the other without at once entering into relations with extraordinary powers that excite him to the point of frenzy. The first is the profane world, the second, that of sacred things.*" (249–50, emphasis added)

The important point in this analysis for understanding Gogol is the connection between the wholeness of the group and the experience of a reality perceived as totally different from everyday life because it is more intense and heightens all aspects of perception. It is from the experience of the group that an individual derives the most powerful sense of an other so absolute that it gives a new context to the lesser otherness of other individuals, putting that alterity into a greater whole, and thus, in some measure, erasing it. The sacred's quality of otherness has the effect of fusing the individual with the group because it provides a completely different ratio of difference: as distinct from each other as fellow human beings may be, they can never be so very removed as the utter difference of the sacred. We therefore enjoy a common bond in the face of so immense an otherness. This sense of the quality of utter difference in the sacred requires the experience of group ritual, a shared activity, whose intensity can then be drawn on as spiritual capital to underwrite the values of the community. It is the contrast between the sacred and profane worlds that models the difference between the individual and the group, permitting a negotiation of the difference that otherwise would be impossible. Thus the wholeness of the community is a function of the sacred's power to fuse individuals into a group, just as the sacred is a function of the group. The two are inseparable. Moreover, understood in this way, i.e., the group as individual subjects whose being is modified by the effects of sharing in rituals, it can be seen that the collective thus constituted by the sacred is the source of all that is new or the highest in the community, since it is the site where individual subjects are taken literally out of themselves.

It is important to see that this essentially religio-sacral explanation for society is one that is not itself religious in the sense that it invokes the sacred as a category in the history of social development. It does not depend on any particular definition of the transcendent (as mediated by particular religions), positing rather the interdependence of the group and the sacred: if there is a god in Durkheim, it is society, the community that shapes values, and without which life becomes literally meaningless.

If, now, we return to Gogol, it will not take much reflection to see that his most characteristic scenes are the most quotidian. What is memorable in them is what in "Old World Landowners" he openly analyzes as habit: the concern for eating and drinking that dominates the existence of his characters in the *Dikanka* and *Mirgorod* cycles, and the obsessive daily rituals of his petty clerk characters in the later tales, such as Akaki Akakievich's need, in *The Overcoat,* to do copying, even when he is not in the office, or Poprishchin's compulsive routines in "Diary of a Madman." What is important about this emphasis on bodily needs and predictable cycles of behavior is that it underscores the depth of immersion of Gogol's characters in the repetitive, unthinking patterns of life, a level of existence not far removed from that of ruminant animals. The absolute lack of any heightened experience in their lives (remember that the sudden disappearance of the old world landowner's cat is enough of a break in the characters' habits to precipitate their decline and death) dramatizes the absence of the sacred and, in Durkheimian terms, of community. Gogol, in other words, is wrestling not the angel of the lord, but the category of the sacred.

How this plays out in a specific example may be seen in the instance of Akaki Akakievich's famous appeal to his fellow workers in *The Overcoat.* The passage, which will be known to most readers by heart, goes like this: after describing several indignities visited on Akaki, the narrator says, "Only if the jokes became too unbearable … would he say, 'Leave me alone, why do you torment me?' There was something strange in these words and the way he said them. His voice had a peculiar sound which made you feel sorry for him, so much so that one clerk who was new to the department, and who was about to follow the example of the others and have a good laugh at him, suddenly stopped dead in his tracks, as though transfixed, and from that time onwards saw everything in a different light. Some kind of supernatural power [*kakaia-to neestestvennaia sila*] alienated him from his colleagues ... And for a long time afterwards, even during his gayest moments, the clerk would see that stooping figure with a bald patch in front, muttering pathetically, 'Leave me alone, why do you have to torment me?' And in these piercing words he could hear the sound of others, 'I am your brother.'"[27]

What we have in this much commentated upon passage is a textbook example of what Durkheim meant when he wrote that "... man is double. There are two beings in him: an individual being which has its foundation in the organism and the circle of whose activities is therefore strictly limited, and a social being which represents the highest reality in the intellectual and moral order that we can know by observation—I mean society."[28]

That is, we are all constituted of differing ratios of polarity between impulses that derive from the unique place in existence we occupy as biological organisms, and other impulses that derive from the model of reality we carry around in our heads, a collective representation that is shaped by the group around us—what Durkheim calls society. In the scene from *The Overcoat,* we see the young clerk as he is first of all moved by the impulses coming from his immediate environment—the company of other young clerks who have harassed Akaki for years. As a new member of this community, the clerk "was about to follow the example of the others," but has a change of heart when he hears Akaki's plea. At that point the clerk hears another voice, which stopped him "dead in his tracks, as though transfixed," and he desists from provoking Akaki. What he hears in that voice is, of course, the judgment of the social, the value which says that it is wrong to make unprovoked attacks on helpless innocents. This voice is stronger than even the voice of the immediate community of clerks, who choose not to hear, or to ignore, the injunction of the larger community. The polarized subject that Durkheim posits is illustrated in the ratio between self and other that the stricken clerk manifests. Between the two external claims to community—on the one hand, with the group of fellow clerks he has just joined with whom he can bond by tormenting Akaki, on the other hand, the larger group of society at large that says to do so would be wrong—there is a subject who makes a choice. It is the necessity of choosing between conflicting social demands and the capacity to do so from a particular subject position that defines what Durkheim means when he says 'man is double.'

Gogol has been criticized for stylistic excess in this passage, and indeed, it is rhetorically overcharged; but I would argue that its very markedness is an indication of the importance Gogol attaches to the scene. Beneath the overdetermined pathos of the passage, under the layers of sentimentality, is the key to the larger phenomenon of homesickness in Gogol, one whose importance cannot be overstated: the ineluctable need to overcome the gap between the individual subject and one or another of the communities into which such a subject is cast. The home for which Akaki, the clerk who desists from mocking him, and Gogol himself all long is the home constituted by a seamless fit with a community, a social space in which decisions are not necessary and—because one has completely fused with the group around him—there is no gap between the self and the other.

"Some kind of supernatural [or unnatural] power" alienates the young clerk who desists tormenting Akaki from his colleagues, and we are in a better position to recognize what that power is: it is the power precisely *not* of nature (*neestestvennaja*), but of its opposite, culture, as it is inscribed into social norms. When the clerk seems to hear Akaki cry out "I am your brother," the voice the stricken young man hears is not Akaki's, but the voice of the social norm. Gogol is quite accurate in assigning to this force a supernatural origin, for it is Durkheim's argument that the voice of the community has been traditionally encoded as the voice of God: religious rites are the means by which humans represent their relations not to a transcendent deity but rather with each other. Values are collective representations that derive from the group's ideas of itself as a group; they lay out the complex space that encloses the individual and his or her community. Such collective representations are by definition extra-individual, making themselves an impersonal force. And this impersonality is then transformed into terms that traditionally have been transcendent, mysterious, religious, with all the explanatory power and supreme authority that attaches to the sacred.

Some of the most typically Gogolian of Gogol's tales can be read almost as parables of the social, dramatized as the breakdown of the community. This breakdown between the individual and the group is then represented as a breakdown in communications: the supernatural voice of social norms can no longer be heard, and thus there can be no shared language. In "Diary of a Madman" Poprishchin's madness is, of course, the vehicle for dramatizing the ripping of threads that connect the individual to the group, but that madness, in turn, is manifested in the form of a breakdown in communications. Not only is Poprishchin unable to share thoughts and values with others, but even the dogs he invents to screen his perceptions are unable to understand each other in their correspondence. At several points in Poprishchin's increasingly incoherent diary, he resorts to the repeated phrase, "Nothing ... nothing ... silence!" These exclamations more often than not burst out when Poprishchin is thinking of women, as in the entry for 8 November, when he writes: "A lot of amusing plays are being written these days. I love going to the theater. As long as I have a penny in my pocket you can't stop me. But these civil servants of ours are such ignoramuses, you'd never catch such peasants as *those* going to the theater, even if you gave them a ticket for free. One of the actresses sang very well. She reminded me of ... ah! Canaille! It's nothing, nothing! Silence!"[29] And in the entry for 11 November, he describes a visit to his superior's office, during which, while passing the boudoir of the director's daughter, he imagines "one peep into her bedroom to see what wonders are there, sheer paradise, more blissful than heaven ... [I can imagine her foot as she climbs out of bed]. And then, over that tiny foot, she begins to pull on her snowy white stocking. It's nothing! Nothing! Silence!"[30] And in the entry for 13 November, when he reads that one of the dogs whose correspondence he has been reading refers to the director's daughter ("My mistress, whom Papa calls Sophie, is passionately fond of me," the dog Madgie writes), he says again, "It's nothing! Nothing! Silence!"[31]

Now, much could be made of the sexual repression that is so patently a part of Poprishchin's use of the *"Nichego, nichego ... molchanie!"* formula, invoked as it is in connection with women. But union with a woman may also be viewed—especially in the context of "Diary of a Madman," where Poprishchin's object of desire is the director's daughter—as a longing for union with the other. The class associations merely underscore the presence of the social. Thus, the silence that desire always occasions in the story figures the absence of communication between Poprishchin and the world around him—the absence of anything that can mediate between his individual place and the community. His condition of eccentricity leads to the depths of madness, a madness, moreover, that is defined by Poprishchin's conception of himself as King of Spain, marking not only his condition as foreigner within his own group, the Russian state, but his imagined bonding with another group: as their king, he is the personification of the Spanish community. But whether in Russia or "Spain," Poprishchin is 'not at home.' He is an outsider because he fails to find a means for negotiating the space between his individuality and the collective, between the forces Durkheim posits as the basis of the sacred/profane distinction. He is another victim of the excluded middle, unable to operate in the arena of the secular, the space of between.

If this melding of social and sacral terms initially appears jarring, consider the even more openly parabolic exploration of that blending in "How Ivan Ivanovich Quarreled with Ivan Nikiforovich." The whole tale is organized around a dizzying catalogue of binary oppositions, beginning with the title. As the story opens, a long series of comparisons and contrasts is laid out: the narrator says, "In spite of their great friendship, there were differences between the two. The best way to find out about their characters is to compare them." But these contrasts and comparisons turn out to be less than meaningful: "Ivan Ivanovich's head is like a radish with the tail pointing downwards; Ivan Nikiforovich's like a radish with the tail sticking up.... Ivan Ivanovich is furious if a fly falls into his borscht. He loses his temper completely, throws the plate up in the air, and heaps abuse on his housekeeper. Ivan Nikiforovich is terribly fond of having a bath, and when the water reaches his neck he has a table and samovar put in the tub, for he loves drinking tea while cooling off. Ivan Ivanovich shaves twice a week, Ivan Nikiforovich once...." [32] And so it goes—similarities and differences that are all equally non-significant: they represent, if you will, a kind of pure opposition, insofar as the only clear-cut distinction between them, for all of the non-signifying details piled up by the narrator, is the simple fact that they are not the same.

They are algebraic markers of sheer difference, a fact of the tale rendered even clearer by the opposition that develops between them after the quarrel that constitutes the main event of the story. The causes of their dispute are as meaningless as the factors adduced by the narrator's catalogue of details as he compares and contrasts them in their pre-quarrel phase. When Ivan Ivanovich is rebuffed in his request for an old gun of Ivan Nikiforovich, words are exchanged; the former is called by the latter a goose, and they part in anger. The ensuing legal dispute becomes a structural metaphor for the impossibility of overcoming the absolute difference between the two Ivans.

Above and beyond the sheer fun of Gogol's brilliant parody of legal language in the depositions that each of the former friends prepares to submit to the local court, there is a disturbing undercurrent of arbitrariness in the separation between the two. These documents are masterpieces of nonsense, with sentence after sentence of clotted meaninglessness. Ronald Wilks gives something of the flavor in his rendering of the letters, from which I adduce the example of Ivan Ivanovich's deposition, part of which reads:

> This same indecent and gross gentleman did attempt to damage my property, inherited by me from my father, Ivan Perepenko, clergyman and son of Onisyev, of blessed memory, in that, in flagrant violation of the law he erected a goose shed directly in front of my porch, with the sole intent of rubbing in the insult for this aforementioned shed up to that time had stood in a very convenient position and was quite strong into the bargain. But the vile motive of the aforementioned gentleman consisted in this alone: viz: to make me a direct witness of indecent action, for it is known to everyone that no man goes into a shed—and least of all a goose shed—for reasons that can be mentioned in polite society. In carrying out this illegal act, the two front posts encroached on my land, handed down to me by my father when he was still alive, Ivan Perepenko, son of Onisyev, of blessed memory, starting from the granary and extending in a straight line to the place where the women wash their pots.[33]

The absurdity of this style makes clear once again that in such highly marked examples of non-communication Gogol is not merely making jokes or dramatizing the low intellectual gifts of his characters, but rather pointing to the void that exists between human beings. That chasm is so wide that the most powerful institutions that society has developed to mediate differences, such as the law, may be seen as having the opposite effect. There is no uniting ideal that holds people together in Gogol's vision, thus all the mechanisms for binding individual subjects are merely that—dehumanizing and mechanical, and thus giving the appearance of an uncontrollable arbitrariness. They are merely conventions, lacking the deep connections that only the shared ideals of a tightly knit community can sustain.

The particular device Gogol chooses to embody the breakdown of community in this tale is one he employs in several others, but nowhere else is it so paradigmatically schematized: it is the device of hyperbolizing the absence of clear-cut value distinctions. This device is at the heart of the original disagreement between the two Ivans, who haggle over the value of the gun that Ivan Ivanovich covets, but which Ivan Nikiforovich refuses to let go. It should first of all be noted that the rifle is old and missing parts, and is neither useful for its purpose of shooting, nor beautiful as an object. Nevertheless, Ivan Ivanovich says, "If you don't want to give me the rifle, how about taking something for it in exchange?" He begins by offering a pig, but his friend refuses: "There's nothing complicated about a rifle, but the devil knows how a pig is going to turn out!" In addition to the pig, two sacks of oats are added, but Ivan Nikiforovich responds, "Who ever heard of swapping a rifle for two sacks of oats?"[34]

suggesting that the value his neighbor has put upon the rifle is wildly off the mark. Now, it could be argued that this dialogue is simply a haggling over price that is perfectly logical: the owner of the desired object is seeking to get as high a price as he can for it. But the whole exchange, which is too long to reproduce here, is characterized by grotesque misunderstandings, culminating in the fateful break between the two.

Let us recapitulate what happens in this scene, since it contains the paradox that accounts for Gogol's homesickness. Schematically reconstructed, the scene is one in which two friends part because they cannot agree on values. This much is obvious. What is less obvious is that the values at issue are invoked in connection with objects that have little or no value in themselves—a useless gun and a brown pig. Where could there be any commensurability? So lacking in intrinsic value, these objects cannot mark clearly defined differences, resulting in a condition in which distinction is therefore arbitrary. Restating the paradox, we might then say that the lack of clear cut differences in value that is highlighted in the haggling bout is precisely that difference between the erstwhile friends that results in the ultimate difference of their separation. In other words, no difference, expressed as the inability of values to define mutually acceptable distinctions, is the very thing that causes the difference that drives the two Ivans to break their bond of friendship.

The two Ivans have trouble agreeing on values that can be shared because they lack connections to a larger community that could dictate the worth of things unambiguously to those who are its members, defined as such precisely by means of the values they share. The larger community that should so be defined is aware that the quarrel between the two men has a significance for it that requires the community to intervene. The social world containing the two Ivans is made present in two ways: in its formal institutions, most spectacularly represented by the legal system; and through the informal attempts of the village to bring the two Ivans together after their quarrel.

The failure of the legal system is dramatized not only in the incomprehensibility of its language (it is a communal resource that produces failed communication as its most spectacular product). Just how unsuccessful the law is as a force for social cohesion is demonstrated when—just as the judges are packing the various bribes of food left by the two Ivans—a pig jumps onto the table and runs off with one of the depositions: "Seizing the document, the brown porker [*xavron'ja*] ran off so fast that not one of the officials could catch it, despite the inkwells and rulers they threw at it."[35] That more than mere comic effect is at issue in this scene is made clear when court officials later visit Ivan Ivanovich, having come to inform him that he has 'broken the law' insofar as it was his pig that ran away with the official document. What is underscored in Gogol's telling is precisely the breakdown of values that the law should be upholding. The confusion of values emerges when Ivan Ivanovich says, "Thanks very much for putting me on the same level as a pig!" (226). The mayor (who learned to write longhand only in his thirtieth year), as the voice of the community, answers, "… the law states 'those guilty of theft …' Listen carefully: *guilty!* No species is mentioned, or sex or rank, therefore an animal can be judged guilty" (227). This amusing point (and one that had been a large issue in medieval law) is most telling in the ambiance of the whole tale as yet another indication of the breakdown of the powers of discrimination due to the absence of clear-cut distinctions. It is not by chance that the pig in question is the very same animal that was originally offered for Ivan Nikiforovich's gun, an incident in which the confusion of values led to the breakdown of exchange that initiated the quarrel that is the main event of Gogol's tale. That the ability of the law to make the community cohere has broken down is emphasized again when the mayor and Ivan Ivanovich work out a deal stipulating that the pig will be killed, but its owner will send the mayor some of the sausages made from the meat of the dead criminal (228).

The second way in which the social world constituting the community around the two Ivans is made present is through the efforts of the village notables to reconcile the combatants. The two Ivans are described as "the pride and glory of Mirgorod" (208). The town recognizes that the division of the two is an affront to the cohesiveness of the community: "They refused to see each other again and broke off all relations, yet these two had been friends who previously had always been thought of as inseparable.… When I heard about the quarrel [says the narrator], I was thunderstruck. For a long time I just couldn't believe it: Ivan Ivanovich had quarreled with Ivan Nikiforovich! Such a fine pair! Is there nothing that endures in this world?" (208). Aware that the rupture in relations between the Ivans is an affront to the shared values of the community, various attempts are made to reconcile the two, culminating in a grand feast. The citizens trick both Ivans into attending the mayor's reception, and for the first time since their break they occupy the same time and space. Each tries to avoid the other, but before either can leave, the assembled guests get behind them and physically push them together: "This is the usual way peace is made in Mirgorod" (239). But just as the two start talking to each other, Ivan Nikiforovich repeats the word 'goose' that had sparked the original split between the two: "Everything went to the devil ... all was at an end!" (240–41). The quarrel escalates in the courts, dragging on for ten years.

Now, I have dwelt at length on the way in which the absence of shared values in the community serves as the obsessively recurring subject of this tale to illustrate my opening thesis, that what is at issue in Gogol is his profound sense of the absence of the sacred, in the specific sense of a group whose cohesion is forged in awareness of something so utterly different from it on a vertical plane, that the troubling differences between individuals on a horizontal plane pale in significance. He is homesick for the wholeness of the sacred, a longing that is clearly present in the conclusion to the story of the community that is broken by the quarrel of the two Ivans.

Immediately after the spectacular failure of the community to bring the combatants together at the mayor's reception, there is a break in the narrative. The narrator appends a kind of coda to the tale, organized around a visit to the town five years after the unsuccessful ritual of reconciliation. The scene could not be more lugubrious: "I had picked a bad time. It was autumn, and the weather was miserable, damp and misty, with mud everywhere. A kind of unnatural green substance—produced by the continuous, tiresome rain—covered the fields with a watery pattern.... The remains of wattle fences and hedges stuck out depressingly" (242).

What is most significant for our analysis is that the narrator's visit coincides with a holiday, which we may assume is a holy day, since he goes first to the church. As depressing as the rain in the foggy, muddy countryside may have been, the church is worse: "The church was almost deserted, and there was almost none there at all. Evidently the mud had frightened away even the most pious. The candles looked strangely unpleasant in that gloomy, rather sickly light. The dark porch had a melancholy look, and the oblong windows with their circular panes were streaming with tears of rain" (243). In this temple of sadness, the narrator encounters the faded figures of the two Ivans, in different parts of the church, of course. Each assures him that the lawsuit will be settled soon, and in his favor. Completely depressed, the narrator retreats to his carriage and drives out of town: "The dampness soaked right through me. The wretched road with its sorry sentry box, with an old soldier sitting there repairing his rifle, slowly passed by. Those fields again ... drenched cows and jackdaws; monotonous rain, and a tearful sky without one ray of sunlight shining through the clouds.... It's a depressing world, gentlemen" (243).

The church, deserted except for the two quarreling opponents, could not be a more paradigmatic example of Gogol's despair: the place of the sacred, the site where community is forged, is empty even on a ritual occasion. Instead of serving to bring individuals together into a larger whole, it is haunted by the wraith-like figures of the two Ivans, the very icons of the community's failure to achieve cohesiveness. The home Gogol longed for, the togetherness of the sacred, looms farther and farther in the distance. Indeed, it is a depressing world.

In closing, let me suggest what some of the immediate implications of such a view might be for contextualizing the figure of Gogol. At the most specific level of close reading, it suggests a connecting link between all those instances in Gogol's texts in which communication fails, such as the infamous exchanges between Akaki Akakievich, his colleagues, and everyone else in *The Overcoat;* the obsession with meaningless difference in the tale of the two Ivans; the hilarious misdirection that clouds conversation between Ivan Shponka and his aunt; or the whole structure of misunderstanding that governs *The Inspector General.*

But at a more general level, the figure of Gogol, as exemplum of the homesickness for God, may be read as a cautionary tale about the dangers of enthymemic thinking as they have dominated the Russian imagination, the insistence on ends and goals being so exalted that they exclude any middle ground between them and those fallen creatures who dream their possibility. Is such longing not central to the fevered dreams of the Nihilists, of the spiritual as well as political Bolsheviks who sought in the Revolution to transform Russia from a land of contingent randomness into a necessary condition of a future that would see the end of time? Is the figure of Gogol not the secret legislator of the twentieth-century avant-garde's contempt for the present, and its impatience to break through to a higher reality?

Gogol's last words as he lay dying were "A ladder, faster, give me a ladder!"[36] Various constructions could be put upon these words, but in light of the secular, it could be said that here was one last call for a means to get from the profane world to the sacred home of oneness for which Gogol always longed.

Notes

1. A text that makes this point with overwhelming erudition is the still useful Zhirmunski volume on Goethe in Russia: V. Zhirmunski, *Gete v russkoj literature* (Leningrad, 1937). See esp. pp. 9–38 and 91–224.
2. Cf. Dmitri Tschizewski, *Hegel in Russland* (Halle, Inaug. Diss., 1934) and the collection the same author edited: *Hegel bei den Slaven,* 2nd ed. (Bad Homburg, 1961). Still of use, too, is Wsewolod Setschkareff's dissertation, *Schelling's Einfluss in russischen Literatur der 20-er und 30-er Jahre des XIX Jahrhunderts* (Berlin, Inaug. Diss., 1939).
3. Nicholas V. Riasanovsky *Nicholas I and Official Nationality in Russia, 1825–1855* (Berkeley, 1969), p. 102.
4. Quoted in G. P. Gooch, *History and Historians in the Nineteenth Century* (Boston, 1968), p. 17.
5. Vladimir Nabokov, *Nikolai Gogol* (New York, 1944), p. 156.
6. The case for this view is made in my ancient article, "The Devil in Mufti: The *Märchenwelt* in Gogol's Short Stories," *PMLA,* vol. 82, no. 5 (October 1967), pp. 352–62.
7. The title, of course, of Eliza Butler's famous book. Cf: E. M. Butler, *The Tyranny of Greece Over Germany: A Study of the Influence Exercised by Greek Art and Poetry Over the Great Writers of the Eighteenth, Nineteenth, and Twentieth Centuries* (New York, 1935).
8. Quoted by Leonid I. Strahovsky, "The Historianism of Gogol," *American Slavic and East European Journal* (October 1953), p. 363.
9. *Duxovnyi put' Gogolia* (Paris, 1934).
10. *Kirche, Liturgie, und Frommigkeit im Schaffen von N. V. Gogol* (Bern, 1986).
11. "Gogol and the Legacy of Pseudo-Dionysius," in *Russianess:Studies on a Nation's Identity* [*Festschrift* for Rufus Mathewson], ed. Robert A. Belknap (Ann Arbor, 1986), pp. 44–55. Mathewson might have found much to support his view in Amberg, whom he does not cite; he also makes the common mistake of assuming that the Pseudo-Dionysius is identical with Dionysius the Areopagite, whom St. Paul is said to have converted in Athens.
12. *Sobranie sochinenii v shesti tomax*, ed. A. Slonimskii (Moscow, 1953), vol. 5, p. 259.
13. "Zapiski sumasshedshego," *Sobranie sochinenii v shesti tomax,* vol. 3, p. 194.
14. Mircea Eliade, *Le sacre et le profane* (Paris, 1970), p. 17.
15. Rudolf Otto, *The Idea of the Holy,* trans. John Harvey (New York, 1967).
16. Emile Durkheim, *The Elementary Forms of Religious Life*, trans. J. W. Swain (New York, 1915), p. 52.
17. Durkheim, *Elementary Forms,* pp. 53–54.
18. René Girard, *Things Hidden Since the Foundation of the World,* trans. Stephen Bann and Michael Meteer (Stanford, 1987).
19. Cesario Bandera, *The Sacred Game: The Role of the Sacred in the Genesis of Modern Literature* (University Park, PA, 1994).
20. Richard Rorty, "Habermas and Lyotard on Post-Modernity," in *Zeitgeist in Babel,* ed. Ingeborg Hoesterey (Bloomington, 1991), p. 91.

21. Samuel P. Huntington, "The Clash of Civilizations?" *Foreign Affairs* 72 (Summer 1993).
22. Paul Johnson, review of *The New Cold War* by Mark Jurgensmeyer, *The New York Times Book Review*, 9 May 1993, p. 7.
23. See especially the pages devoted to a critique of 'apriorism,' as on pp. 26–31. Another relevant text here is Durkheim's *Sociology and Philosophy*, trans. D. F. Pocock (New York, 1974), esp. section 2 of Part III, "Individual Reason and Moral Reality."
24. Durkheim, *Elementary Forms*, p. 95.
25. Benedict Anderson, *Imagined Communities: Reflections on the Origins and Spread of Nationalism* (London, 1983), p. 19.
26. Durkheim, *Elementary Forms*, p. 249.
27. *Sobranie sochinenii v shesti tomax*, vol. 3, p. 132.
28. Durkheim, *Elementary Forms*, p. 29.
29. *Sobranie sochinenii v shesti tomax*, vol. 3, p. 179.
30. Ibid., p. 180.
31. Ibid., p. 182.
32. Ibid., pp. 196–97.
33. *Diary of a Madman and Other Stories*, trans. Ronald Wilks (Harmondsworth, 1972), p. 134; *Sobranie sochinenii v shesti tomax*, vol. 3, p. 218.
34. *Sobranie sochinenii v shesti tomax*, vol. 3, pp. 205–6.
35. Ibid., p. 223.
36. Vsevelod Setchkarev, *Gogol: His Life and Works*, trans. Robert Kramer (New York, 1965), p. 91.

Chapter 6

The Tyrannical "Knout" of World History

Russia in the Writings of Heinrich Heine

Roger F. Cook

In Heine's writing it is common to find in association with a person, nation, or event a recurring topos that fixes and often reduces the signification of its referent to a particular thematic context. In the case of Russia, Heine, who frequently employed animal figures for such purposes, made surprisingly little use of the wolf or the bear to represent the dangers that the Russian expanses or its rugged inhabitants might hold for Germany and its European neighbors. Instead, he chose the knout as a historical and a linguistic figure to convey another aspect of Russia. A loan word from Russian in both German (*Knute*) and English (but originally of Scandinavian origin and related to English *knot*), it denotes the traditional Russian shepherding or flogging whip that features a lash of leather thongs twisted with wire. Along with its literal meaning, in German it is also used to denote oppression or tyranny. This second meaning stems apparently in part from Heine's frequent topical use of the word to denote the Russian czar's tyrannical exercise of power.[1]

However, as my title suggests, when Heine employed the knout as a symbol for absolutist Russian authority, he usually did so in a context that imparted a more universal meaning to it. The power of the knout to suppress individual liberty refers not only to absolutist political rule but is transferred metonymically to the scheme of world history that frames Heine's depiction of nineteenth-century czarist Russia. From the late 1820s on, Russia figures in Heine's writings almost exclusively as the great European empire to the east that was playing an important part in world history in two ways. On the one hand, through the initiative of the Romanov line of rulers it was acting to whip the coarse peoples inhabiting its vast territories into members of a progressive, modern European civilization. On the other hand, the absolutist principles and tyrannical methods of the Romanov czars were diametrically opposed to the cause of freedom and individual rights championed by Heine. His comments on these two aspects of nineteenth-century Russia reflect an essential ambivalence that pervades all of his thinking on the political events of his day and their larger historical significance.

This essay traces the course of changes in this depiction of Russia from a rather inauspicious passage in his 1829 Italian travel journal to the early 1840s, when Heine adhered to a dialectical scheme of history. As he became influenced more strongly by early socialist and Left Hegelian thinking, his evaluation of czarist Russia shifted accordingly. Ultimately, the knout topos disappeared almost altogether as his own unique version of the historical dialectic gave way to a new understanding of the dynamic between individual and collective history. The surprisingly sparse references to Russia in his later writings, if examined alone, would present a lopsided picture of Heine's thoughts on history. However, a contextual analysis of the significance Russia played over the whole period indicates how an ambivalence about teleological vision tempered Heine's concept of world history even in the moments of his strongest faith in dialectical logic. Moreover, it suggests that in the end he did not merely resign himself to the pessimistic belief that the poet can only engage history in a melancholy, consolatory vision, but rather viewed the dynamic of history in a fundamentally different manner.

* * * *

Perhaps the most glaring mistake Heine ever made in his assessment of contemporary political affairs occurred in the second piece of his *Travel Pictures, Journey from Munich to Genoa*, in which he acclaims Nicholas I the standard-bearer of freedom (*der Gonfaloniere der Freiheit*).[2] He writes that all true friends of the Revolution should recognize that their hopes lie with Russia, both in its war with the Ottoman Turks and in its differences with the other Restoration powers of Europe. Furthermore, he proclaims the new culture of freedom emerging in Russia as the true heir to the liberal principles of the French Revolution, and juxtaposes it with the civil liberties achieved in England, which he says are actually rooted in medieval institutions. He calls Russia a democratic state, even a *Christian* state in the most cosmopolitan sense of the word—one that openly contests the power of the church and the nobility. In response to those "alarmists" who fear its size and the absolutist reign of the czar, Heine offers that he would gladly endure a little "knout" as long as he knows that the enemy is receiving its dose as well. After all, he adds in his typical style of critical jest, a little more or a little less oppression will hardly be noticed in the German states.[3]

In certain respects, this defense of Nicholas and Russian imperialism by one of the most outspoken liberal opponents of the reactionary Metternich regime is not as surprising as one might first think.[4] In contrast to his fellow liberals throughout Europe, Heine did not necessarily support the establishment of a German republic or even a constitutional government. He did, however, persistently defend the equal rights of all, not only to the basic necessities of life but also to its pleasures. The first order of political business, he contended, was to secure these rights for all Europeans; only then should the discussion ensue as to which form of state could best guarantee them. Both in his private correspondence and in his political commentaries, he admitted quite unashamedly that he was not even opposed to absolutism per se. In a letter to Heinrich Laube of 23 November 1835, he declares that "the political form of the state is only a means: monarchy or republic, democratic or aristocratic institutions are matters of indifference as long as the outcome of the fight for the basic principles of life, for the very idea of life itself, has not been decided."[5]

Nonetheless, shortly after it appeared in 1830 his glowing support for Nicholas proved particularly embarrassing. This was due in part to the shift in political climate, for the July Revolution in France had revived Heine's belief in the French Revolution and convinced him that sweeping social change would spread from France to Germany and beyond. Then, when the Russian emperor's troops quashed the Polish uprising of 1830–31 and brutally forced Russian political and cultural hegemony on the Poles, Heine quickly reversed his assessment of Nicholas. In the 1831 introduction to a collection of letters by a liberal critic attacking the German nobility ("Kahldorf on the Aristocracy"), he explains how Nicholas, once the *Gonfaloniere* of the liberals, has become the European aristocracy's leader in their resistance to the July Revolution. He reiterates his earlier claim that the Russian monarchy grants equality to all its citizens on the basis of their loyalty to the state, but now Heine concedes that czarist absolutism is ultimately incompatible with a constitutional freedom that protects the rights of citizens. What disturbs him most about the Polish episode is the alliance between the Russian royal government and the Prussian aristocracy. He calls the Prussian backers of Nicholas casuistic "Knoutologists," charging that they depict the czar as the new defender of church and country, falsely comparing him to their Russian allies in the 1813 War of Liberation against Napoleon. He warns, however, that Nicholas is a Russian wolf who has donned the moth-infested purple robe of the old Byzantine Empire, the time-worn imperial breeches of the Holy Roman Empire, and the diamond-studded old-Frankish crown of Charlemagne—and one day he will throw off this grandmotherly attire of the old empires and devour the Little Red Riding Hood of Freedom.[6]

Heine expresses his revised view of Nicholas in more subtle fashion in another text of the same year. In *French Painters*, his collection of articles on the 1831 Paris Salon art exhibition, he writes that while he was viewing the paintings throngs of armed Parisians were roaming through the streets crying "Warsaw has fallen! Our avant-garde has fallen!" This reminder of the turbulent political situation throughout Europe disrupts his solitary reflections on art, and even causes the paintings to take on a new face. While he is studying Paul Delaroche's painting of Cromwell lifting the coffin lid to observe the corpse of Charles I, he contemplates the role of revolutions in history. Suddenly, Cromwell is transformed into the figure of Nicholas, and the corpse of Charles Stuart becomes the body of defeated Poland. As he imagines the fate of his fellow soldiers in the struggle for freedom, he envisions Nicholas together with the pope and the victorious aristocracy standing triumphantly over the coffin.[7]

Once again Heine distances himself from his earlier praise of Russian absolutism and allies himself with the popular uprisings against aristocratic rule that were spreading throughout Europe in the wake of the July Revolution. Yet his other thoughts on Delaroche's painting indicate how his enthusiasm for constitutional freedom might easily revert to support for Nicholas. He compares the painting of Cromwell with the one hanging next to it by Leopold Robert, a pastoral scene of haycutters returning from the fields with their harvest. As he contrasts these two works, Heine encapsulates quite elegantly an ambivalence that permeates all his thinking on history. He writes that the painting of Cromwell represents "world history": "That history … that rolls around idiotically in dirt and blood, often remains standing foolishly for centuries, and then in haste takes off again clumsily to rage about without direction."[8] The painting next to it by Robert depicts "the history of humanity": "A history without beginning and end that repeats itself endlessly and is as simple as the ocean, as the sky, as the seasons; a sacred history that the poet narrates and whose archive can be found in every human heart."[9] At a time when Heine was enthused by fresh hopes for a revolution spreading from France to Germany and all of Europe, and bolstered by his own loosely Hegelian view of world history, this other, human side to history is relegated to providing comfort for those caught in its great wheels: "Alas, it is surely necessary that the wonderful, indestructible, melodious history of humanity comfort our soul amidst the dissonant racket of world history."[10]

In 1831, the first of the twenty-five years of self-chosen exile in Paris that lasted until his death in 1856, Heine had not yet assumed the more radical Hegelian position he would take a decade later. At this point in time, he saw himself as a poet who had sacrificed his beloved Romantic poetry in order to engage his pen in the service of a great revolution. He reiterated his own now well-known proclamation of three years earlier, which declared that the *Kunstperiode*, the great blossoming period of German culture epitomized by Goethe, was coming to an end. On the one hand, he was saddened by this, lamenting that the times would not allow him to focus on "the wonderful, indestructible, melodious history of humanity," as Goethe had done in Weimar for the last forty years. Yet, he resolved that it was not possible to ponder the aesthetic qualities of the Salon paintings in quiet reflection while a whole nation (Poland) lay vanquished in the boulevards of Europe.[11]

Over the course of the next decade Heine envisioned ways to overcome this either/or dilemma of art. Before returning to this question and examining how the figural depiction of Russia fits into his late concept of poetry as a mediating discourse, it is necessary to trace changes in his thinking during the same period with regard to its historico-political significance. After the harsh criticism of Nicholas's role in the Polish uprising, Heine abstained from attributing any specific blame to Russia, focusing rather on its place in the dialectical scheme of world history. Actually, after he recanted his praise of Nicholas in 1831, he discussed Russia only rarely both in his published writings and in his private correspondence, despite the fact that his younger brother Maximilian served as an officer and doctor in the Russian military in St. Petersburg. Nonetheless, there is a marked shift in his representation of Russia, one that corresponds with his changing views on history and the role art can play in determining its course.

The next occasion for addressing the question of Russia arose in his 1833 essay on the German Romantic School, in which Heine discusses briefly Ernst Raupach, a German writer who had lived for eighteen years in Russia and began his literary career there before taking a position as playwright for the Prussian royal family in the 1820s. Heine uses this occasion to caution once again those Germans who see in Nicholas an ally for their own political interests. While the main thrust of his criticism is directed at the Prussian aristocracy, this passage also reveals his more fundamental concerns about the Russian Empire. Heine explains that Raupach was initiated into poetry by the Muscovite muse, whom he describes, in an allusion to the fiery poet Denis Davydov, as a snub-nosed, brandy-swilling figure cloaked in sable, who wears a saber and carries the "knout of tragedy."[12] Here the knout of tragedy suggests the dangers associated with a fanatical Russian patriotism that could induce the people to fight avidly even for an oppressive monarch. Later, in one of his articles for the *Augsburger Allgemeine Zeitung*, Germany's largest and best-known newspaper of the time, Heine warns of this fanaticism more directly, claiming that the Russian people's blind obedience to the holy, infallible power of their czar is actually a vehicle for their own self-glorification.[13]

In the following years, when Heine occasionally mentioned the threat Russia posed to Germany and the rest of Europe, his concern was counterbalanced by a changing view of world history. This is clearly evident when he looks back on the Polish revolution in his 1840 book on the German liberal writer Ludwig Börne. Heine designed the book as a response to those fervent German Republicans, including Börne, who questioned his loyalty to the liberal movement. In what turned out to be a severe blow to his reputation among many of his former allies, he ridiculed the recently deceased Börne as a narrow-minded, Nazarene moralist who was unable to view the political events of the day from a more detached, grander perspective. His account of Börne's responses to the Polish revolution was supposed to serve both as an apologia and a political primer that would warn the Germans about rash, impassioned attempts at revolution. Quoting frequently from Börne's Parisian letters at the time of the Polish uprising, he contends that his fellow exiled liberal became caught up in the storm of political emotions around 1830 and ignored all restraints of reason in his support of Poland.[14] Now, eight years later and influenced by the Left Hegelian concept of revolution, Heine is even more critical of the Poles than of Nicholas. Although sympathetic to their plight, he shows disdain for their romantic devotion to antiquated chivalrous ideals, charging that they were unable to pursue their freedom in a clear-headed, pragmatic manner. He claims that, blinded by quixotic idealism, they committed the most foolhardy, self-destructive political blunders. Of course, he deplores the wounds inflicted on them and reassures his reader that he remains the same champion of freedom who condemned the Russian emperor at the time; but he also warns that sentimentality has no place in the writings or actions of those intellectual giants who will guide Germany through the tumultuous events of the coming years.[15]

In the years following the Börne book, Heine adhered quite closely to the radical Left Hegelian view of a world historical revolution. In an 1842 article for the *Allgemeine Zeitung* he speculates about the role Russia will play in the momentous upheavals awaiting Europe. Anticipating widespread, all-out war, he foresees the end of Germany and France as autonomous political states, while predicting that Russia and England will be able to survive even if they suffer devastating defeats on the battlefield. Revealing his familiarity with the radical French socialists and the budding communist movement, he maintains that the second act in this dramatic European phase of world history will pit the disenfranchised masses against the wealthy and landed elite. It will be a transnational class struggle, he writes, in which neither the seas nor the vast continental expanses will save the old order in England and Russia.[16]

As he began to interpret history more strictly in terms of the Hegelian dialectic, Heine returned to almost the same positive view of czarist Russia that he had expressed before the Polish revolution. In an unpublished fragment from the early 1840s, he proclaims that, like France, Russia too is infused with the spirit of the revolution: in France this spirit resides in the masses, while in Russia it is imposed upon the people from above by the monarch. In both cases, he asserts, it is fueled by a scorn for the past and an overriding drive to mold out of the remnants of the old feudal order a modern nation-state. In this context he even welcomes Nicholas's efforts to turn the chivalrous Poles into Russians, to assimilate the Jews into modern European culture, and to use the knout freely to transform the Russian people from two-legged animals into civilized modern beings.[17]

For an unknown reason, Heine decided against including this fragment in the series of articles he submitted to the *Allgemeine Zeitung*. It was apparently written in 1843,[18] at a time when he was most strongly convinced that a comprehensive and universal social revolution was inevitable. It brings him back almost full circle to the position expressed in the travel journal of 1829. And it must have called to mind his own strong affiliation with the cause of the Polish revolutionaries and reawakened his deep-seated ambivalence about world history. Perhaps he even held it back after recalling how once before he had regretted condoning the Russian knout as a tool for the advance of humanity.

Whether this played a role in his decision or not, his misgivings about such an Olympian view of history were still the very ones he had expressed most persuasively in the same Italian journal passage that praised Nicholas. There the idea of Nicholas as the *Gonfaloniere* of freedom occurred to him as he looked out over the battlefield at Marengo, the site of Napoleon's important victory over the Austrians in 1800. As he observes the graves of fallen soldiers, he asks whether the cost in human lives was worth the victory in the name of liberty: "Alas, every inch humanity moves forward costs streams of blood; and is that not a bit too dear of a price? Is the life of the individual not just as valuable as that of the whole human race? Every individual human being is, after all, a world in its own right that is born with that person and dies with that person; underneath every tombstone lies a world history."[19] The narrator of the travel journal immediately quells these doubts and affirms all the sacrifices made for freedom: "Enough of such talk; that is what those who fell here in battle would say. We, on the other hand, are still alive and must carry on the hallowed struggle for freedom."

These voices questioning our grand narratives of history were never completely quelled within Heine himself. Ultimately, he granted the demands of each individual just as much validity as those of humanity as a whole. This unresolved ambivalence perhaps explains why after fifteen years Heine's figurative use of Russia as the tyrannical knout of world history came to an end. However, not only does this topos not resurface in his works in the final twelve years of his life, but there are astonishingly few references to Russia at all. Where one does occur, it is invariably dictated by the topic at hand and gives only minimal attention to Russia as a theme in its own right.

This absence suggests more than just the return of this ambivalence about world history. When he abandoned the Hegelian view of history in the late 1840s, Heine questioned as well the dichotomy he had established in *French Painters* between art that depicts world history and art that investigates a human history common to every individual. Even there he had argued that a successful alliance of these two functions is not impossible in and of itself, only that it was not possible in his lifetime. Taking the ancient Greek and Renaissance Florentine masters as examples, he claims that at other times the artist did not have to lead "an egotistical, isolated life."[20] By contrast, Heine saw his own age as a transitional period of history in which this unison was not possible. In *French Painters* he looks forward to a new age in which such cultural articulation will again be possible; but until it arrives, he counsels, the artist should pursue "a most self-intoxicated subjectivity, an individuality that is free of all worldly concerns."[21] Even so, he had left the ultimate question of human history open, wondering whether the frenzied subjectivity of Romanticism was an eerie premonition of renewal or the final convulsions of a dying culture.

As he moved closer to a pure Hegelian idea of history, Heine modified his view of the dichotomy between these two modes of history. He began to fancy that the poet attuned to the spiritual advance of world history could do more than merely await the future in feverish anticipation. To coordinate this new notion of poetry with his standard view of the poet's calling, he took up a famous metaphor he had formulated some fifteen years earlier. In the *Baths of Lucca* (1829) the narrator laments that a great tear in the world runs straight through the heart of the Romantic poet and that his psychic wounds are also those of the collective soul of alienated modern humanity. As his own sense of self-importance became linked to a teleological view of history, Heine came to see his poetic articulation of this collective wound as an integral part of the drumbeat driving world history on to revolution and a new wholeness. Eventually, he regarded himself as the herald of a glorious future and proclaimed that his poetic revelation of the historical process would usher in the great changes he intuited.

When his faith in the dialectical progression of history dissipated, Heine renounced this self-aggrandizing notion of his poetic mission. However, he did not merely embrace the idea of a comforting, poetic history of humanity, nor did he reject outright the idea of historical progress. On the contrary, his renunciation of the Hegelian dialectic, sealed by his drastic physical collapse in 1848, resulted in a radically new conception of a poetry that can negotiate between the needs of the individual and the demands of humanity as a whole. In contrast to the poet-herald of historical change, the poet as portrayed by Heine in his late poetry trains his faculties on the deep past, restoring to consciousness latent memories, dreams, myths, legends, sagas, and visions. This unearthing of buried poetic treasure is not restricted to the realm of the imaginary, but is a *writing* of history. Thus Heine assigned an even greater significance to what he had called human history and to the poet's role in opening up its archive in the human heart. No longer restricted to comforting the masses as they were ground down by the wheels of progress, in this late view the poet resonated with the voice of the people in writing an alternative history.[22]

In sketching these central deliberations on the function of poetry, my intention is not to portray Heine's changing perception of Russia as any kind of precipitous moment in the major turns his thinking on history took over a volatile twenty-year period. On the contrary, it was only one, relatively minor piece in a shifting overall vision of world history. Appropriately, however, there is a fitting touch of irony associated with his late views on poetry and the corresponding disappearance of the knout topos as a figurative representation of Russia, for when Heine finally abandoned the Hegelian vision of historical progress, the result was his own dialectical advance to a new conception of *poésie engagée.*

Notes

1. Jacob Grimm and Wilhelm Grimm, *Deutsches Worterbuch*, vol. 14 (Munich, 1984), facsimile of 1st ed. (Leipzig, 1854–1971), pp. 1028–29.
2. Heinrich Heine, *Sämtliche Schriften*, ed. Klaus Briegleb et al., vol. 3 (Munich, 1968–76), p. 379.
3. Ibid., p. 380.
4. In his commentary on Heine's letters, Friedrich Hirth (Heinrich Heine, *Briefe*, ed. Friedrich Hirth [Mainz, 1950–51] vol. 4, pp. 185–86) contends that in his praise of Nicholas I, Heine was influenced by the Russian poet and diplomat Fyodor Tiutchev with whom he was friends in Munich (see also Klaus Briegleb's commentary in *Sämtliche Schriften*, vol. 4, pp. 864–65). There is, however, no documentation to support this suggestion. Moreover, at the time Heine referred to Nicholas as the champion of liberal causes there was considerable support for Russia throughout the liberal camp in Germany. In the journals of the most prominent liberal publisher, Johann Friedrich von Cotta, articles appeared praising Russia for carrying out the will of its people and those of Europe in putting down the crescent of the Orient (i.e., the Ottoman Turks). The liberal playwright at the Bavarian royal court, Harro Harring, even joined the Russian army as a volunteer in its fight against the Turks. Thus, Heine's characterization of Nicholas reflects a widely held view of Russia's heroic fight in defense of European culture against warrior hordes from the East. See Heinrich Heine, *Historisch-kritische Gesamtausgabe der Werke*, Düsseldorfer Ausgabe. ed Manfred Windfuhr et al., vol. 7/2 (Hamburg, 1973–), p. 917.
5. Heinrich Heine, *Säkularausgabe. Werke, Briefwechsel, Lebenszeugnisse*, ed. Nationale Forschungs- und Gedenkstätten der klassischen deutschen Literatur in Weimar and Centre National de la Recherche Scientifique in Paris, vol. 21 (Berlin and Paris, 1970–), p. 125; see also Heinrich Heine, *Historisch-kritische Gesamtausgabe*, vol. 13/2, p. 1352.
6. Heine, *Sämtliche Schriften*, vol. 2, p. 666.
7. Ibid., vol. 3, pp. 69–70.
8. Ibid., p. 68.
9. Ibid.
10. Ibid., p. 69.

11. Ibid., pp. 70–71.
12. Ibid., p. 482.
13. Ibid., vol. 5, p. 344.
14. Ibid., vol. 4, pp. 75–77.
15. Ibid., pp. 78–81.
16. Ibid., vol. 5, p. 406; see Gerhard Höhn, *Heine-Handbuch: Zeit, Person, Werk* (Stuttgart, 1987), p. 394.
17. Heine, *Sämtliche Schriften*, vol. 6/1, pp. 634–35; Heine, *Historisch-kritische Gesamtausgabe*, vol. 13/1, pp. 331–32.
18. Heine, *Historisch-kritische Gesamtausgabe*, vol. 13/2, pp. 1348–49.
19. Heine, *Sämtliche Schriften*, vol. 2, p. 378.
20. Ibid., vol. 3, p. 72.
21. Ibid., pp. 72–73.
22. For an extensive analysis of the changes in Heine's thinking on history and poetry after 1848, see my book on his late writings: Roger F. Cook, *By the Rivers of Babylon: Heinrich Heine's Late Songs and Reflections* (Detroit, 1998).

Chapter 7

PUSHKIN AS "UNIVERSAL" POET

Varnhagen von Ense and Dostoevsky

Marina Kanevskaya

This essay presents a comparative analysis of K. A. Varnhagen von Ense's[1] article "The Works of A. Pushkin" (1839) and F. M. Dostoevsky's "Pushkin Speech," which was delivered during the celebrations to commemorate the unveiling of the Pushkin monument in 1880 in Moscow.[2] Although no one knows for sure whether Dostoevsky was familiar with Varnhagen's article, circumstantial evidence suggests that he might have been: in an article by his close friend and colleague N. N. Strakhov, "The Main Treasure of Our Literature" (1867), which was dedicated to Pushkin, Varnhagen's opinion is quoted at length and used as an example for Russian critics. Building on the similarities between Varnhagen's and Dostoevsky's texts, I will demonstrate how, on the one hand, the Hegelian approach dominated art and poetic personality in nineteenth-century Russia[3] and how, on the other hand, the myth of Russian national superiority, built on these concepts, emphasized independence from European cultural sources.

The comparison between Varnhagen and Dostoevsky would sound reductive if one did not bear in mind other prominent representatives of German objective idealism, such as Herder, Fichte, Schelling, and Schiller, who, along with Hegel, significantly influenced the formation of Dostoevsky's aesthetic views. Schelling was probably the single most important influence on the "Native Soil" movement, certainly as far as aesthetic thought was concerned. Schelling's idea of irrational organicism, with its underlying principle of the "revelation in the native soil"

(*Boden* = *pochva*), gave the direction and, indeed, the name for this movement. Dostoevsky's colleague Apollon Grigor'ev was the main advocate for "Native Soil" (*pochvennichestvo*); he was thoroughly familiar with Schellingian teaching.[4] However, even before Grigor'ev and "Native Soil," Schelling had had many followers in Russia since the 1820s; some of them would come to represent the "older generation" of the Slavophile movement.[5] The members of Stankevich's circle (see below), among whom we find K. Aksakov, V. Belinsky, M. Bakunin, and others, were all at certain periods of their life devoted to Schelling's philosophy. In his later writings, Schelling presents a distinct aesthetic theory:

> [A] belief in the cognitive and prophetic powers of poetry; a conception of poetry as a microcosm symbolic of the macrocosm of the world; the notion that poetry rivals rather than imitates nature; a polarized model of the work of art, which is seen as a fusion of the 'ideal' and the 'real,' the 'infinite' and the 'finite,' etc.; an emphasis on the poet's 'inner vision' in the creative process; an organic and historical conception of the relationship between art and social life.[6]

This theory is refined to a definitive religious formula in which God reveals Himself to humanity, with Christianity as the ultimate stage of this emanation of the deity.

Schelling's transcendental idealism had great importance for the formation of Dostoevsky's approach to Pushkin. However, in this chapter I will focus on the Hegelian elements in his "Pushkin Speech." Dostoevsky's keen interest in the journalistic polemic, which he saw as the only manifestation of Russian political and social life, is essential for my thesis. Varnhagen's article remained in the annals of Russian literary criticism as an unusual approach to Pushkin's legacy, precipitating significant debate in the press.[7]

Written to commemorate the publication of the first posthumous collection of Pushkin's writings, Varnhagen's article presents a critical overview of the poet's works. The article, published in translation in the journal *Syn Otechestva* in 1839,[8] demonstrates that the German critic was well-versed in the Russian polemics of the time concerning Pushkin's works and Russian literature in general. Besides being familiar with the German research on Russian literature (such as Heinrich König's book *The Literary Scene in Russia* [1837], written in close cooperation with the Russian philosopher N. A. Mel'gunov),[9] Varnhagen had connections with the Moscow circle of N. V. Stankevich.[10] One of Stankevich's closest friends, Iakov Neverov, translated Varnhagen's article for publication in Russian.

The importance of the Stankevich circle for Russian cultural history is hard to overstate. Its participants came from a variety of backgrounds, ranging from the Slavophile K. S. Aksakov, to the Romantic revolutionary M. A. Bakunin, to the socialist-minded V. G. Belinsky. Among the people close to the circle were A. I. Herzen and T. N. Granovsky. An inevitable split among the members of the circle occurred at the very beginning of the 1840s, after Stankevich's death. Aksakov rose to become the Slavophiles' main religious ideologue. Bakunin emigrated to Europe, only to be handed back over to Russia, where he spent thirteen years (1848–61) in prison and Siberian exile for his political activities; he eventually escaped to America. Belinsky, after completing his ideological "wanderings," which ranged from Hegelianism to the denunciation of Christianity and a superficial socialism, became a recognized authority on Russian literary criticism.[11]

Through these important figures, Hegelian philosophy became part of both idealist and materialist aesthetics in Russia. The Hegelian vision of the divine order as the foundation of society was generally shared by all Russian religious thinkers. At the same time, it was translated by Belinsky's socialist-oriented followers into a positivist system. According to Hegelian aesthetics, each individual poet or artist was inevitably the product of his age. Consequently, any noteworthy creation by an individual was no more and no less than the embodiment of the "Spirit" of that age, expressed in verbal symbols or images. In the later, positivist version of Hegelianism represented in Russia by "Realist" critics (such as N. G. Chernyshevsky, N. A. Dobroliubov, D. I. Pisarev, etc.), the notion of the Spirit was replaced by that of "Society." As the last Russian "Hegelianist," Strakhov wrote in one of his polemical articles, "the theory of social environment (*sreda*) is also inspired by Hegel but wrongly understood."[12] Indeed, the famous debate between "Pushkinian" and "Gogolian" movements (*napravleniia*) in Russian literary criticism during the 1850s and 1860s[13] can be understood as a controversy between idealist and positivist interpretations of Hegel. Both in its religious and in its positivist version, the Hegelian approach remained dominant in Russian aesthetic theory until the end of the 1890s. However, toward the end of the 1850s, the positivist approach prevailed in Russian thinking. As Chizhevsky puts it, Russian materialists did not advocate a new philosophical system but rather struggled against the very idea of philosophical thinking.[14] At the beginning of the 1860s, when anti-positivist or idealist philosophers such as B. N. Chicherin, N. G. Debol'sky, and S. S. Gogotsky were literally excluded from public discussions in the universities and in the press, Strakhov, Grigor'ev, and Dostoevsky remained the sole followers of the Romantic tradition of the 1840s who dared to advocate their views publicly.

In the course of the 1860s, controversies centered around the "Native Soil" movement (which consisted of M. M. and F. M. Dostoevsky, Grigor'ev, and Strakhov, who propagated their aesthetic and social agenda in the journals *Vremia* and *Epokha* [1861–64]). The old Slavophile Iurii Samarin wrote that the "Native Soil" example proved that "Russian thought" could exist only on the basis of Hegelian philosophy. Samarin's appreciation shows how close Dostoevsky's world-view was to that of "older" Slavophiles. (However, the relation between Dostoevsky's thought and that of the Slavophiles will not be examined in this essay as it deserves special and extensive study.) Toward the end of the 1870s, idealist thinking was considered to be backward and even laughable; it was not so much criticized seriously as ridiculed in the main popular journals of the socialist left (such as *Delo* and *Otechestvennye zapiski*), as well as of the liberal center (such as *Vestnik Evropy*). The public expression of non-positivist views was marginalized and relegated to the conservative press (such as the journal *Russkii vestnik* or the newspapers *Moskovskie vedomosti* and *Varshavskii dnevnik*) or the journals that represented particular groups, such as the Slavophile *Russkaia mysl'*. Naturally, in his *Diary of a Writer* (1873–81), Dostoevsky adhered to his deeply held religious and idealist views. Thus the *Diary* appealed only to a limited segment of the "reading society," much narrower than the circle of Dostoevsky's fiction readers.

At the Pushkin festival—the most far-reaching and liberal public celebration imaginable in the Russia of the 1880s—Dostoevsky gave voice to his idealistic opinion, eliciting hysterical sobs and applause from an audience that had been mocking his views for twenty years. In fact, Dostoevsky told his positivist-minded listeners that their philosophy of life, although seemingly consistent and pragmatic, had already brought mankind to the brink of catastrophe. Moreover, he pointed to idealism as the only approach that could provide humanity with hope for universal salvation. I have analyzed this speech and its historical context elsewhere, with the objective of explaining its success at the moment of its presentation and its lasting impact on Russian culture.[15] The present essay sets for itself a considerably narrower goal—that of identifying the Hegelian elements in Dostoevsky's "Pushkin Speech" by using Varnhagen's critique of Pushkin as an immediate precedent.

* * * *

In his article, Varnhagen writes: "A short time has passed since the untimely and tragic death of Russia's poet. Alexander Pushkin's death, as the death of a man struck down by fate, has aroused sympathy even among those who failed to appreciate his poetic qualities.... This occasion has brought about the realization that Pushkin was a bona fide poet. Certainly, the ecstatic praise of his fellow countrymen, the unanimous acknowledgment of his genius, and the deep feeling with which people of all social strata met his works, can serve as inarguable proof of that." It can be seen here that at the very outset of the article Varnhagen states that public appreciation is a key criterion by which true poetic genius can be measured.

Dostoevsky begins his speech with a similar statement, using Gogol's authority—since Gogol was revered in both "materialist" and "idealist" circles—to bolster his own pronouncement: "'Pushkin is an extraordinary, and perhaps unique, phenomenon of the Russian spirit,' said Gogol...."[16] Expanding on this statement, Dostoevsky points out Pushkin's significance for all Russians, echoing the idea that a great poet must be appreciated by his people because he incarnates its Spirit: "Yes, in his appearance, to all us Russians, there is something indisputably prophetic."

In the Hegelian system, acknowledging a poet's genius and viewing him as a prophet are congruent concepts. In that sense, Dostoevsky and Varnhagen agree, both leaning on Hegel's "expressionist" theory. According to the tenets of this theory, poets and artists whose individual ambitions coincide with the interests of national self-expression are filled with a sense of mission. They intuitively sense the importance of what they are doing, and are the first to realize and give expression to what must be the next stage in the people's development. Dostoevsky put this idea in the following words: "Pushkin appeared precisely at the very inception of our genuine self-consciousness, which was just then coming into being and which originated in our society after a whole century following Peter's reform; and his appearance greatly helped to illuminate our obscure path with a new guiding light. It is in this sense that Pushkin is a prophecy and a revelation." In the passages quoted earlier, Varnhagen stresses the nation's appreciation as the true criterion for identifying poetic genius; for Dostoevsky, the idea of prophecy was more important.

According to Varnhagen, the European nations—which he considered the leading nations of modern civilization—had begun to pay closer attention to Slavic languages, and Russian in particular. The reason for this increased attention was a recognition of Russia's new role in European history, "a new flight which surpassed all expectations." At the same time, Varnhagen registered the growth of Russian national self-consciousness and confidence in its strength. In turn, Dostoevsky attributes a similar constellation of proud and hopeful feelings to Pushkin's poetry: "Had there been no Pushkin, perhaps our faith in our Russian individuality and our belief in our future independent mission in the family of the European nations, would not have manifested itself with so unyielding a force ..." Significantly, Varnhagen dated the beginning of Russia's new role to the era of the Napoleonic wars; Dostoevsky pinpoints this period as "the very inception of our true self-consciousness," the exact moment when "Pushkin appeared."

Varnhagen's idea that Pushkin was the one who attracted all of Europe's attention to Russia finds an echo in the last and most important passage of Dostoevsky's "Pushkin Speech." Stating that Russia's participation in the Napoleonic wars was a sign of its initiation into pan-European life, Varnhagen illustrated his thought with Pushkin's poems "The Hero," "The Anniversary of the Battle of Borodino," and "To the Slanderers of Russia." According to Varnhagen, Pushkin realized that Russia's highest aspirations lay in its evolving and central role among the Slavs. Varnhagen interpreted these poems as evidence of the poet's blaming Western Europe for its failure to understand Russia's new role in the world. Thus, with the help of Hegelian teaching on the creative genius, Varnhagen managed to interpret Pushkin's pronounced imperialism and nationalism positively.

Given the somewhat scandalous reputation of these poems as servile and conservative, Dostoevsky does not discuss them in front of the liberal audience gathered for the Pushkin festival. Nevertheless, the idea is implicit in his approach when he notes: "Had [Pushkin] lived longer, perhaps he might have revealed great and immortal images of the Russian soul which would be intelligible to our European brethren; he might have attracted them to us much more than they are attracted at present; perhaps he might have explained to them the whole truth of our aspirations, and thus they would comprehend us better than at present and might foresee our destiny; they would cease to look upon us as suspiciously and haughtily as they still do."

The nature of Russia's participation in European life is of great importance for Dostoevsky. Many earlier Russian thinkers, such as P. Ia. Chaadaev, I. V. Kireevsky, V. F. Odoevsky, and Pushkin himself, thought that Russia's separation from Europe was part of a divine plan that included its eventual great mission among the nations of the world.[17] Following this thinking, Dostoevsky qualified the nature of Russia's participation in European life since the time of Peter the Great as "service" (*sluzhenie*). Here, "service" must be understood in the clerical meaning of the term, as in a priest's delivery of the "service" to his parish. Dostoevsky suggested:

> If you analyze our history after Peter's reform ... what else has Russia been doing in her policies, during these two centuries, if not serving Europe much more than herself?... Oh, the peoples of Europe have no idea how dear they are to us!... To become a genuine Russian means to seek finally to reconcile all European controversies ... in our all-humanitarian and all-unifying Russian soul, to embrace in it with brotherly love all our brethren, and finally, perhaps, to utter the ultimate word of great, universal harmony, of the brotherly accord of all nations abiding by the law of Christ's Gospel!

In essence, Dostoevsky's vision of unifying love based on religious faith—Russian Orthodox Christianity—presents a Russified version of the Hegelian concept of universal love representing the ultimate stage of progress and self-expression. Varnhagen voiced this idea as well, though naturally without such Russocentric fervor, in his analysis of Pushkin's "Feast of Peter the Great": "The poet speaks here with a high patriotic and pan-human goal, in the sense of universal forgiveness and all-inclusive reconciliation." Varnhagen declared that if the Russian poet had reached this all-loving and all-embracing idea, then "Russian poetry without reservations can appear at the same level with any other poetry."

Because of the multinational composition of the Russian Empire, only a Russian poet, Varnhagen asserts, could have delivered such a message to Europe:

> To Pushkin, things usually separated by distance in space and heterogeneous in spirit were presented naturally as a whole within the boundaries of his country. Everything is equally known to him—South and North, Europe and Asia, savagery and refinement, the ancient and the modern; by portraying the most heterogeneous things he portrayed his Fatherland.... In the world of an old rural custom and in the mundane world of fashion, in the palaces of the rich and under a Gypsy's tent—everywhere he is in his native element, everywhere his poetry takes root.

Here Varnhagen applied the Hegelian concept of the subordination of the "subjective principle" (the individual) to the "objective principle" (relating broadly to the state). Thus, the power of the state, which in the case of the Russian Empire happened to be multinational, gave a protean nature to the genius of its poet.

Dostoevsky favored the Russocentric policy of the Russian government, especially the idea of Russia's domination of the rest of the Slavs.[18] At the Pushkin celebrations, Dostoevsky attempted to avoid confrontation—without, however, abandoning his own political convictions. In his speech, he expressed his views on the superiority of the Russian nation over the rest of Europe using spiritual terms such as "the Spirit of the People," "the People's Predestination," "the People's Faith," etc., instead of the more provocative (for his liberal and radical audience) notion of "the Russian State."

In doing so, Dostoevsky, unlike Varnhagen, avoided the sensitive issue of Russia's policy toward the other Slavic peoples and the Empire's minorities. Instead, he focused on the depiction of Europeans in Pushkin's works. Dostoevsky invited his audience to consider such works as "A Scene from Faust," *The Covetous Knight*, "Once There Lived a Poor Knight" (as he calls the ballad in *The Scenes from the Times of Knighthood*), *Don Juan* (as he referred to *Stone Guest*), *A Feast During the Plague,* "The Wanderer," and *The Egyptian Nights.* Each of them, he maintained, illustrated Pushkin's exceptional ability to perceive and convey "the spirit of the age … history itself," and his "universal responsiveness … that reincarnation in his spirit of the spirit of foreign nations, an almost complete, and therefore miraculous, reincarnation."

Varnhagen focused on the expressiveness of Pushkin's language. Here again, he followed Hegelian theory in seeing language as a vessel of the "People's Spirit" (*Volksgeist*), a notion based in turn on Herder's idea of language as the unifying medium of the nation.[19] Dostoevsky did not speak about language directly; here he referred instead to poetic imagery, which in Hegel's aesthetic hierarchy is the superior stage in a language's existence and hence most appropriate for a nation's self-expression. For example, in Pushkin's image of the monk in *Boris Godunov,* Dostoevsky sees "the potent spirit of Russian national life.… Therefore, the spirit of the people that conceived it must exist, and it is great and boundless.…"

According to both authors, the expressiveness of a language manifests itself in its ability to create art. It serves as objective, sensual proof of a nation's degree of development. This expressive ability was given to the Russians by Pushkin, along with the "prophetic sign" of their new mission in the world—something half-consciously recognized by foreigners and by a limited number of Russians. Pushkin's genius and Russia's mission manifested themselves in the attempt to merge with the spirit (genius) of the European nations. Dostoevsky claimed this "miraculous" power of incarnation to be an exclusive trait of Russian character. As has been demonstrated, in his conclusions Dostoevsky goes much further than Varnhagen, even though the latter was essentially moving in the same direction.

Conclusion

The central idea put forth in both of the texts discussed here is Pushkin's universality. This term—universality—which most of Dostoevsky's critics found meaningless in 1880, was Hegelian in essence. First of all, when applied to a poet, "universality" suggests that the poet so evaluated is a truly "national" poet. This means that the poet's art represents the highest stage of his nation's cultural or—as Dostoevsky would have it—spiritual development, and the highest level of a nation's self-perception. Such a poet not only utilizes the language—the materialized self-expression of the people—but also transfigures this language into art, which, according to Hegelian aesthetics, is viewed as the highest embodiment of *Volksgeist.* This represents the nation's growing self-esteem, while simultaneously evoking the greater appreciation with which other, more advanced nations view it.

Both Varnhagen and Dostoevsky cite the period of the Napoleonic wars as the time of "awakening" for the Russian people. At this particular moment, Russians were sent a providential sign, or, as Dostoevsky puts it, a prophesy and revelation: Pushkin revealed to the Russians their destiny in the form of art. Varnhagen sees this moment as coinciding with the general development of European civilization. Dostoevsky, from the perspective of Russian nationalism, sees it entirely as a manifestation of Russian self-development. He denotes the relations between Russia and Europe since the Westernizing reforms of Peter the Great with the term "service." Understood in the religious sense, this notion identifies Russia's mission as the spiritual leader of the European nations. It is here and not in their philosophical approach as such, that the essential difference between Varnhagen and Dostoevsky lies. Both men viewed Pushkin as the figure who successfully articulated his nation's mission. On this basis, Varnhagen believed that Russian poetry as epitomized by Pushkin was entitled to take its place among the most developed literatures in the world. In contrast, Dostoevsky's idea aimed at Russia's leadership of the whole Aryan world;[20] first, on the grounds of having the only true Christian faith (Orthodoxy), and second, because of his belief that the Russian ability to embrace and to assimilate the spirit (or genius) of foreign nations was a providential sign sent to mankind through the phenomenon of Pushkin. Dostoevsky interpreted this sign as meaning that Russians could overcome any alienation among individuals and nations and eventually succeed in unifying these nations in a single community. This vision of universal community and new spiritual order is fully congruent with the Hegelian vision of the divine as freedom in Spirit.

Notes

1. Karl August Varnhagen von Ense (1785–1858) was a literary critic, poet, translator, and diplomat. He lived most of his life in Berlin, where his wife Rachel Varnhagen hosted a literary salon. In 1827, Varnhagen began to contribute to Hegel's literary annual *Jahrbücher für wissenschaftliche Kritik*, where the article on Pushkin was originally published in October 1838.
2. F. M. Dostoevskii. *Polnoe sobranie sochinenii v 30-ti tomakh* (Leningrad, 1972–88) 25, pp. 136–49. For quotations the following translation with some modifications was used: F. M. Dostoevsky. *A Writer's Diary.* Translation and annotations by Kenneth Lantz (Evanston, Ill., 1993) 2, pp. 1281–96.
3. There are two books on the aesthetics of Russian literary criticism from the point of view of Hegelian philosophy: Dmitry Chizhevsky's *Hegel in Russia* (*Gegel' v Rossii*, 1939) and Victor Terras's *Belinskij and Russian Literary Criticism* (1974). Chizhevsky devotes several chapters of his book to the dominating role that Hegel's aesthetics played in Russian literary criticism of the nineteenth century. The Hegelian "organicist" approach to a work of art was broadly accepted in Russia, practiced most consistently by the "Native Soil" critics—a movement to which Dostoevsky belonged (see also Wayne Dowler, *Dostoevsky, Grigor'ev, and Native Soil Conservatism* [Toronto, Buffalo, London, 1982], pp. 40–60, 116–29). Chizhevsky makes the specific point that Dostoevsky was exposed to Hegelian philosophy through the influence of his friends Apollon Grigor'ev and Nikolai Strakhov, if not by reading Hegel himself. Oddly, later in the same passage, the author states that in Dostoevsky's writing one cannot find any "observable trace of Hegel" (D. I. Chizhevskii, *Gegel' v Rossii* [Paris, 1939], p. 321). In his book, Terras also emphasized the continuity of the Hegelian approach to art and literature throughout Russia in the nineteenth and twentieth centuries, well into the Soviet period (Victor Terras, *Belinskij and Russian Literary Criticism: The Heritage of Organic Aesthetics* [Madison, Wis., 1974], pp. 3–9).
4. Terras, *Belinskij and Russian Literary Criticism,* pp. 214–15.
5. Ibid., 20–21.
6. Ibid., 21–22.
7. N. N. Strakhov, "Glavnoe sokrovishche nashei literatury," *Otechestvennye zapiski* (1867) no. 12 (reprinted in N.N. Strakhov, *Zametki o Pushkine i drugikh poetakh* [The Hague, 1967]); see also *Pushkin. Itogi i problemy izucheniia*, ed. B. P. Gorodetskii, N. V. Izmailov, B. S. Meilakh, et al. [Moscow-Leningrad, 1966], p. 35.
8. In Russian literary criticism, Varnhagen's article (Karl August Varnhagen von Ense, "Sochineniia A. Pushkina," *Syn Otechestva* [1839], vol. VII, part IV, pp. 1–37. Reprinted in V. Zelinskii, *Russkaia kriticheskaia literatura o proizvedeniiakh A.S. Pushkina. Khronologicheskii sbornik kritiko-bibliograficheskikh statei* [Moscow, 1897], part 1, pp. 105–26) is rarely mentioned. There is a review of it—barely more than a summary—in a book by S. S. Trubachev called *Pushkin in Russian Criticism* (Pushkin v russkoi kritike) (St. Petersburg, 1889), pp. 279–80, 285–86. In 1897, V. Zelinsky included a reprint of the article in the collection *Russian Critical Literature about A. S. Pushkin's Writings* (Russkaia

kriticheskaia literatura) mentioned above. Some basic information about the article is given in a collective monograph: *Pushkin. Itogi i problemy izucheniia,* pp. 34–35). There are a few other studies in which the article is briefly mentioned. For example, in his article "Velimir Khlebnikov" (1941–42), N. Ia. Berkovsky gives an extremely idiosyncratic interpretation of Varnhagen's statement on the multinational phenomenon of the Russian Empire (N. Ia. Berkovskii, *Mir, sozdavaemyi literaturoi* [Moscow, 1989], pp. 148–57, 152–53). In Iurii Mann's book *Russian Philosophical Aesthetics,* a single sentence is dedicated to Varnhagen's interest in Pushkin (Iu. Mann, *Russkaia filosofskaia estetika (1820–1830)* [Moscow: 1969], p. 232).

9. Heinrich König (1790–1869), *Literarische Bilder aus Russland,* Herausgegeben von H. König; mit den Bildnissen von Derschawin und Puschkin (Stuttgart und Tübingen, 1837).

   In 1839, in the supplement to the journal *Otechestvennye Zapiski* (no. 4, Supplement to vol. 3, pp. 3–30, first pagination), N. A. Mel'gunov published a controversial article entitled "A History of One Book" in which he advocated the originality of König's book and the general importance of the European recognition of Russian literature. Mel'gunov wrote with great respect about Varnhagen's views on Pushkin.

   Mel'gunov's article was followed by a second publication of Varnhagen's article on Pushkin, in M. N. Katkov's translation, under the title "A Foreigner's Opinion on Pushkin" (*Otechestvennye Zapiski* [1839], no. 4, Supplement to vol. 3, pp. 1–36, second pagination). In the introduction to this publication, Katkov writes: "We firmly believe and clearly comprehend that Pushkin is not a poet of a single epoch but of the whole of mankind, not of a single country but of the whole world ... a great poet of felicity and inner harmony.... The People of Russia are not inferior to any other people; likewise, Pushkin is not inferior to any other poet in the world." Katkov informs the Russian readers that Varnhagen's article first appeared in the journal that Hegel founded. Katkov continues: "In this journal, the foreigner has uttered the full and solemn recognition of our motherland's greatness, he has articulated an adequate appreciation of our Pushkin." Katkov insists that this foreigner, Varnhagen, is Hegel's worthy pupil and speaks on his behalf. "On Hegel's behalf, Germany itself outstretches its hand to us, and on behalf of Germany, all of Europe and mankind.... They do not call us students and imitators anymore.... Our teachers address us as their equals.... Shame! We have been surpassed in the appreciation of our Pushkin!" (p. 4).

10. About V. N. Stankevich (1813–1840) and his circle, see *Byloe i Dumy* (chapter 25) by A. I. Gertsen. *Polnoe sobranie sochinenii v 30-ti tomakh* (Moscow, 1956), IX, pp. 9–46; Chizhevskii, *Gegel' v Rossii,* pp. 50–75; V.V. Zen'kovskii. *Istoriia russkoi filosofii v 4-kh tomakh* (Paris, 1948), vol. I, part 2, pp. 39–72; Edward J. Brown. *Stankevich and His Moscow Circle 1830–1840.* (Stanford, 1966); Iu. Mann, *Russkaia filosofskaia estetika,* pp. 191–248. Stankevich met K. A. Varnhagen in 1839, in Berlin, at the Russian literary salon of the Frolovs (see Chizhevskii, *Gegel' v Rossii,* p. 75; Mann, *Russkaia filosofskaia estetika,* pp. 231–32).

11. For studies on Belinsky's world-view, see Ivanov-Razumnik. *Velikie iskaniia* (Sankt-Peterburgh, 1911), vol. III (*Vissarion Grigor'evich Belinskii*); see also: Chizhevskii, *Gegel' v Rossii*, pp. 113–43; Terras, *Belinskij and Russian Literary Criticism*, pp. 33–38.

    In his article, Varnhagen calls Pushkin's travels around Russia "wanderings" and suggests that it was "beneficial to his poetry." Here Varnhagen introduces the connotation of *Bildung* into the notion of "wanderings"—the general Romantic concept, which had a very strong influence on Dostoevsky's perception of "wanderer" (*skitalets*) and "wanderings" (*skitaniia*).
12. Chizhevskii, *Gegel' v Rossii*, p. 277.
13. On the polemics between the "Gogolian" and "Pushkinian" movements in literary criticism in the context of the history of ideas in Russia, see B. F. Egorov, *Bor'ba esteticheskikh idei v Rossii serediny XIX veka* (Leningrad, 1982).
14. Chizhevskii, *Gegel' v Rossii*, p. 266.
15. M. Kanevskaya, "F. M. Dostoevsky's Pushkin Speech in a Cultural Context" (Ph.D. diss., Bloomington, 1997).
16. At first glance, Dostoevsky endeavors to overcome the controversy between Pushkin's admirers and opponents in his "Pushkin Speech." He begins with a quotation from Gogol, an authority for the entire audience. This friction between the "Pushkinian" and "Gogolian" movements still existed in 1880. For example, at the celebrations, on the day before Dostoevsky's speech, I.S. Turgenev expressed the opinion of the liberal intelligentsia in his speech. Turgenev narrowed the criterion of "People's poet" (*narodnyi poet*) to the meaning "comprehensible to common people," and discarded it as nonsensical. Then he expressed doubt about Pushkin's right to be named a "national poet in the sense of Shakespeare, Goethe, etc." Turgenev preferred to "leave this question open" (I. S. Turgenev, *Polnoe sobranie sochinenii i pisem v 28-mi tomakh* [Moscow-Leningrad, 1960–68], vol. XV, pp. 68–69).
17. On the significance of Russia's isolation from European civilization to the course of Russian thought, see Zen'kovskii, *Istoriia russkoi filosofii*, vol. I, pt. 1, pp. 115–219.
18. Dostoevsky expressed these political views many times—for example, in his *Diary of a Writer* under the general term "the Eastern question." His views clashed bitterly with the opinions of the liberal majority of the Russian intelligentsia. These circles also disapproved of Pushkin's nationalistic predilections. See for example V. Vodovozov, "Pushkin's Political and Social Views in the Last Period of His Life" (Politicheskie i obshchestvennye vzgliady Pushkina v poslednii period ego zhizni), *Pushkin [Sobranie sochinenii]*, ed. S. A. Vengerov (Petrograd, 1907–15), vol. VI, pp. 368–89.
19. Charles Taylor, *Hegel* (Cambridge, London, New York, Melbourne, 1975), pp. 80–94.
20. In his speech, Dostoevsky equated "pan-human unification" with the unification of "all tribes of the great Aryan clan." The notion "Aryan," it should be noted, had undergone numerous interpretations in the previous two centuries and needs a moment of commentary. In the nineteenth century, the term referred to Europeans or the peoples speaking Indo-European languages. The idea of language taxonomy was first introduced in 1786 by the British Orientalist Sir

William Jones. It was based on the hypothesis of a generic relationship between certain European languages and their derivation from a common source. The term "Aryan languages" was proposed by Max Friedrich Müller, the German Orientalist and historian of religion, at the beginning of the nineteenth century. According to his theory, three great races inhabited Asia and Europe: the Turanians (including Uralic and Altaic peoples), the Semites, and the Aryans, with three corresponding large language families. Jones's assertions were confirmed by such contemporary scholars as the Dane Rasmus Rask and the German Jacob Grimm. The last full-scale treatment of the Indo-European language family is found in the multivolume work of the Neo-grammarian Karl Brugmann titled *Grundriss der verleichenden Grammatik der indogermanischen Sprachen* (Outline of Comparative Indo-European Grammar, 1897–1916). Thus, the term "Aryan tribe" was familiar to Dostoevsky's audience. Toward the middle of the nineteenth century, this notion, as defined in the writings of Herder, Fichte, Schiller, Renan, and others, circulated widely in Russia. However, under the influence of J. A. Gobineau's book *Essai sur l'inégalité des races humaines* (Essay on the Inequality of Human Races, 4 volumes, 1853–55), this notion had already begun to imply the idea of the superiority of the Germanic peoples, who were seen as incarnating the summit of civilization. Characteristic in this respect is the prominent populist journalist N. K. Mikhailovsky's response, in which he directly accuses Dostoevsky of nationalism and of anti-Semitism (see N. K. Mikhailovskii, "Literary notes," *Otechestvennye zapiski* [1880], no. 9, II, p. 131). Nevertheless, the majority of Dostoevsky's audience realized that he meant "Aryans" as a Christian conglomerate. This view of Christianity as the religion of the highest order, as an absolute religion of complete spirituality and freedom, is derived from Hegelian philosophy (specifically from his *Lectures on the Philosophy of Religion*, 1832). For a complete classification of the nationalist European theories of the nineteenth century, see William Zebina Ripley, *The Races of Europe: A Social Study* (1899). For a modern perspective on the same subject, see Ivan Hannaford, *Race: The History of an Idea in the West* (Baltimore, London, 1996), especially pp. 235–324.

Chapter 8

CHURCH AND STATE

Dostoevsky and Kant

Alexander von Schönborn

The title of this essay is the conjunction of what one might view as two truisms. The first of these is that Dostoevsky's main concern throughout his life was with the nature of social order and, more particularly, with its specifications as church and state. The large secondary literature follows Dostoevsky himself in viewing the "Legend of the Grand Inquisitor" chapter in *The Brothers Karamazov* (1879–80) as the culmination of his lifelong pertinent ruminations. Thus, to pick but one example, Bruce K. Ward[1] writes in regard to this chapter: "This short writing, considered by him to be the 'culminating point' of *The Brothers Karamazov*, can be regarded as the culmination also of his religious and political thought—his 'final statement' concerning the question of human order." The key phrases here—"culminating point" and "final statement"—are Dostoevsky's own, cited by Ward from a letter to N. A. Liubimov dated 10 May 1879.[2] The second truism leading to the title of this short essay concerns the conjoining of Dostoevsky and Kant. This chapter presents a dispute in regard to the issue of whether freedom or happiness ought to be the highest principle in human affairs. Kant initially raised this question, which has ever since vexed ethicists and thoughtful people generally, reflecting on the standards of their conduct.

I want therefore to look at Dostoevsky's "Grand Inquisitor" chapter in the light of Kant's philosophy. In particular, I seek to ascertain how the Cardinal's rejection of freedom in favor of happiness might be understood in terms of the conceptual resources provided by Kant. The result of this effort is to provide various possibilities of interpretation between which a reader of the rest of the novel and of Dostoevsky's other works will have to decide.

Unfortunately, the secondary literature linking Dostoevsky and Kant is exceedingly sparse. What little there is does not prove helpful. H. Tencler, in his book on the problem of freedom in *The Brothers Karamazov*, mentions Kant once to indicate that what Dostoevsky means by "soul" is what Kant calls "personality."[3] In terms of more extended treatments, there is the work by Ia. Golosovker.[4] Since we know that on 22 February 1854—over a quarter century before the publication of *The Brothers Karamazov*—Dostoevsky wrote to his brother requesting a copy of Kant's *Critique of Pure Reason* (1781), Golosovker plausibly assumes that Dostoevsky read this work. He then treats the relation between Dostoevsky and Kant in terms of the doctrine of "the Antinomy of Pure Reason" as put forth in this work by Kant. Focusing specifically on the Third Antinomy, which is concerned with the theoretical issue of freedom and determinism, Golosovker argues that Dostoevsky rejects Kant's belief that the antinomies are insoluble and uses the character of Ivan in *The Brothers Karamazov* to argue against the critical philosophy. But this reading is not helpful. Not only does Kant not consider the antinomies insoluble, he considers it the singular merit of the critical philosophy that it alone solves them. Moreover, the Third Antinomy is concerned with freedom as a theoretical issue only; it is in Kant's moral and religious texts that freedom is treated as a practical principle, i.e., as a principle of human conduct, and Dostoevsky is clearly interested, in *The Brothers Karamazov*, in the question of how we ought to conduct ourselves. In this arena, Dostoevsky clearly is with Kant, not against him.

To anticipate how I think Dostoevsky sides with Kant, let me summarize my reading of "The Grand Inquisitor" chapter in slogan form. In the past, people adhered to the "Divine Command" theory of moral obligation. That theory has, for good reasons, been abandoned and is historically passé. It has been replaced by two more humanistically focused theories of moral obligation: the Kantian account, which defines freedom as autonomy, and the various teleological or consequentialist accounts that make happiness, however defined, the highest principle of morality. The Grand Inquisitor's defense of the latter principle leads to the realization that, when pushed to its extreme in politicized fashion (so characteristic for Dostoevsky), such a consequentialist principle manifests itself as but a secularized version of the Divine Command theory in all its unsavory aspects. This leaves but one viable account, that of freedom as the highest moral principle.

I assume, of course, that Dostoevsky was familiar with Kant's account, an assumption that seems quite reasonable not only in terms of the content of "The Grand Inquisitor" chapter, but in view of the following facts. As a very young man, Dostoevsky had already become acquainted with Kant's moral thought through N. Karamzin's (for whom Kant himself apparently summed up the main points of his ethics) *Letters of a Russian Traveler, 1789/90.*[5] Dostoevsky was also a member of the "Society of the Lovers of Wisdom" whose meetings "were concerned almost exclusively with the reading and discussion of modern German philosophers such as Kant, Fichte and Schelling" (17). Finally, various Russian translations of Kant's moral and religious writings were in wide circulation during the second half of Dostoevsky's life.

But let me start at the beginning—the Preface to Ivan's prose poem—which immediately raises a series of questions. Ivan begins by noting that if his story, set in the sixteenth century, had been written at the time of its setting, it would have required a preface "to bring down heavenly powers on earth."[6] Such a preface is then provided. It contains, as its center, a paraphrase of the poem "The Wanderings of Our Lady Through Hell." In it, Our Lady, shocked by what she observes, "bids all the saints, all the martyrs, all the angels and archangels to fall down with her and pray for mercy on all without distinction" (20). She obtains from God a respite from suffering, at which point the sinners cry thankfully from hell, "Thou art *just,* O Lord, in this judgment." With that cry the paraphrase of the poem ends, and Ivan immediately adds again that his "poem would have been of that kind if it had been written at that time" (20), which, of course, it was not. This raises some obvious questions: Why is the Preface there at all, especially in light of the fact that it is clearly historical in a way that no sixteenth-century preface ever was? What is the meaning of that strange cry—strange, given what one might imagine oneself saying under those circumstances—emanating from hell?

Parenthetically, I might note the unfortunate fact that the various major treatments of this famous chapter make no attempt to deal with the Preface. This strikes me as unfortunate because of Dostoevsky's contorted efforts to have it in place and the issues of content it raises. Connected with these omissions is the absence of an adequate account why a sixteenth-century setting was chosen for the story. The typical explanation—a number of examples can be found in a collection edited by J. S. Wasserman[7]—alludes to Dostoevsky's desire to castigate both socialism and Roman Catholicism for attempting the compulsory organization of human happiness and for claims to infallibility thereby required. This seems true enough. But the choice of a sixteenth–century setting makes reference to socialism by anything more than allusions and formulas obviously difficult. This difficulty could easily have been removed by setting the plot in the present, which would have had the added advantage of possible reference to the pronouncements of the First Vatican Council—*The Brothers Karamazov* was, after all, written in the decade following that Council's proclamation of papal infallibility. Be that as it may, I think that an appropriate consideration of both the Preface and the setting of the Legend provides an answer to R. L. Cox's plaint: "Some, notably Berdyaev, identify Dostoevsky with Christ rather than with the Inquisitor, but they point to no cogent and specific reasons within the chapter itself for doing so."[8]

But how might one interpret that thankful "Thou art *just*, O Lord, in this judgment"? Surely God was not unjust before granting the temporary reprieve from suffering, nor would He have been so without heeding Mary's supplication. I would suggest that the only view of justice on which this cry, under the conditions of its utterance, makes sense is the Divine Command theory of moral obligation. According to that view, to which even such late medieval thinkers as John Duns Scotus and William of Ockham were still partial,[9] "what ultimately makes an action right or wrong is simply its being commanded or forbidden by God and nothing else."[10] If God decrees no respite, then He is just; and if He decrees a respite, then He is just as well, because it is His decree that ipso facto makes something just. Such decrees, like any legal positivism, are accompanied by external sanctions and coercion, as indeed they must be: temporal punishment in this life and in purgatory, eternal reward or punishment in the life to come. Dostoevsky alludes to this view and does so in the distinctive fashion I have indicated to show that this view is no longer a live option. Very few Western philosophers of note have maintained it since late medieval times. Kant in his *Groundwork of the Metaphysic of Morals* (1785)—the most widely read of his ethical works—dismisses it as the very worst of moral theories because it forms "the basis for a moral system which would be in direct opposition to morality."[11]

Dostoevsky is, of course, not interested in the theory of obligation per se but rather in its historical realization, as both the historical character of the Preface and the later, nostalgic allusion by the Inquisitor to "the rigid, ancient law" prior to Christ's initial appearance on earth make clear.[12] It is worthwhile to look for a moment at the Israel of the Old Testament through Kantian eyes, for this community is clearly an instance of the theory I have mentioned above.

Kant distinguishes between state and church on the basis that the former is a civil society bound by coercive laws, while the latter is a civil society bound by noncoercive laws, i.e., moral laws. Legality and morality are not to be conflated. The principle of freedom as governing our maxims—the intentional rules prompting our actions—is the principle of morality. As normative for external legislation, for laws which prescribe actions rather than maxims of actions, the principle of freedom is the highest principle of the state and of its specification of civil rights. In regard to religion, Kant distinguishes between an ecclesiastic faith—convictions resting on statutory laws and on putative historical facts concerning the founding of a church—and a religious faith that is grounded in, and is an extension of, moral practice.

In terms of these distinctions, one would have to say that Israel had no church, but only a theocratically organized state, and that in this state there was no possibility of genuine religious faith, but only of an ecclesiastical faith. True religion was thus not possible in that historical context, given its conception of moral obligation. According to Kant, what God's becoming incarnate symbolically represents, regardless of historical beliefs that may have prompted this insight, is the possibility of genuine morality and hence of true religion. This, in turn, allows for an appropriate distinction between, and a coexistence of, church and state. But if so, then this Israelite realization of the Divine Command theory is as passé historically as the underlying theory of Divine Command. Both have thus yielded the stage—and thus serve quite literally as preface—to the contemporary options discussed by the Grand Inquisitor.

As mentioned, these options are freedom and happiness. They are respectively championed in our chapter by Christ and by the Grand Inquisitor, who is the servant of Satan. Why do freedom and happiness have these particular champions, given that, for example, we do not normally associate happiness with Satan or, more generally, with evil?

Here again Kant is highly instructive. In many of his works, Kant argues that the moral law (formulated, for example, as "Act in such a way that you always treat humanity, whether in your own person or in the person of any other, never simply as a means, but always at the same time as an end"[13] or, as I might put it very simply, "Respect every human being equally") is the expression of our autonomy; we are the source not only of free or self-initiated actions, but also, through practical reason, of the law to which these voluntary actions ought to conform and for the sake of which they ought to be undertaken. The moral law, he further argues, commands that all of our intentions and dispositions conform to it. This entails, since complete fitness of all our intentions to the moral law is defined by him as "holiness," that the moral law requires us to attain holiness. In his *Religion Within the Limits of Reason Alone* (1793), Kant goes on to add that since holiness is originally the trait of divinity, the moral necessity of such holiness in human beings requires us to think of God as uniting Himself with humanity: God made Flesh. Christ is thus the exemplification and instantiation of the ideal of moral perfection, of a fully autonomous human being. We experience the actuality of this incarnation in moral obligation; how this is possible we do not and cannot understand, though we must believe in the possibility of this incarnation as a postulate of the rationality of our moral life. Religious belief, therefore, is generally an outgrowth of our practice of morality, of the right exercise of our autonomy. As such, genuine religious faith cannot countenance any aspect of ecclesiastical faith that has a destructive impact on practical rationality as the source of our autonomy or on freedom as the genuine foundation of religious faith. On the other hand, any human act not springing from autonomy is morally wrong, even if, viewed externally, it appears to be in accord with the moral law; its principle is one of heteronomy rather than autonomy. When we make a principle of heteronomy the basic reason for our action, we subordinate our autonomous freedom to that principle. Such subordination is the nature of moral and religious evil. Since happiness, as the putative ground of our conduct, is for Kant the paradigmatic instance of an heteronomous principle when this happiness is made the ultimate justification of that conduct, it is easy to see why Satan's man should be cast in the role of the champion of happiness.

Let me now turn to the dialogue itself or, rather, to the Inquisitor's monologue. What are the Inquisitor's charges? I quote:

> I swear, man is weaker and baser by nature than Thou hast believed him! Can he, can he do what Thou didst? By showing him so much respect, Thou didst, as it were, cease to feel for him, for Thou didst ask far too much from him—Thou who hast loved him more than Thyself! Respecting him less, Thou wouldst have asked less of him.[14]

Again, in altered form:

> Thou wouldst not enslave man by a miracle, and didst crave faith given freely, not based on miracle.[15]

> There are three powers, three powers alone, able to conquer and to hold captive forever the conscience of these impotent rebels [i.e., most of

> humankind] for their happiness—those forces are miracle, mystery, and authority. Thou hast rejected all three and hast set the example for doing so.[16]

In order to understand the Legend, it is important not to cede "miracle, mystery, and authority" to the Inquisitor; after all, the story begins with Christ performing a miracle. As Cox has noted, "the Inquisitor reverses the real situation—he is the one who rejects miracle, mystery, and authority, and proposes to meet man's needs by *magic, mystification,* and *tyranny.*"[17] These Christ does reject. As the Inquisitor notes, "Thou wouldst not deprive man of freedom and didst reject the offer, thinking, what is that freedom worth, if obedience is bought with bread?"[18]

"Bread," in the remark last cited, is the Inquisitor's symbol for happiness, so that his question becomes "What is freedom worth if acting for its sake is motivated by and hence subordinated to happiness?" The answer from Christ's perspective is that, morally and religiously speaking, it is worth nothing at all. But why do I think that the Grand Inquisitor uses "bread" as a synonymous expression for "happiness"? The antonym to "bread" in the story is "hunger." Like Kant, the Inquisitor sees human finitude in terms of empirical needs, deprivations, and the desires for their respective assuaging. The satisfaction of our desires, to the extent that they fit into a coherent whole, is generally the modern Western conception of happiness.

For Kant and Dostoevsky, of course, this is only part of the human condition. A human being is also a being of reason or, in Dostoevsky's terms, a being with a soul, and practical reason—or, more colloquially, will—specifies human ends that set limits qua morality to the intended satisfaction of desire. Our autonomy dictates what we morally ought or ought not to do. Moreover, for Kant, though not for Dostoevsky, autonomy entails as external freedom a body of rights, of civil liberties. Dostoevsky restricts the principle of freedom to internal freedom—having a conscience—and views civil rights only as a historically associated co-product of what he calls in *The Raw Youth* the "Geneva" idea.

Kant takes the unconditional character of autonomy's demand on us to mean that we can satisfy it. In his famous dictum, "Ought implies can." It is in regard to that "can" that the Inquisitor, on the one hand, and Dostoevsky (with the restriction just mentioned) and Kant, on the other, part company. For the latter, that "can" characterizes every human being as a moral subject, as a "person"; for the former, it characterizes only the strong, the scriptural "elect." Put in terms of the basic human dispositions pertinent to will, which Kant analyzed at the start of his

Religion Within the Limits of Reason Alone,[19] the Inquisitor believes that all human animals are capable of attaining "humanity"—theoretical rationality, which can also serve desire—but only very few are capable of the exercise of "personality," of being capable of moral accountability or of that respect toward the law of the will that is a sufficient incentive for our power of choice. Put differently, according to the Inquisitor we are all humans, but very few of us are persons in fact. He vacillates about how this fact is to be construed. At times he suggests that the majority of human beings will become premoral children still using the notions of right and wrong and those of heaven and hell as well—even atheists such as the Inquisitor or Robespierre recognize the necessity of the latter notions for social order in situations where the coercive sanctions of positive law are not effective. However, being unable to derive the notions of right and wrong from their own theoretical but also premoral rationality, these "children" will have to receive them heteronomously. At other times, the Inquisitor suggests that all human beings have moral autonomy—they all have a genuine conscience—but the vast majority choose to give its exercise away without culpability owing to weakness of the will, self-deception, and deception by the state. For example, if being able to ascertain the facts relevant to a moral decision is a necessary condition for the exercise of autonomy, they would lack this condition (à la George Orwell's *1984*).

What are the implications of this profoundly ambiguous characterization of human beings? The Inquisitor has no quarrel with freedom as the highest internal principle in the lives of persons, a relatively tiny subclass of humans. But for the vast majority of humans, freedom in the sense of autonomy cannot be the highest principle of their inner and outer conduct. It is in reference to this human multitude that the Inquisitor declares: "Dost Thou know that the ages will pass, and humanity will proclaim by the lips of their sages that there is no crime, and therefore no sin; there is only hunger?"[20] Universal happiness must, for most of us, take the place of autonomy whose demands we cannot or will not meet.

To say that, for most of us, universal happiness must be the highest principle does not mean, to the Inquisitor, that this principle suffices. As he sagely notes, "without a stable conception of the object of life, man would not consent to go on living, and would rather destroy himself than remain on earth, though he had bread in abundance" (32). Given the modern growth in consciousness of freedom, it will thus be necessary to persuade the human majority that "they will only become free when they renounce their freedom to us and submit to us" (38), so that "people are more persuaded than ever that they have perfect freedom, yet they have brought their freedom to us and laid it humbly at our feet" (26). The freedom thus relinquished cannot be that of autonomy; it must refer to such manifestations of freedom as civil liberties, as the preferential choices of prudence, and as what the Inquisitor himself calls "free thought" (38). The Inquisitor does not mention the devices, beyond simple deception, whereby this persuasion is accomplished. Orwell's *1984* can once again suitably fill in the blanks.

According to the Inquisitor, the quest for universal happiness will initially turn to science for the satisfaction of hunger. But science will prove disappointing. In meeting needs, it generates new ones without a limiting principle so that "freedom and bread enough for all are inconceivable together, for never, never will they be able to share between them!" (38).

It is here that Dostoevsky and Kant part company. Writing the better part of a century after Kant, Dostoevsky might have sought to expand the modern doctrine of civil rights through a deeper understanding of the notion of autonomy than Kant's. Indeed, given his considerable sensitivity to profound economic inequality and inequity, he might have pointed ahead to the next century by seeing whether a doctrine of economic rights might be anchored in the principle of freedom as autonomy. Instead and as mentioned, Dostoevsky equates the principle of freedom with inner freedom. This forces him, in effect, into ceding the political arena to the Inquisitor. Dostoevsky seems convinced that the individual pursuit of happiness within the juridical framework of a liberal state grounded in a conception of freedom, as envisioned by Kant, can in principle neither satisfy nor contain within moral limits the demand for "the universal happiness of man" (37). More and more state intervention—"socialism"—will be required for building the requisite institutions and, given the disparity of wealth among nations, will finally have to take the form of a single world–state. Moreover, this socialist or communist program will have to be melded with, in Kant's terms, an ecclesiastic faith ascribing omniscience or infallibility, paternalistic benevolence, and beneficence to its governing group. For only such can be trusted to distribute bread equitably and to make continuous progress in its maximal availability. (We need only recall here the role of the chocolate rations in Orwell's *1984*.) Only such an ecclesiastical state will be "able to conquer and to hold captive forever the conscience of these impotent rebels for their happiness" (33).

Having relinquished the political realm, Dostoevsky envisions his alternative social order as a form of Christianity. Specifically, he seems to envision the church as the community of religious faith based on the principle of freedom—Luther's and Kant's "invisible church"[21]—with some monks of the Russian Orthodox community and the Russian commune forming, again in Kant's terms, the visible and true church, representing the invisible to the extent possible within experience through such marks as freedom, holiness, universality, and depth of redemptive suffering. Dostoevsky calls this visible community "Russian socialism."[22] As Father Zossima urges elsewhere in the novel, the state must become the church! This is not a demand for a true theocracy in opposition to the Inquisitor's illusory one; it is not a demand for a current version of ancient Israel. What Zossima demands is "church" as specified above. Dostoevsky thus seems implicitly committed to what, from a political perspective, might be called a Christian anarchism. (Perhaps, given Russia under Alexander II's late rule, this provides an additional reason for Christ's silence in the "Grand Inquisitor" chapter.)

But let me return to the Inquisitor's infallible state. What Dostoevsky means to show through its characterization by the Inquisitor is that the quest for universal happiness must culminate in a state of affairs that can only be described as a secularized institutionalization of the Divine Command theory, a humanistic refurbishing of a discredited view and of the practices it entails. The Inquisitor himself comes to this realization when, upon having fully articulated his view, he is nevertheless kissed by Christ in a sign of love and respect. Freedom is the right moral principle, but its demands are, in the Inquisitor's enduring judgment, too stringent for the majority of human beings to meet. In short, he continues to see no alternative except for the Elect. "The kiss glows in his heart, but the old man adheres to his idea."[23]

Notes

1. B. K. Ward, *Dostoyevsky's Critique of the West* (Waterloo, 1986), p. 101.
2. Ibid.
3. H. Tencler, *Das Freiheitsproblem in Brüder Karamazov von Dostoevski* (Riga, 1928), p. 26.
4. Ia. E. Golosovker, *Dostoevsky and Kant* (Moscow, 1963).
5. Ward, *Dostoyevsky's Critique of the West*, p. 16, n. 33.
6. F. Dostoevsky, *The Grand Inquisitor on the Nature of Man* (Indianapolis, 1948), p. 19.
7. F. Dostoevsky, *The Grand Inquisitor*, ed. J. S. Wasserman (Columbus, 1970).

8. R. L. Cox, "Dostoevsky's Grand Inquisitor," in F. Dostoevsky, *The Grand Inquisitor*, pp. 77–92, here p. 77.
9. A. Fagothey, *Right and Reason* (St. Louis, 1953), p. 111.
10. W. K. Frankena, *Ethics* (Englewood Cliffs, N. J., 1973), p. 28.
11. I. Kant, *Groundwork of the Metaphysics of Morals* (New York, 1956), p. 111.
12. Dostoevsky, *The Grand Inquisitor on the Nature of Man*, pp. 32–33.
13. Kant, *Groundwork of the Metaphysics of Morals*, p. 96.
14. Dostoevsky, *The Grand Inquisitor on the Nature of Man*, p. 35.
15. Ibid., p. 34.
16. Ibid., p. 33.
17. Cox, "Dostoevsky's Grand Inquisitor," p. 79.
18. Dostoevsky, *The Grand Inquisitor on the Nature of Man*, p. 29.
19. I. Kant, *Religion Within the Limits of Reason Alone* (New York, 1934), pp. 21–23.
20. Dostoevsky, *The Grand Inquisitor on the Nature of Man*, p. 29.
21. Tencler, *Das Freiheitsproblem in Brüder Karamazov von Dostoevski*, p. 36.
22. Ward, *Dostoyevsky's Critique of the West*, pp. 179–81.
23. Dostoevsky, *The Grand Inquisitor on the Nature of Man*, p. 45.

Chapter 9

The Shape of Russian Idealism

From Kant and Hegel to Dostoevsky to the Russian Religious Renaissance

Steven Cassedy

Dostoevsky

To judge from a letter Dostoevsky wrote in 1854, just after his release from prison camp, access to the works of Kant and Hegel was practically a matter of life and death. The young writer implored his brother to send him Kant's *Critique of Pure Reason* and Hegel's *History of Philosophy*, adding urgently that his entire future depended on this.[1] So far, no one has presented any hard evidence to show that Dostoevsky ever studied (or perhaps ever even perused) the two books that were so vital to his future or that he read any other works by Kant or Hegel. One could certainly argue that German idealism was so much a part of the intellectual culture in which Dostoevsky passed his early adult years that no reading, writing, and thinking person could resist its lure. In Dostoevsky's writings, there is no mistaking a set of concerns similar to those that the two German philosophers addressed, nor is there any mistaking the same set of concerns in Vladimir Soloviev (1853–1900) and Pavel Florensky (1880–1937), true philosophers who were inspired by Dostoevsky's fiction but who had a firsthand acquaintance with Kantian and Hegelian thought.

Dostoevsky's world is a world of barriers and limits. The thinking subject in this world knows of the existence of a higher reality but is constantly reminded that there can never be any true access to it. He (it is almost always male characters in Dostoevsky's works that reflect on these barriers and limits) knows that there is a perfect form of conduct (actually both a perfectly good and a perfectly bad form of conduct), but he is reminded that he can never achieve it. He knows that there is a perfect form of belief in an inaccessible higher reality (just as there is a perfect absence of belief), but he is reminded that he can never attain this, either. He knows that there is a perfect way to love his fellow human beings, but he is reminded that it is impossible ever to realize perfect love. He knows that there is a truth that can reveal the causes of all the world's injustices, but he is reminded that such a truth is "not of this world" and that he will never know it. He knows that there is a perfect brotherhood, founded in the perfect practice of a perfect faith, but he is reminded that such a brotherhood is merely a distant and probably unattainable goal.

Barriers and limits present themselves in two ways in Dostoevsky's world, one static and one dynamic. In the first, the subject finds himself captive within "this world," confronting an immutable and impenetrable barrier on the other side of which lies an ideal quality (Truth, God, perfect love). In this static conception, there are two worlds: we inhabit one, the ideal quality inhabits the other, and we cannot cross the barrier between them. In the second, dynamic conception, an ideal exists at the end of a process; the subject finds himself on a continuum that leads to the ideal end, though he knows that he can never reach it. The process can be a personal quest, as when a saintly man devotes his life to fulfilling God's commands or an individual searches for perfect, unselfish love; or it can be something much larger, like the movement of world history toward an absolute end.

I will limit myself to one example—the diary entry Dostoevsky wrote as he kept vigil over the body of Maria Dmitrievna, his first wife, in April 1864. In a haunting combination of the personal and the philosophical, the grieving writer notes that "Masha is lying on the table," asks "Will she and I see one another again?" and then launches into a prolonged meditation introduced by this thought: "To love a person *as oneself*, according to Christ's commandment, is impossible. The law that there must be such a thing as an individual person is binding on earth."[2] In other words, true, ideal Christian love is impossible on earth since such love, if total, would bring about a complete merging of the self with the other and hence the annihilation of the self. There are two worlds: the material world, where physical reality prevents the ideal fusion of two individuals, and a transcendent world, where such fusion is possible.

But as Dostoevsky's meditation continues, it becomes clear that the "two worlds" model does not exhaust his thinking on this subject, for he soon begins to write of a process of development (*razvitie*) by which we earthly subjects can come ever closer to the ideal (without, of course, actually reaching it). Only Christ could fulfill His own commandment, Dostoevsky notes,

> but Christ was an eternal ideal toward which man strives and, by a law of nature, must strive. After the appearance of Christ as the *ideal of man in the flesh*, however, it became as clear as day that the highest, the ultimate development [*razvitie*] of the individual person must reach the point (at the very end of the development, at the very point of attainment of the goal) where man can find out, recognize [*soznal*], and, with all the force of his nature, be convinced that the highest use he can make of his individual person, of the fullness of the development of his self, is, as it were, to annihilate this self.[3]

Life on earth is characterized by the "law of the development of the individual person," while existence on the other side (true Christian love, the nature of God) is "synthetically full, eternally pleasurable and filled"; it is an existence where "there shall be time no longer," where we find a "merging of the full self [*slitie polnogo* ia] ... *with everything*."[4] So in this formulation, it is not a barrier that separates us from the ideal, but a continuum that we must, by a law of nature, follow to an ideal "end."

Thus Dostoevsky's writings reflect both a static conception of the ideal (God, perfect love, etc.) as transcendent object and a dynamic conception of the ideal as limit (or limit-of-development). One might say that the first conception corresponds, *mutatis mutandis*, to Kant's theoretical reason and the second, to Kant's practical reason. For theoretical reason, such "ideas" as God, freedom, and immortality are simply transcendent, and the human subject, whose ordinary understanding is limited to the operation of the categories, cannot escape a static position from which those ideas can never be reached.[5] For practical reason, on the other hand, the object that is sought is the highest good, and as Kant tells us in the *Critique of Practical Reason* (1788) and *Religion Within the Limits of Reason Alone* (1793), we approach that infinitely distant object dynamically, by a "progress *ad infinitum*," or "a continuous progress from the deficient good to the better *ad infinitum*."[6]

But Dostoevsky's dynamic conception also evokes the language of his Hegelian contemporaries, who were inclined to view the human spirit and human history as a process of development toward Absolute Knowledge or the Absolute Idea. At the end of the *Logic* section of the *Encyclopedia*, Hegel describes the Absolute Idea as the "unity of the subjective and objective idea," "the absolute and entire Truth," "the unity of the theoretical and the practical idea." "Until now," he says in a textual "addition" (*Zusatz*), "we have had as our object, throughout, the Idea in its development through its various stages; now, however, the Idea is objective in itself."[7] And in the final section of the *Phenomenology of Spirit*, devoted to Absolute Knowing, Hegel speaks of a moment when Spirit extinguishes time, and the self "cancels out" or "supersedes" (*hebt auf*) its "time-form" (*Zeitform*).[8] Though Dostoevsky employs a mixture of scriptural and popular theological language to express his sentiments, terms like "development" (*razvitie*) and "ideal," together with references to progress toward an ideal goal where differences are resolved and the separation between subject and object is erased or narrowed, are a fairly reliable indication that a representative of the Russian intelligentsia in this era had Hegel or one of Hegel's followers in mind. It is hardly surprising to see Dostoevsky use such terms at this time, since this appears to be the period of his greatest exposure to Hegelian philosophy through his friendship and collaboration with the philosopher Nikolai Strakhov.[9]

Philosophers such as Vladimir Soloviev and Pavel Florensky drew heavily on their direct knowledge of the Western philosophical tradition, and so the classic words and phrases of German idealism find an easier application in a discussion of their works. Where Dostoevsky speaks informally of an ideal of love that is unrealizable on earth (thus placing a wall between our material existence and the Christian ideal), Soloviev or Florensky might well speak of the impossibility of directly apprehending the underlying mechanical causes of experience or the inadequacy of intellectual knowledge (thus placing a Kantian wall between our cognitive faculties and the thing-in-itself). Where Dostoevsky speaks informally of "an eternal ideal toward which man strives" and "ultimate development," Soloviev or Florensky might well outline a dialectical process of history that strongly resembles what we find in Hegel.

Soloviev

Dostoevsky and Soloviev were close friends during the time when Soloviev was writing his early and enormously influential *Lectures on Godmanhood* (1877–81) and Dostoevsky was writing *The Brothers Karamazov* (1879–80). It appears impossible to sort out precisely the manner in which the two influenced each other intellectually in this relationship, though much has been written on the subject.[10] It is unlikely that Dostoevsky had much of anything to offer his much younger friend in the way of formal, speculative philosophy. And yet one cannot deny the possibility that Dostoevsky suggested certain ways of looking at things, a certain "shape of thought" for which Soloviev provided the sort of framework and terminology that we would expect in a professional philosopher. And whether Dostoevsky's particular ways of looking at things ultimately stemmed from his youthful encounter with a culture devoted to German idealism and Soloviev subsequently learned this directly from his older friend, or whether Soloviev simply took certain "idea-feelings" and gave them a framework that had its own origins partially in German idealism, there is no denying that Dostoevsky's cast of thought bears a strong resemblance to Soloviev's.

No work helped to establish Soloviev's reputation as much as the *Lectures on Godmanhood*, which the young philosopher delivered in public in the early months of 1878. Dostoevsky was among those in attendance. The first few of the *Lectures* present theories of experience and cognition that are essentially Kantian. This is clearest in the opening pages of the third *Lecture*, where Soloviev describes the "unconditional principle" of all that exists, something that he had mentioned at the very beginning of the first *Lecture*. "It is completely beyond doubt," he writes,

> that the reality of the unconditional principle, as something existing in itself and independently of us, the reality of God (like the independent reality in general of any other being besides ourselves) cannot be deduced from pure reason, cannot be proved purely logically. The necessity of the unconditional principle for the higher interests of man, its necessity for the will and for moral activity, for reason and truthful knowledge, for feeling and creation—this necessity makes the actual existence of the divine principle merely probable in the highest degree; a full and unconditional conviction in it can be given only by faith: and this applies ... not only to the existence of the unconditional principle but also to the existence of any object and the entire external world.[11]

Though Kant himself shuns the notion of probability in connection with a belief in the reality of such ideas as the existence of a supreme being and though he uses the term "faith" almost exclusively in connection with moral truths and with the ideas of God, freedom, and immortality, the claims Soloviev makes here are fully consistent with Kantian critical thinking.[12] In fact, in the sentence following the passage just quoted, Soloviev shows his debt to Kant even more clearly by writing that we know the world only through our own sensations and that any inferences we draw about it are merely probable.

In the fourth *Lecture*, Soloviev turns to the relation between the world as our representation and the underlying causes of that representation, once again drawing a distinction between the true, physical basis of experience and the experience itself. "If what we see is merely our representation, then it does not follow from this that this representation has no causes that are independent from us and that we do not see. The obligatory character of this representation makes the assumption of these causes necessary."[13] Later in this *Lecture*, after a brief discussion of Platonism, Soloviev asserts the existence of a "true reality," which he calls a "realm of ideas," and then describes an ascension through progressive levels of ideas to the broadest and most general of all, namely love (55–57). Thus Soloviev makes two distinctions. One is the epistemological distinction between the "true" basis of knowledge (what Kant would call the thing-in-itself) and subjective knowledge. The other is between the sensible world (without regard for how we apprehend it) and a supra-sensible world of ideas. Presented in this way, both distinctions are ultimately static.

It soon becomes clear, however, that such static notions do not reveal the whole picture. As Soloviev shows early in the *Lectures on Godmanhood*—and in almost everything he wrote—the world at all levels is characterized by *process*. Ordinary life is a "natural, material-conditional process, subject to the laws of physical necessity" (27). In fact, this observation is what leads Soloviev to assert the existence of the unconditional principle. That principle, too, is characterized by process. In the tenth *Lecture*, Soloviev speaks of a "world soul" (*mirovaia dusha*). "The eternal idea of the absolute organism," he says, "must be gradually realized, and the striving toward this realization, the striving toward the incarnation of the divinity in the world, the striving that is general and single in all and that consequently passes beyond the limits of each, this striving, which constitutes the inner life and the principle of movement in everything that exists, is the world soul." He goes on to assert that the world soul contains within itself the divine principle, which he now defines as the "active, forming, and determining principle of the world process" (144–45).

Soloviev here introduces the terms "Logos" and "incarnation." Though the incarnation can be seen as a static bridge between the two worlds in a static conception, Soloviev shows here that it is a dynamic process. In fact, he explicitly mentions the contrast between the static and the dynamic views of the incarnation: "The divine principle in and of itself is the eternal all-unified [*vechnoe vse-edinoe*], abiding in absolute peace and immutability; but in relation to the plurality of finite being that issues from it, the divine principle appears as an *active* force of unity—a Logos *ad extra*.... Thus the divine principle appears here (in the world process) as an active force of the absolute idea, striving to realize itself or incarnate itself in a chaos of disparate elements" (145). The end of the process that Soloviev describes here is the ideal union of the world soul with the divine principle, a union that represents the incarnation of the idea (146).

Elsewhere in Soloviev's works, the principle of process has an even more obviously Hegelian cast. As it happens, this tendency is particularly evident in the "Three Discourses in Memory of Dostoevsky" (1881–85), in which Soloviev finds and celebrates his own theology in Dostoevsky's art. For example, in the "Second Discourse," he presents a vision that he claims to have shared with the late novelist. It is a vision of an ideal society that is composed of a free, pan-human (*vsechelovecheskoe*) unity, a universal brotherhood in the name of Christ. If a true society is to be founded in Christianity, however, we must remember that Christianity exists in three stages: temple (*khramovoe*) Christianity, focused on the external forms of worship; domestic (*domashnee*) Christianity, focused on private, individual worship; and universal (*vselenskoe*) Christianity, focused on all humanity. The first two stages of Christianity exist in reality as mere "facts"; that is, they are grounded in physical reality. The third does not exist in reality; it is a task (*zadacha*) (200). "The true Church, which Dostoevsky preached," Soloviev says, "is pan-human, above all in the sense that in it the division of humanity into tribes and nations striving with each other and hostile to each other must disappear at the end" (210). The final stage in this vision is clearly an ideal that exists at the end of a process, like Dostoevsky's concept of pure Christian love. But what is especially curious about the vision, which Soloviev attributes to Dostoevsky, is how closely it follows a dialectic in the style of Hegel. Soloviev's process moves toward a final stage that is both universal, as his label for it explicitly shows, and concrete, since at this stage Christianity, like the incarnation, "must spread itself over all mankind and all human affairs." Along the way to this final stage, Christianity has incorporated within itself, and transcended, elements of the individual and the particular, since, as Soloviev explains, at the domestic stage Christianity is concerned largely with the inward, individual life and the private affairs of man (200).

This incorporation and transcendence, incidentally, shows one sense in which Soloviev departs from Dostoevsky, even where he thinks he sees his own image in the works of his late friend. Dostoevsky's ideal is one that is positioned well beyond all individuation. For him, the synthesis implicit in ideal, Christian love and the principle of individuation in the finite, human world are mutually exclusive, whereas for Soloviev, the ideal represents a Hegelian *Aufhebung*, not only a cancellation or negation, that is, but also an incorporation and a transcendence of a lower level (in this instance, individuation). In a series of essays called "The Meaning of Love," written a little over a decade after Dostoevsky's death, Soloviev insists on this *Aufhebung* in human love in general, and sexual love in particular.[14] "The meaning of human love in general," he writes, "is *the justification and salvation of individuality through the sacrifice of egoism*" (where "egoism" means the tendency of an individual to esteem himself too highly and to give himself "an unconditional significance and infinite dignity") (16, italics in the original). All love, Soloviev goes on to say in the same essay, represents an attempt to overcome egoism to protect individuality. The ultimate joining (*soedinenie*) of two beings affirms individuality not only because the act of love itself affirms the individuality of each partner, but also because the act of sexual love in particular *creates* a new individual (19). For Soloviev, the ideal is thus a union, even an "inseparable joining," of two individual beings (22), not the absolute dissolution of an individual into the universe.

In the utopian *Russia and the Universal Church* (1889), Soloviev further develops his conception of the ideal church. He shows here what might very well be a textual debt to Hegel. In speaking of the Trinity, Soloviev explains how, in the third person of the Trinity (the Holy Spirit), "the divine being, by an interior doubling … arrives at the manifestation of its absolute unity, returns to itself … possesses itself, and enjoys itself in the fullness of its consciousness."[15] Similarly, in the final sentence of the *Encyclopedia*, Hegel discusses philosophy, "which has self-knowing reason, the absolute universal, as its center, which divides itself into spirit and nature." "The eternal Idea," he continues, "existing in and for itself, acts as absolute Spirit, engenders itself, and enjoys itself."[16]

Dostoevsky certainly never used language like that employed by Soloviev and Hegel in these passages. But, as Soloviev explained, even though the novelist had left behind no formal theory or system, he was still able to elevate his fundamental goal and idea to "an unprecedented height."[17] It is clear that the two writers agreed to a considerable extent not only on the essence of the fundamental questions of human existence, but also on how to conceptualize those questions. They appear to have agreed on a certain "shape" for describing those questions. They agreed that we start with a static, two-world configuration, in which the human subject finds itself on one side of an impenetrable barrier beyond which lie certain intangible qualities, and that we move from this configuration to the notion of progress toward an unattainable goal, whether that goal is selfless Christian love, as Dostoevsky saw it, the Universal Church, according to Soloviev, or Absolute Spirit (Absolute Knowing, Absolute Idea), as in Hegel.

Florensky

Florensky is another thinker who appears to have taken a static, two-world model as a logical point of departure and worked from that model to a dynamic, progressive model. Much of his principal work, *Stolp i Utverzhdenie Istiny* (The Pillar and Ground of the Truth, 1914), is devoted to duality (and the problem of its reconciliation with Trinitarian doctrine). In the first section, titled "To the Reader," Florensky introduces the term *tserkovnost'* (ecclesiality) and, after noting the difficulty of defining this term, asks if it is not simply equivalent to "*life*, particular, new life, given to people but, like every life, inaccessible to the intellect [*razsudok*]." In a lengthy footnote to this remark, he mentions as proponents of the view that life is inaccessible to the intellect the Slavophiles Khomiakov and Kireevsky, in addition to Henri Bergson and a number of other contemporary thinkers.[18]

But Florensky most dramatically presents a Kantian, dualistic worldview in the first of the twelve "letters" that make up his work. "Two Worlds" is the title of this letter, which Florensky presents in the form of an intimate, personal letter to a beloved but unnamed friend. The letter is organized around a series of separations: the author's separation from his friend, the separation between spring and autumn, between life as an amorphous rhapsody of inchoate impressions and the "Pillar and Ground of the Truth," between intellectual (*razsudochnoe*) and spiritual knowledge. Oddly (though characteristically for the Russian tradition he continues), Florensky presents in Christian terms what is clearly a Kantian epistemological dualism. He refers to Matthew 11, in which Christ preaches in the cities of His disciples. Among Christ's pronouncements are these:

> I thank Thee, O Father, Lord of heaven and earth, because Thou hast hid these things from the wise and prudent, and hast revealed them unto babes. Even so, Father: for so it seemed good in Thy sight. All things are delivered unto Me of My Father: and no man knoweth the Son, but the Father; neither knoweth any man the Father, save the Son, and he to whomsoever the Son will reveal Him. Come unto Me, all ye that labour and are heavy laden, and I will give you rest. Take My yoke upon you, and learn of Me; for I am meek and lowly in heart: and ye shall find rest unto your souls. For My yoke is easy, and My burden is light. (Matthew 11:25– 30)

In Florensky's view, this passage has a "principally cognitive" or even "theoretical-cognitive, gnoseological" significance. In fact, the subject of all of Matthew 11, as Florensky sees it, is the question of knowledge (*poznanie*), the "inadequacy of intellectual knowledge and the necessity of spiritual knowledge." Jesus, after all, says to God, "Thou hast hid these things from the wise and prudent, and hast revealed them unto babes." This statement, Florensky thinks, demonstrates that "intellectual [*umstvenn*oe] 'infancy' [the babes unto whom these things are revealed] … might very well turn out to be a condition for the attainment of spiritual knowledge." Jesus is actually emphasizing the difference between the "easy yoke" and "light burden" that He has to offer, on the one side, and the "cruel yoke and heavy, awkward burden of science," on the other. Florensky concludes this series of reflections by addressing the following remarks, reminiscent of Dostoevsky's diary passage, to his friend: "And if there were even one soul who felt I was talking to him not with my mouth and not into his ears, I could want nothing more. I know you will accept me, because you are the one who can burst the bounds of my egoism."[19]

The next letter in Florensky's work is titled "Doubt," and the following one is called "Triunity." The issue in both letters is Truth (*Istina*) and the necessity of conceiving of it in Trinitarian terms. Doubt is a dualistic concept—the Russian word, *somnenie* (literally, "co-thinking"), like the equivalent words in many other languages, suggests duality—and therefore inadequate to the full meaning of Truth. Florensky's discussion of doubt focuses on the truth claims of identity statements (A = A), which he sees as both dualistic (the second *A* in the equation is, for Florensky, in actuality a kind of "not-*A*") and empty. Genuine Truth, he shows in the letter on triunity, emerges only in a Trinitarian structure, which must be understood in terms of love. An escape from the emptiness of identity requires the presence of an Other, and through its relation with this Other, the *I* objectifies itself and becomes a *He*. This, in turn, suggests the Trinity: "Truth is the contemplation of Oneself through an Other in a Third: Father, Son, Spirit" (48). Truth is thus, like the Trinity, "one substance in three hypostases" (49).

It is also the vehicle by means of which we can escape the confines of individuality (which, we have seen, implies duality for Florensky). If knowledge (*poznanie*, the standard Russian term for ordinary cognition) cannot give us access to Truth, it is because only the interrelated powers of faith, hope, and love can do so. "I cannot *know* whether or not there is Truth," Florensky says. "But I love it—love it more than everything that exists." It is for the sake of Truth that we renounce doubt. "Through the threefold action of faith, hope, and love, the stagnation of the law of identity is overcome. I cease to be 'I,' and my thought ceases to be my thought; through an incomprehensible act, I renounce the self-affirmation of '*I* = *I*'" (67–68).

There is thus a tension in Florensky, just as there had been in Soloviev, between an ideal that cancels or negates individuality and one that transcends and incorporates it. As he shows later, in a discussion of friendship, the ideal both transcends and incorporates individuality, as it did for Soloviev. "On one hand," Florensky says of friendship, "the separate individual is everything; but on the other, it is something that exists only where there are 'two or three.' 'Two or three' is something qualitatively higher than 'one,' though it was precisely Christianity that created the idea of the absolute value of the separate individual. The individual can be absolutely valuable in no other way than in absolutely valuable communion" (419).

By the time Florensky gets to his fifth letter, "The Comforter," he has established that knowledge of Truth occurs through the grace of the Holy Spirit. "The Comforter" (*Uteshitel'* in Russian, *Paraklētos* in Greek) is an alternate name given to the Holy Spirit in John 14:26: "But the Comforter, which is the Holy Ghost, Whom the Father will send in My name, He shall teach you all things, and bring all things to your remembrance, whatsoever I have said unto you." A life in Truth, Florensky says, is directed by the Holy Spirit (110).

In his early discussion of Truth, Florensky had provided the foundation for a dynamic, processive model that would allow him to break free from static duality. Florensky showed Truth to be Trinitarian in a passage reminiscent of the section of the *Logic* in which Hegel derives the idea of "becoming" from the opposition between *being* and *nothing.* "The truth of being, like that of nothing, is thus the *unity* of both; this unity is *becoming* [*das Werden*]," Hegel wrote. "*Becoming* is the true expression of the result of being and nothing, as their unity; it is not only the *unity* of being and nothing, but *unrest* in itself—unity that exists within and against itself not just motionlessly as relation-to-itself, but through the difference between being and nothing that exists in becoming."[20] In a supplementary note to this passage, Hegel goes on to explain that "becoming is the first concrete thought and therefore the first concept, while being and nothing are empty abstractions," adding that life and spirit themselves are higher forms of becoming (192–93). To be sure, Hegel's purpose in this section of the *Logic* is different from Florensky's in the letters on doubt and Truth. But both show the clash of opposites (being and nothing in Hegel, *I* and not-*I* in Florensky) as the point of inception of a process (logical becoming in Hegel, the path toward Truth in Florensky).

Given what appears to be the irresistible logic of Florensky's dialectic model, it is not surprising that he soon begins to speak of a "path." In "The Comforter," he introduces the notion of progress, time, and history. A true knowledge of the Holy Spirit would bring about a complete deification of the created world, and that would represent the end of the "path," the end of history, the moment at the boundary of history when the Comforter reveals Himself and there is time no longer.[21] Our march to Truth must, then, be conceived as a progression through the persons of the Trinity. Florensky views this progression in historical rather than in purely individual terms. To each of the three hypostases there corresponds a stage in the development of the human spirit. The first hypostasis corresponds to the religion and life of antiquity and to its organic worldview; the second corresponds to the religion and life of the new era and the logical worldview of that era; the third corresponds to the subsequent, modern era, the striving toward beauty, the renunciation of "scientificity," the "joyful, light inspiration of the coming 'gay science,'" Florensky says in an acknowledged reference to Nietzsche (127–28).

Florensky finds a precedent for his views in the works of the fourth-century Church Father St. Gregory of Nazianzus, also known as St. Gregory the Theologian (c. 330–c. 389). In his *Fifth Theological Oration*, St. Gregory presents his doctrine of the gradual revelation of the Trinity, according to which the Father is revealed in the Old Testament, the Son is revealed in the Gospels, and the Holy Spirit is revealed in the subsequent life of "us" (members of the Church). The context of St. Gregory's remarks is a discussion of what he refers to as time's two notable "transformations" (*metatheseis* in the Greek, *preobrazovaniia* in Florensky's translation), to which he also refers as Testaments (*Diathēkai*) and "quakes" (*seismoi*). The first led from the period of idols to the period of the law (the Old Testament), and the second led from the period of the law to the period of the Gospel (the New Testament). St. Gregory proceeds to preach the coming of a third "quake," one that will move from this place or world to the other place or world, to a place of the immutable and unshakeable.[22]

Though Florensky cites a patristic precedent for his theory of the progress toward Truth, he gives his thought a distinctly modern, Hegelian cast. St. Gregory, of course, did not have a modern theory of history. When he refers, for example, to the two great "transformations," he uses the phrase "of all time" (*ek tou pantos aiōnos*) (160) but lacks the conceptual apparatus to refer explicitly to anything like the modern, Hegelian notion of historical progress. Florensky, however, easily reads

St. Gregory in modern terms. After a long quoted passage from the Church Father, Florensky offers this comment:

> St. Gregory the Theologian is affirming the gradualness [*postepennost'*] of the historical appearance [*iavlenie*] of the Spirit; but it is necessary to think of still another side, too—the interruptedness [*preryvnost'*] of His [the Spirit's] trans-historical revelation, just as the Kingdom of God has both a gradual-historical and an interrupted-eschatological appearance, neither of which is reducible to the other. Otherwise it is incomprehensible just how the final condition, the enlightening of God's creatures, the casting out of death, in a word, "the coming age," can be distinguished from the preliminary, waiting condition of "his age," in which death still reigns.[23]

In a footnote, Florensky refers us to a later section of his book, in which he speaks of Sophia, the Kingdom of God, and the possibility that both, which can be seen as ends of a process, preexist the world (332–33).

Dostoevsky proposed a model of process that leads to a condition outside of time, a model that bears a resemblance to Florensky's secular history as it approaches a condition in which there is "time no longer." This in turn bears a resemblance to the process by which Hegel's human spirit approaches its final, absolute stages. In the next-to-last section of the *Encyclopedia*, "Revealed Religion," Hegel presents his own Trinitarian structure of the next-to-highest stage of spirit, Christianity, which represents the highest form of religion. Hegel describes three different stages, each corresponding to a person of the Trinity. Curiously, once we reach absolute Spirit, stages like these are no longer temporal or historical. Hegel emphasizes the presupposed nature (*das Vorausgesetzte*) of the several persons of the Trinity, even though these persons are part of a stage of development that might be described as historical. In this way, incidentally, he follows a tradition in Christian thought that ascribes to Christ the paradoxical status of being both eternally preexisting and born at a determined historical moment.

As Hegel presents the Trinity, it is thus not proper to conceive of the Father as existing first and then engendering a Son and a Holy Spirit. Instead the Persons of the Trinity must be seen as proceeding *logically* and *necessarily* in a way that transcends temporality. This is because the three Persons correspond to the three moments of the Hegelian Notion (*Begriff*), namely the Universal, the Particular, and the Individual, and they are bound together by the logic of the Notion rather than by temporal sequence. Thus the Father (the Universal) begets a Son who is a distinct being, but who nonetheless remains in a condition of "original unity" with the Father.[24] This is because the Son, having become finite and particular, dies (the sign of His finitude and particularity) and then resumes the character of the Father. We then reach the moment of the Holy Spirit, "the idea of the eternal, but *living* Spirit present in the world" (376), where all contradictions are resolved. Thus even though in Hegel's view the highest stages of the human spirit come at the end of a dialectical or historical process, they are themselves, like the final moments of the process in Dostoevsky and Florensky, outside of time; in fact, they can be seen as even preexisting the dialectical or historical process at the end of which they occur. The very last paragraph of the *Encyclopedia* cited earlier shows the same tension between progressive development and extratemporality. If Absolute Spirit, as the highest moment of human spirit, exists eternally in and for itself, then it would appear to exist outside the process that leads to it, just as the Son of God exists outside the temporal process that led to His physical birth and just as the existence that Dostoevsky evokes as "synthetically full, eternally pleasurable and filled" is one in which "there shall be time no longer."[25]

Florensky had his own avowed reasons for admiring Dostoevsky. A few years after he published *The Pillar and Ground of the Truth*, in a set of autobiographical reminiscences he wrote of the manner in which Dostoevsky, in the spirit of the antinomies of the Apostle Paul, had revealed "the salvific nature [*spasitel'nost'*] of the fall and the blessedness of sin."[26] The antinomian character of Dostoevsky's thought that Florensky notices is no doubt a key to a deeper affinity between the two writers—the obsession with overcoming duality.

* * * *

Because of the difficulties facing anyone who attempts to uncover a clear philosophical lineage for a writer as unscholarly as Dostoevsky, one can only speculate on the reasons for the set of concerns that Dostoevsky shared with representatives of the Russian religious renaissance. One thing, however, is certain: Dostoevsky, Soloviev, and Florensky all confronted the same problem—namely, reconciling the duality of Kant's theoretical reason with the notion of temporal or spiritual process. For the three Russian writers, the path to an ideal end was itself the means by which duality was overcome, whether the ideal end was "the highest synthesis of life" (when there shall be "time no longer"), the establishment of the Universal Church, the reign of the Holy Spirit, or a number of other millennial moments.

Notes

1. F. M. Dostoevsky, *Polnoe sobranie sochinenii v tridtsati tomakh* (Leningrad, 1972–90), 28/I:173.
2. Dostoevsky, *Polnoe sobranie sochinenii,* 20:172. Italics in original.
3. Ibid., p. 172.
4. Ibid., pp. 173–74
5. Kant uses the term "ideas" for God, freedom, and immortality in the *Critique of Pure Reason,* B 395. See Immanuel Kant, *Werkausgabe,* ed. Wilhelm Weischedel (Wiesbaden, 1956), 1:338.
6. For the passage in the *Critique of Practical Reason,* see Kant, *Werkausgabe,* 7:259. For the passage in *Religion Within the Limits of Reason Alone,* see ibid., 8:720.
7. G. W. F. Hegel, *Werke in zwanzig Bänden,* ed. Eva Moldenhauer and Karl Markus Michel (Frankfurt am Main, 1970), 8:388.
8. Ibid., 3:584.
9. On Strakhov's possible role in Dostoevsky's Hegelian education, see Malcolm Jones, "Some Echoes of Hegel in Dostoyevsky," *Slavonic and East European Review* 49 (1971), pp. 500–20. Strakhov worked with Dostoevsky on the journals *Vremia* and *Epokha* between 1861 and 1865.
10. Marina Kostalevsky, in a recent book on the subject, has provided a detailed account of the friendship between Dostoevsky and Soloviev and has shown important resemblances in their work, but she has prudently avoided explicitly identifying intellectual debts where such debts are difficult to prove. See Marina Kostalevsky, *Dostoevsky and Soloviev: The Art of Integral Vision* (New Haven, 1997).
11. V. S. Soloviev, *Sobranie sochinenii,* ed. S. M Soloviev and E. L. Radlov (St. Petersburg, 1911–14), 3:32–33.
12. See the *Critique of Pure Reason,* A 775/B 803 (Kant, *Werkausgabe,* 2:656), for probability, and see B xxx (ibid., 1:33), for the mention of faith in connection with God, freedom, and immortality. Kant uses the word *Glaube* (belief, faith) in several slightly different senses in the final chapters of the *Critique of Pure Reason,* but what he calls "doctrinal belief" (*der doktrinale Glaube*) is the only one that applies to the doctrine of the existence of God. See A826/B 854 (ibid., 2:691).
13. Soloviev, *Sobranie sochinenii,* 3:51.
14. Soloviev, "Smysl liubvi" (The Meaning of Love), *Sobranie sochinenii,* 7:3–60.
15. V. S. Soloviev, *La Russie et l'église universelle* (Paris, 1922), p. 217.
16. Hegel, *Werke,* 10:394.
17. Soloviev, *Sobranie sochinenii,* 3:204.
18. Pavel Florensky, *Stolp i utverzhdenie istiny: Opyt pravoslavnoi feoditsei v dvenadtsati pis'makh* (The Pillar and Ground of the Truth: Essay in Orthodox Theodicy in Twelve Letters) (Moscow, 1914), pp. 5, 608–609n2. Florensky's monumental work has recently been published in English translation. See Boris Jakim, trans., *The Pillar and Ground of the Truth: An Essay in Orthodox Theodicy in Twelve Letters* (Princeton, 1997).

19. Ibid., pp. 12–13.
20. Hegel, *Werke*, 8:188, 191.
21. Florensky, *Stolp*, p. 112.
22. J.-P. Migne, ed., *Patrologiae Cursus Completus: Series Graeca* (Paris, 1857–66), 36:160–61. Much of St. Gregory's discussion paraphrases passages from Heb. 12.
23. Florensky, *Stolp*, p. 136.
24. Hegel, *Werke*, 10: 375.
25. The phrase (which is not always understood as describing an extratemporal state) is from Rev. 10:6.
26. Pavel Florensky, "O literature" (On Literature), *Voprosy literatury* 1 (1988), p. 159.

Chapter 10

Bakhtin, Benjamin, and Historical Representation of the Holocaust

A Juxtaposition

Michael Bernard-Donals

There is no evidence of which I am aware that Mikhail Bakhtin knew the work of Walter Benjamin. It is doubtful, also, that Benjamin could have known Bakhtin's work, given that the latter had been internally exiled and his writings had been effectively suppressed by 1929. In this essay, however, I propose to suggest that their work has a great deal in common, particularly as it addresses the problem of historical time and the possibility that the critic's and the historian's task is to rupture the language of narrative and of history in order to allow chronology to yield to the events themselves. There is a certain irony to the fact that the Russian-German juxtaposition in this case is so historically muted—their philosophical groundings are diametrically opposed (Bakhtin's in neo-Kantian philosophy, Benjamin's post-Kantian; Bakhtin was anti-Marxist while Benjamin turned to Marxism after flirting with Jewish mysticism) and their historical circumstances do not meet at all. Nevertheless, their philosophical positions find a meeting place not in history itself but in a tear in the fabric of history. They meet not in the most significant German-Russian juxtaposition of the twentieth century (i.e., World War II) but in a museum in the United States, built at the end of the century, commemorating the Jewish Holocaust.

Because of the nature of this collection, this essay can only be a brief analysis in which the work of the two thinkers is set side by side. The locus of the juxtaposition—the representation of the Holocaust—is important because the events that comprise the Holocaust trouble history to the point where we are forced to ask how or whether such an event can be represented to make it resonate in the contemporary moment. I can also analyze only a capsule of a single representation of the event, in this case a small portion of the permanent exhibit in the United States Holocaust Memorial Museum. But it is a moment, I think, that illustrates well the connections, and the importance of them, between Bakhtin and Benjamin and their theorization of an excess of history that may well be connected to what I will call the Divine.

* * * *

Walter Benjamin, a thoroughly assimilated upper-middle class Jew who was as adept with Kant and Hegel as he was with Kabbala, died by suicide at the French-Spanish border after he was denied entry to Spain while trying to flee the Nazis four months after their occupation of France. Benjamin was no historian, but what he was concerned with throughout his life was the nature of truth and the means by which truth, something which seemed at once Divine in origin and yet mutable in experience, could be glimpsed if not attained. Specifically with regard to chronological time, he was interested in the relation between the historical, momentary, and palpable objectivity of the object (or moment) of history and the suprahistorical meaningfulness of that same object that the Divine nature of truth reveals. What one learns from historical objects—in this case, objects found in a museum devoted to the disaster that led to Benjamin's suicide, such as a desecrated Torah ark or historical narratives of survivors (or, for that matter, perpetrators)—is the event's material content; what one learns from one's encounter with such objects or narratives is its truth content. Beyond these two moments, or in the relation between them, is the possibility of recovering the eventness of the event itself (for Benjamin, its Divine origin).

Benjamin's thinking about history and its relation to the actual or to the originary events that spawn history was profoundly affected by Jewish mystical texts collectively known as the Kabbala. At the moment of Adam's fall, the world—which had, according to these writings, been governed by the Tree of Life in which the Divine and the mutable were in communion—becomes differentiated in its elements. As a result, humans resort to the Tree of Knowledge, which enables them to distinguish between elements such as good and evil, the holy and unholy, the truth and the lie. The primary means by which humans do this is through the world of the senses and through language. Though there is an implicit link between sense experience and experience of the Divine, and between human language and Divine language, that link has been rendered invisible by language's (and sense-experience's) tendency to naturalize existence and the name. Though God created the world through an act of Divine language, "the Fall marks the birth of the human word, in which the name no longer lives intact, and which has stepped out of the language of names ... the word must communicate something (other than itself)."[1] Yet it was still possible to glimpse the originary link between the heterogenous human language and the Divine Ursprache. Revealing the shape of this link, if not the link itself, was the task of redemptive criticism.

But of course, all this work must take place in the realm of the profane: historical time. Our narratives of ourselves and our places in the world have so naturalized human language as to stand full force in the way of redemption: common language prevents the recovery of the Divine, of the originary, structurally determinate moment in which material content and truth content engage in relation. In order to engage in the historical work of redemption, the historian sees his job as "rescuing the few unique visions of transcendence that grace the continuum of history, the now-times (*Jetztzeiten*), from the fate of oblivion which incessantly threatens to consume them."[2] In the historical materialist language that characterizes the later phases of Benjamin's writings, the historian "takes cognizance of [the historical object] in order to blast a specific era out of the homogenous course of history—blasting a specific life out of the era or a specific work out of the lifework."[3] Though he never quite lays out for the historian what this would look like in practice, he does make some telling suggestions in his 1923 essay, "The Task of the Translator." Redemptive critique is "a work of remembrance: it is a process of preserving the truth content or idea of a work [or an object] from the ever-threatening forces of social amnesia to which humanity has over the ages become inured."[4] Translation is in effect the piecing together of language. A translator does not try to match a translation to a work in the original language (trying, in effect, to match historical narrative to the event it purports to narrate), nor to write a translation that sounds as if it had been written originally in the language of the translation (trying, in effect, to match historical narrative—by means of contextualizing it in the language of the narrative itself—to an interpretive "truth"). Instead, the translator—and by extension, the historian—must "constellate" the original and the translation so that their superimposition releases "that pure language which is under the spell of another."[5] In a passage from the same essay, Benjamin states that the translator, as one who has before him "fragments of a vessel which are to be glued together" so as to match one another, must "incorporate the original's mode of signification, thus making both the original and the translation recognizable as fragments of a greater language, just as fragments are part of a vessel."[6]

There are two moments to translation, and by implication to historical work, that must be constellated: the moment as it is embedded in historical time (its interpreted significance, its retelling as testimony and as evidence), and the moment as it has significance as "now-time," as an inseparable part of the observer's own life-world, the observer's own language and memory. The redemptive philosophy of history is not interested in the possibility of recovering the events that, in Vidal-Naquet's terms, "precede discourse" if by that we mean getting the historical moment right. The events of history are heterogenous because they are not easily plotted, and they are not easily plotted because their interpretive significance and their effect upon the participant's and the observer's lived lives are by definition incommensurable. But they most definitely have an effect, an effect that is palpable, and capable of being mapped as it takes place in the impasse engaging spoken language, survivor testimony, a desecrated Torah scroll, and the lived life of the religious Jew or the Lithuanian émigré. And it is this gap, this impasse—"the strait gate" in Kabbalistic terms—"through which the Messiah might enter"[7] or through which, in the terms a historian might use, we have a sense, though not evidence, of event as event, prior to discourse.

* * * *

Let me say a word now about a portion of the permanent exhibit at the U.S. Holocaust Memorial Museum (USHMM). In a glass case about eight feet in height, in the middle of the floor, stands the wooden arch that was the front of a Torah ark from the synagogue in Essen, Germany. Immediately next to it is a six-by-six-foot photograph of the interior of the synagogue on the morning of 10 November 1938. Both of these items take as their context the Nuremberg laws, established in 1935, and the ever-increasing legal ostracism of the Jews in Europe from 1935 through 1939. Along the wall behind the photograph and ark are framed identification passes carried by Jews. Nearby are photographs, taken by members of the Düsseldorf police in the 1930s, of Jehovah's Witnesses, homosexuals, and others who were rounded up along with prominent Jewish businessmen during the suppression of those years. In a glass case in the middle of the floor, some paces removed from the ark, one can see desecrated Torah scrolls. Clearly, this section of the museum is intended to lead visitors to understand the series of events that culminated in the violence of *Kristallnacht* and the sanction the German state provided for individual acts of persecution of Jews and others covered under the Nuremberg laws. The narrative of this section of the museum might be as follows: laws were passed ostracizing Jews, and others, in Germany and the territories; this sanction led to spontaneous and state-led destruction of Jews' personal property, synagogues, and places of business; and this sanction in turn led to deportations from Germany and the territories and, later, full-scale removal and extermination of Jews and others from the newly conquered lands in Poland and the east. The evidence of this narrative is on display in the museum: the text of the Nuremberg laws is presented in translation, photographs of Jews lined up for exit visas and identification cards are hung on the wall like living-room paintings, and the recovered remains of synagogue interiors are exhibited like wounds.

But what this teaches us is far less complete and less compelling than accounts some visitors may know from Hilberg's or Dawidowicz's books on the trajectory of the Holocaust in Europe.[8] The artifacts may make the train of events more accessible to visitors who do not know those works, but they are presented as a narrative under a traditional view of the purpose of museums. If you understand the ark and the photographs of the Essen synagogue as interruptions in the documentary evidence of the deportation of European Jews or state-sanctioned antisemitism, then this section of the U.S. Holocaust Memorial Museum does not exactly teach, if by teaching we mean providing rational access or a name for an event or object; instead, it provides access to the effect of the event itself. The ark acts as an interruption of the narrative itself: for non-Jews, it may require something of an explanation to say why the ark is important in a synagogue in the first place; for Jews, coming upon a sacred object in the midst of legalese and the bureaucratic trappings of a government bent on suppression is a bit disconcerting. But the immediate effect of the ark—its Hebrew letters, seen by Nazis on the night of 8 to 9 November 1938 as evidence of the profane, effaced by an ax—is to draw attention away from the bureaucratic and to the place of the ark itself. The paint on the ark is unblemished, smooth as the day it was painted in blues and greens, and there is only a molding missing over the opening from which the Torah scrolls would have been removed. What has been damaged, what has been hacked away, is the part of the ark that marks the point of connection between the congregants in Essen and the individual, particularly the Jewish individual, museum visitor: the words that mark the moment of the opening of the ark and the removal of the Torah scrolls on a Saturday morning: "Be mindful of that before which you stand."

There are two narratives here—both of which are supported with historical evidence, historical artifacts, and testimonial evidence of those who were there (from the Wexner Learning Center)—that are placed in relation with one another. Several items displayed here might be taken for the "material content" of the display: the desecrated ark, the photograph of the interior of the synagogue, the documentary and tangible evidence of the ostracism and deportation of urban Jews in the late 1930s. Several other elements of the display can be taken as its "truth content": the significance of the ark and the Torah to Jewish daily life, the significance of carrying government-issued documents that identify individuals by "race." But whatever access is available to the historical event of *Kristallnacht*, or deportation, or even such an event as something called "the Holocaust," it is available between truth content and material content.

To speak only for myself, what I experienced in the moment that immediately followed viewing the hacked-away words on an ark that had been the central focus of Jewish life in the city of Essen in 1938 and for many years before it was something like the shock of recognition. This was not recognition of my identity as a Jew, or of a central revelation involving the history of violence against Jews in National Socialist Germany, or even of a semblance of the significance of the moment in which those marks had been made on the ark itself. What had occurred was the denaturalization of both the material content and the truth content of the display in such a way as to allow the observer (or at least me) to understand the historical event as heterogenous, immediate, prior to significance, and—in the aftermath of the moment—potentially redemptive. In the same way that, as I explained to my father-in-law during a visit to the USHMM, I saw the faces of my daughters, and of my mother, and of my father-in-law in the photographs of shtetl life on a two-story wall in a walkway of the museum, I recognized a moment of a life—mine, perhaps, but perhaps not mine but another's—that came closer to the historical moment emplotted along the USHMM's display than perhaps was planned. Adorno's concern[9] that the objects of interest in museums had been stripped from their contexts and reified as objects of representation, and had become used by viewers as objects of pleasure rather than understanding, seems here called into question. Though the ark—and the shtetl photographs that I will describe later—have not exactly been stripped of their context, they have nevertheless been recontextualized out of narrative time in such a way that they can potentially reintegrate the lived life of the observer with the traces of lived life to be found in the artifacts themselves.

* * * *

It is precisely this dissonance that Bakhtin discusses in his book on Rabelais. There Bakhtin is interested in the way Rabelais's images "have a certain undestroyable nonofficial nature;... These images are opposed to all that is finished and polished."[10] This unfinishedness has to do with the conundrum of discourse, of the impossibility of naming once and for all an other or a moment. When in his discussion of remnants of the grotesque during the Renaissance Bakhtin uses the term "interior infinite," he is referring to what he calls the "important discovery" of a subjectivity that has "depth, complexity, and inexhaustible resources."[11] No name, no story, fixes a subject, because the person creating the name cannot occupy the place of the other, and in fact cannot fully understand the complexities of one's own location either. The result is what Bakhtin called an "excess of vision," the result of an attempt to write a coherent story that conscientiously ignores those pieces of the puzzle by which we are anguished and which simply do not fit.

If, as Bakhtin suggests, the act of speaking produces an excess of language or an unspeakable sign, then the act of speaking is unfinished in the sense that the unspeakable, ambient sign can itself be perhaps not heard but certainly felt in its effect upon any subsequent utterance. Because it is not part of the "official version" of the utterance, it destabilizes it in measurable ways. Language itself excludes and marginalizes aspects of being that return to haunt that official version both politically and aesthetically, and it is this shadow that acts as a wedge that potentially opens up a space in which we are apt to catch a glimpse of excess and, perhaps, of the Divine.

We investigate this aspect of any utterance through history, but history viewed not as sequence, as pearls upon a string, but as the panoply of lived life. For Bakhtin, lived life includes not only what we speak and name, but also what we cannot speak (or refuse to speak). Lived life returns in some other form that is nevertheless recognizable as a distinct portion of itself. As history plunges into the present, these moments come to be associated with "crisis time," "moments of death and revival" "of breaking points in the cycle of nature or in the life of society and man."[12] Recognizing the crisis associated with the excessive and unspoken is important if we are to reconsider the possibility of recognizing and understanding the voice of the other and perhaps the voice of God. It is important because it forces us to be aware of the critical and disruptive potential of particularly traumatic events, such as the death of a family member or, in the case of the artifacts and photographs of Jewish life in Eastern Europe found at the United States Holocaust Museum, of objects that refract the material conditions of the mundane acts of living and working and praying and laughing in villages and shtetls in Lithuania and Poland in the years prior to 1939. In the case of the USHMM, as I will suggest in a moment, crisis time is met with a narrative of the everyday—of life and death, of outings in the country, and of men and women in love—but it is a narrative that is so upended by the disaster that it becomes interrupted, shattered by what cannot be told at all. One is left with neither history nor the disaster but the voice of history disrupted by the unspoken presence of the unseen and inaccessible center of the event itself.

The disruption of history assumes a view of time that is nonlinear, and (at least in Bakhtin's view) cyclical. There is much to be suspicious about in such a romantic notion of time and its connection to the material, but one thing about it is not suspect at all: if we attempt to name the everyday events of a lived life from the perspective of historical time, those named images nevertheless "preserve their peculiar nature, entirely different from ready-made, completed" being and historical time.[13] The effect of trying to understand the marginal or everyday by means of the historical is rupture, what (in another context) Homi Bhabha recognizes as aporia, ellipsis, the dislocation of the grotesque, or perhaps marginal, certainly subaltern ("not possessing one's own hegemonic position" from which to speak).[14] This rupture expresses itself in the blurring of the difference in perception of past and future, the possible and the impossible. "The author of the '[prophetic] riddle'" of events of dread and wonder also foretells the "complete destruction of the established hierarchy, social, political, and domestic"[15] and, I would add, generic and discursive. Bakhtin's crisis time may well parallel the moment—in Benjamin's scheme—blasted out of the continuum of history, the moment in which the figural Angelus Novus moves forward through time with his back to the future and with his eyes on the disaster of the present that propel him.[16] If so, Bakhtin and Benjamin provide us a way to understand not the language of history but the circumstances of crisis that increase the likelihood of an "authorial exchange" whose anxious attempts to name the other provide radical excess. They provide a means to understand the character of prophecy and the future in the present through which the language of that other might be expressed or through which it may have an effect. Such an effect uncannily rings in a reader's future in a way that forces the reader to reconsider the name of (aspects of) the other in him- or herself.

Bakhtin is interested in the way in which the destructive potential for parody and debasement may lead to a radical otherness, one that—as I will imply in the conclusion—is connected to sublimity and the Divine. Such a state of being has effects that can be measured in certain instances, instances that potentially change human ethical activity from that point forward. This is what Bakhtin has in mind when he ascribes to the crisis leading to carnival a "peculiar logic of the 'inside out' (*à l'envers*), of the 'turnabout,' of a continual shifting from top to bottom, from front to rear" which "permit[s] no distance between those who came in contact with each other and [destroys] norms of etiquette and decency...."[17] This inversion has an "indissoluble and essential relation to freedom" from the law (both conventional and natural), and from orthodox notions of the Divine.[18] If we turn to Bakhtin's earlier book on Dostoevsky,[19] it is an inversion that comes from a feeling of ambivalence concerning the question of which is the "serious" word in a given utterance and which word is meant to undercut or destabilize that utterance.

Bakhtin's understanding of the overturning logic of crisis is connected to a historical time that is not linear but, in Benjamin's terms, shot through with other time. Such an understanding re-situates not only the subject (the viewer in the museum or a reader of a text), but space and time—the event of history—itself. The identification by that subject of an imagined, named place and time (in the case of the photographs in the museum, rural Lithuania 1938) allows him or her to identify with the individual figures or characters in those photographs (a Pole, a Jew, a citizen, a parent, or a child); but it also (by dint of what cannot be said) forces the subject to realize that he or she is a stranger to those aspects of nation and culture (the Final Solution, Auschwitz, living life as an ostracized member of a community) that cannot be brought full force to the subject by appending to them a name. The state, the location of utterance, is an originary point of destruction, destabilization, and potential trauma, one that is irrepressible and yet one that announces the weight of the historical *grands récits* by the ruptures that cannot help but be seen in that larger narrative. It is this originary point through which, in Benjamin's terms, the Messiah—in all of His glory and destruction—will come.

* * * *

It is this originary point that may be found in certain places in the permanent exhibit at the USHMM and that makes the museum capable of providing the visitor a sense of the Holocaust as a structured event. Of course, the directors of the USHMM have been quite clear about the purpose of laying out the museum's "tour" as a historical narrative: it marks a point of departure for the visitor to come into contact with the events that mark the Holocaust and the evidence of those events as they unfold in time. Visitors enter an elevator on the first floor of the museum, which takes them to the fourth floor of the exhibit. The visitors travel from that floor down through the third and second floors of the museum, each one devoted to a historical period of the events in Europe: from the time leading up to Hindenburg's handing over the German chancellorship to Hitler in 1933 to the beginning of the War (on the fourth floor); from the beginnings of the move to exterminate Jews in the pale in Poland and the Ukraine in 1939 to the end of the systematic killing of the European Jews and the liberation of the camps in 1945 (on the third floor); and the aftermath of the Holocaust in Europe, the United States, Israel, and elsewhere since the end of the War (on the second floor). Though the visitors to the museum are free to walk around in any sequence on each of the floors, by my observation in the museum they follow either the people, usually strangers, standing in front of them, or along the walls of the exhibit. As described by Weinberg and Elieli, the purpose of the narrative arrangement of the building is to enable visitors to "project themselves into the story and thus experience it like insiders while at the same time remaining at a distance.... Drawn into the flow of the narrative, visitors view the display with their senses tuned to sequence, coherence, and transformation."[20] Later on, they argue that it is the "plot" of the narrative that maintains the interest of the visitor.

But this really is only a point of departure for the museum, and I would argue that what makes the exhibits function historically—what enables the viewer to have access not to the events of the Holocaust but to the set of material and ideological markers upon which meaning is founded—are those interruptions that Weinberg and Elieli (and, before them, the planners and architects of the museum) take pains to re-contextualize, such as the glass walkways between sections of the exhibit floors, and the watchtower rooms, one of which contains the photographs collected from the Lithuanian shtetl of Ejziszky destroyed in the 1940s. It is these interruptions that most clearly mark the point—the constellation of the historical events or objects—at which the visitor's sense of history-as-narrative is ripped apart.

Where, one might ask, is God in the recognition of a life not one's own, a life that is only a glimpse represented by a photograph or set of photos—or, in other places, by testimony, artifacts, or displays, all arranged to produce an effect, though different, perhaps, from the one intended? Is it the God the museum's designers suggest by calling the museum's memorial component "awe-inspiring ... [provoking an experience] similar to that experienced by a believer in a holy place"?[21] I would argue otherwise. It is telling that the museum designers liken the museum's Hall of Remembrance—a large and empty hexagonal space near the exit to the permanent exhibit—to a cathedral, implying an overwhelming sense of greatness, one that makes "the individual feel small and insignificant."[22] But I think what occurs here is quite the reverse: the individual's relation to the moment of trauma is opened up rather than closed down in the face of the terrible or all-powerful.

There is another way to understand the moment of interruption in terms of what might be called "Divine," or aesthetic, or exterior to language or the ability to name. It is tempting, in such a view, to understand the rhythms of daily life as something like the manifestation of God in human life. The walks to work, the work itself, prayer, watching one's children grow up—it is very easy to say that this lifeworld is a reflection, in some small way, of the Divine in one's life. As for the lives that are opened up in the artifacts that remain from them in the museum, either in the posed photographs of yeshiva students or in the faces of people strolling to no place in particular, one can imagine the points of connection between the intention one believes one sees in a person's gait and a similar intention when that person smiles for photographs at weddings or family reunions. The horrible accidents of life—as the death of a child, the destruction of dearly held belongings—are the reverse of this reflection, and there are innumerable instances of such a reflection in the museum: the piles of shoes taken from victims at Majdanek, or the photograph of a ten-year-old boy, his hands raised in surrender at the point of a gun during the liquidation of the Warsaw ghetto. But this kind of understanding of one's lived life in effect ties together the Divine and the human or historical in such a way as to enable us to see one in the other. The phenomenal world of life and death and of stories and laws and experiences, and conceptual aspect of life that provides for our capacity to describe it as such, and the noumenal world that we cannot see but can only think—all become oddly commensurable. We come to make an easy equation between them.

The relation that is established at the wall of photographs is something else: this particular representation of the Holocaust reinserts the division between the noumenal and the phenomenal, between our capacity to render or name the event and the event as it is understood by us as an event. Far from uniting us with those whose lives appear to be made accessible to us through the photographs taken by family members or friends, far from linking an imagined horror in our own lives with the unseen and unimaginable horror that we know only through the liquidation of a shtetl, the relation established by that moment of interruption rips a seam in the fabric that could so easily be seen to weave together those moments. The opening that it provides, violent and traumatic enough to reduce visitors to a public museum to tears, also manages to provide warmth to others. It does so by providing an experience that cannot be named in terms of the Divine, that cannot be explained in the name of God. This is where the traditional lamentations and questions—"where was God?" "God has died on the gallows in Auschwitz"—are instructive: the person of God would not allow such a thing to happen. But to think in these terms is to wonder aloud why God is not as good a person as I am. The theological point is that God was never absent in the Holocaust.

There is a point in the third *Critique* where Kant suggests that "the most sublime passage in the Jewish law is the commandment: thou shalt not make unto thee any graven image, or any likeness of any thing that is in heaven or on the earth."[23] In Hebrew, God is referred to as "hashem" (the name). Many historians now refer to the Final Solution as "the event." In each instance, the refusal to name is intended to forestall representation as that which tends to close off the effect of the event upon our neatly packaged lives. The resistance to the name, the univocal, is the very thing that Kant's third *Critique* points to as the essential quality of art, but also of the human capacity for knowledge.

This is not the vastly oversimplified Kant whose categorical imperative is read as the subsumption of individualist ethics to a universal morality, or the Kant whose third *Critique* is read as the bridge between the pure and practical reason by way of the suspension of the conceptual in the aesthetic. For this Kant, the aesthetic (and more so the aesthetic's underside, the sublime) provides the ability to think the supersensible—what might be called by psychoanalysis the real, or by historical materialism as the unmediated relations of human subjects—through our inability to know the object of representation. Kant relegated the idea, and also God, to the supersensible because it must be there in order to provide structure to what we know; it intervenes in its effects, in the phenomenal world. The interruption that marks the moment of the sublime is not the positive, freeing connection of the laws of reason that find their fullness in a sense of freedom. Instead, it is a negative and terrible sense that there is a point in our knowledge and experience and history beyond which we cannot go, but it is a point that gives a shape and a name to where we can. The interruption that marks the moment of the beautiful is not the positive, freeing connection of the laws of nature that find their fullness in a sense of order, but a negative sense that whatever similarities we might like to see between our own and others' situations are those of our own making and not of a broader order that we cannot hope to know. These interruptions, which might be called Divine, announce that which precedes our desire to order something—to call it the Holocaust, or to rationalize trauma or recognition by providing it with a narrative, or by investing it with meaning. It is the jolt of incommensurability that urges upon us the desire to commensurate the mundane act of an eastern European family going on a picnic with the mass extermination of a culture thousands of years old and with our own place in a lifeworld and a culture that are connected only because we say they are.

Representations of the Holocaust qualify as traumatic because they potentially defy the easy commensuration between the Divine and the experiential, the phenomenal and the supra-rational, the event and the name. The Holocaust is paradigmatically traumatic because it is, in its effect, more resistant to explanation than other experiences in the lifeworld. Using Kant's terminology, the magnitude, the force of the event itself is such that it defies concept and language. It is sublime, and representations of the event are characterized by their effect as ruptures. The event can only be explained, using Lyotard's term, as a *differend*, an incommensurability of concept and expression.[24] But I would go further: I would suggest that representations of the Holocaust, particularly those, like the wall of photographs in the USHMM, that function inasmuch as they insert the mundane and the recognizable into the indescribable, represent an incommensurability of concept, of expression, and of the thing itself. This latter incommensurability is not simply a matter of fact (or philosophy) but is, writ much smaller, an integral part of our more run-of-the-mill explanations of our day-to-day lives. (The popularity of American pragmatism, with its convenient forgetting of the Divine, of being, of that archaic term "truth," is the manifestation, fifty years after the Holocaust, of the fact that such an incommensurability is easily ignored or rationalized away for the sake of political and philosophical expediency.)

What we see clearly, in the oddly eclectic but significant views of history and its "other" that Bakhtin and Benjamin patched together, is that representations of the Holocaust potentially mark the point at which human activity and the capacity to understand it part company. But because one requires a capacity to render narrative sense to activity before one actually renders it, that capacity—the Divine, the noumenal—and the narrative itself are intimately related, but not at the level of narrative. They are related at the level of excess, at the point where it is impossible to commensurate my sense of my daughter and my imagined view of life in a shtetl in Lithuania, on the one hand, and my knowledge of the events comprising the Holocaust, on the other, because they produce an effect that works beyond reason, but which effect has everything to do with my ability—or maybe in this case my inability—to process all of this.

Geoffrey Hartman is right: one way to understand the relation between Holocaust and representation is at the level of the sublime, but a sublime that recognizes the necessary mismatch between imagination and reason.[25] The Holocaust, by reintroducing a relation between the experiential and that which exceeds it, is a single terrible example of the way in which any human act or utterance bears a relation to our (in)-capacity to act or to speak. Because it is a relation that can only be said to be a relation of incommensurability, it reflects back upon human action and speech. It is a relation that lays out in front of the speaker, the viewer, the reader, the distance between speech and event and the intimate closeness of them as they take effect. If we take representations of the Holocaust as acts of witness—as narratives that work to render the impossible by shattering our tendency to reduce the event to a name by providing the event and other events that surround it with characteristics that we imagine can simply not be spoken together—they do not bear witness to history, if what we mean by history is the logical or temporal connection between one figure and another, between one event and another.

To put this in terms of the book of which this essay is a part, no representation (neither aesthetic nor academic) can logically connect the lives of two thinkers—one German, one Russian—by examining their views of history. As I said at the outset, such a connection is illogical, unhistorical, and potentially uninteresting. Such a connection, however, may be presented as one that obliterates the historical connection one would want to produce, and I would argue that this very essay examines a historical ellipsis rather than a historical juxtaposition. More to the point, representations of the Shoah—an event that confounds logical consistency and historical reason—bear witness, in terms of which neither Bakhtin nor Benjamin would disapprove, to the Divine. But the Divine is not God. The Divine is the human capacity to understand that the incommensurable—the horrible, the traumatic—has definable characteristics, characteristics that look remarkably like those you grant yourself and your neighbors; but that those characteristics produce an effect that goes beyond what you can name. The Divine is understanding that the fragments of events, not the events themselves, form potential wholes, and that the map of the whole, if we are to repair it, does not come from God, but from the faces in a synagogue in Essen and their effect upon us as we stand in front of the ark of the covenant before which they also stood.

Notes

1. Walter Benjamin, *Reflections*, trans. E. Jephcott (New York, 1978), p. 327.
2. Richard Wolin, *Walter Benjamin: An Aesthetic of Redemption* (Berkeley, 1994), p. 48.
3. Walter Benjamin, *Illuminations*, trans. H. Zohn (New York, 1968), p. 263.
4. Wolin, p. 45.
5. Benjamin, *Illuminations*, p. 80.
6. Ibid., p. 78.
7. Ibid., p. 264.
8. Raul Hilberg, *The Destruction of the European Jews* (Chicago, 1965); Lucy Dawidowicz, *The War Against the Jews* (New York, 1975).
9. Theodor Adorno, *Prisms*, trans. S. Weber and S. Weber (Cambridge, MA, 1981), pp. 175-85.
10. Mikhail Bakhtin, *Rabelais and His World*, trans. H. Iswolsky (Bloomington, IN, 1968), p. 3.
11. Ibid., p. 44.
12. Ibid., p. 9.
13. Ibid., p. 11.
14. Homi Bhabha, *The Location of Culture* (New York, 1994), p. 59.
15. Bakhtin, *Rabelais and His World*, p. 237.
16. Benjamin, *Illuminations*, pp. 262, 257-58.
17. Bakhtin, *Rabelais and His World*, pp. 11, 10.
18. Ibid., p. 89.
19. Mikhail Bakhtin, *Problems of Dostoevsky's Poetics*, trans. C. Emerson (Minneapolis, 1984), pp. 185ff.
20. Jesjahu Weinberg and Rina Elieli, *The Holocaust Museum in Washington* (New York, 1992), p. 49.

21. Ibid., p. 26.
22. Ibid., p. 26.
23. Immanuel Kant, *Critique of Judgment*, trans. W. Pluhar (Indianapolis, 1987), p. 135.
24. Jean-Francois Lyotard, *The Differend: Phrases in Dispute* (Minneapolis, 1988).
25. Geoffrey Hartman, "The Book of Destruction," in *Probing the Limits of Representation*, ed. Saul Friedlander (Cambridge, MA, 1992), pp. 321-38.

Chapter 11

MITTELEUROPA TO MOSCOW

Germanic Links with Russian Architecture

William Craft Brumfield

Of the many European sources that have enriched Russian architecture, Central Europe—and more specifically Germany—has played a surprisingly modest role. During the medieval period there are occasional hints of a German presence, as in the case of the great twelfth-century limestone Cathedral of the Dormition in Vladimir, for whose construction the Laurentian chronicle mentions the bringing of masters from "all lands." In the same chronicle there is later mention of *Nemtsy*, or "Germans"—a term broadly used. In fact, it has been proposed, on slim evidence, that artisans for the Dormition Cathedral were sent by Frederick I Barbarossa to Andrei Bogoliubskii, grand prince in Vladimir.[1] And in the northwest, Novgorod's connections with the Hanseatic League have led to suggestions that the German lands may have served as a conduit for possible Romanesque elements in that city's church architecture during the same period.

But there is no doubt as to the source for the revival of Russian architecture, under the guidance of Moscow, in the late fifteenth century, as the central Russian lands emerged from the debilitating effects of the Mongol invasion (1237–41) and two centuries of internecine warfare. It was the Italian peninsula that provided Russia with its best architects and engineers for the five decades between the 1470s and the 1520s.[2] And after Moscow's recovery from the interregnum during the "Time of Troubles" at the beginning of the seventeenth century, Poland served as a source for early baroque motifs in Muscovite architecture.

With the founding of St. Petersburg in 1703, French influence was predominant at the beginning of the eighteenth century, although Peter I also hired architects from northern Europe, including the prominent Prussian master Andreas Schlüter, who died soon after his arrival in the northern Palmyra in 1714. During the reign of Catherine II, a notable role in the development of St. Petersburg's architecture was played by Georg Friedrich Veldten, whose father Mattias had occupied the important position of head steward to the St. Petersburg Academy of Sciences. After his father's death in 1736, Veldten returned to the German lands, where he completed his education at Tübingen University and continued his apprenticeship through work on the ducal palace in Stuttgart.[3]

Undoubtedly, Veldten's observations of the rich variety of German architecture, from the medieval to the late baroque and early neoclassical, provided him with much material for his subsequent work in St. Petersburg. With the departure of the great baroque master Bartolomeo Francesco Rastrelli from St. Petersburg in 1764, Veldten became one of the capital's most prolific architects, designing not only the South Pavilion of the Small Hermitage and the Chesme Palace (on the southern fringes of St. Petersburg), but also a number of pavilions and interiors at imperial estates such as Tsarskoe Selo. In addition, Veldten directed much of the original work on the city's magnificent granite quays, which provide such an imposing pedestal for its architecture.

During the first half of the nineteenth century, the most notable German presence in the architecture of imperial Petersburg came with the design of the New Hermitage (1839–51) by Leo von Klenze, renowned for his design of the Grecian-style Glyptothek, or museum of sculpture, in Munich. For the New Hermitage, Klenze adopted the Renaissance style, with classical elements that reflected the building's purpose as a museum. Klenze made yearly visits to St. Petersburg to inspect the work, which was supervised by the Russian architect Vasilii Stasov.[4] Another German who contributed to the imperial design during the reign of the Teutonophile Nicholas I was the Prussian architect Karl-Friedrich Schinkel. Although known primarily as a great interpreter of neoclassicism, Schinkel turned to the Gothic revival for his 1829 design of the Capella at Alexandria Park in Peterhof.[5] Schinkel's design was actually implemented by Adam Menelaws, a Scottish architect who designed a number of other Gothic revival structures for the imperial estates.

Although by no means insignificant, these isolated examples of German presence in imperial Russian architecture suggest that the German contribution to the imperial design throughout the eighteenth and early nineteenth centuries remained secondary, despite the proximity of Russia to Prussia and the Austrian Empire. During the nineteenth century, however, the rapid growth of German industry and its related technology created a favorable climate for urban development, with its essential components of architecture and civil engineering. German institutions of higher technical education (including architecture) became increasingly important, and we know that at least a few Russian architects studied in Germany (e.g., Ludwig Bohnstedt, at the Bauakademie in Berlin; Viktor Schröter, at the Berlin Academy of Arts; and Harald Bosse, in Darmstadt), as opposed to the more frequent choice of Paris. And the German states were often on the itineraries of Russian architectural students after their graduation from institutions such as the Academy of Fine Arts in St. Petersburg.[6]

In addition, the establishment of professional architectural societies in Russia's two major cities—most notably the Imperial Society of Architects in St. Petersburg—stimulated further interest in German developments as part of a hoped-for integration of Russian building practices with those of the rest of Europe. The journal *Zodchii*, published by the Society of Architects beginning in 1871, served as a particularly effective conduit for news of contemporary European architecture and design, with attention to German developments, including systems of German technical education.

Perhaps the most important aspect of German influence on Russian architecture at the turn of the twentieth century was in the area of theoretical polemic. During this period Russian architectural publications devoted much attention to the varieties of European modernism, such as art nouveau, but German and Austrian architects were the most prolific and persuasive in advancing arguments for a modernist aesthetic. Architecture, it would seem, had become dependent on the word as an essential component of self-definition. Buildings in and of themselves were no longer sufficient as evidence of architectural culture; their presence had to be justified by a critical text.

Victor Hugo had already predicted this relation between building and the printed text in his *Notre Dame de Paris.*[7] But in architecture it was the Germans who seized the theoretical and textual initiative on behalf of modernism in architecture, and Russian critics followed intently. Indeed, it can be argued that the remarkable spread of *style moderne* buildings in Russia's major cities at the beginning of the twentieth century would have been unlikely without the spate of critical articles in professional publications. And what they had to say—if at times awkwardly—about modernism and its German supporters would also contribute to the extraordinary development in Russian architectural theory later in the century.

One of the earliest and most detailed reports on the new European architecture appeared at the end of 1899 in *Stroitel'* under the heading "Conversations of a Builder." Written by G. Ravich, the report surveyed developments in Belgium, Germany, Austria, and France and included illustrations of interiors and exteriors by Otto Wagner, Victor Horta, Joseph Olbrich, Rudolf Tropsch, and Hector Guimard. Ravich enthusiastically portrayed the new style as a welcome development that would allow architects "to express the inner content of our life" and permit architecture to join her "sisters," poetry and painting, in the quest for new forms.[8]

The first extensive critical assessment of the new style appeared in the February 1902 issues of *Zodchii*, in which Pavel Makarov disputed those who had linked the new architecture to the questionable term *dekadentstvo*, which more properly belonged to literature. Makarov, who had graduated from the Institute of Civil Engineering in 1898, rapidly became one of the most vocal supporters of the *style moderne.* He defined the qualities expected from a revived architectural aesthetic: "Freshness, simplicity, lack of pretension, and a complete rejection of old forms, together with a striving to change the basic principles of the former style and move architecture along the path to rationality, which so clearly appears in the Gothic and was so completely forgotten during the Renaissance."[9]

Makarov labeled contemporary German and Austrian architecture "moderne" (another early use of the term in Russia), a word he seems to have taken from Otto Wagner. He concluded his survey of Europe with a commentary on Austria and the Vienna Secession, giving Otto Wagner pride of place for his buildings as well as his project drawings, which were published in German periodicals. He discussed the work of the Secession, particularly that of Olbrich, although no German or Austrian buildings were included in the illustrations.[10] In the 9 June 1902 issue of *Zodchii*, Makarov[11] published an adaptation of an article by F. Knopf entitled "The Architect and Designer Josef Hoffmann" (originally published in 1901 in Makarov's favorite journal, *The Studio*). Although only two pages long, Makarov's adaptation—photographs and text—was highly flattering to Hoffmann's designs for commercial and domestic interiors.

More significant than the article on Hoffmann was Makarov's report in the August 1902 issues of *Zodchii* on the Darmstadt Exhibition, from May through October 1901—a report that atoned for his omission of Matildenhöhe (an artists' community and architectural showcase near Darmstadt) from his survey of modern architecture the preceding year. The two articles focused on Olbrich, who was responsible for the general design of the colony and in 1899–1901 built houses for himself and most of the other residents, including Hans Christiansen, Julius Glükert, and Ludwig Habich. Makarov described the work of resident artists such as the sculptor Habich, as well as the design of the houses.[12] (The noted architect Peter Behrens built his own house at the colony, although Makarov did not discuss this structure; he commented instead on Behrens's activity as a designer.) He also gave an account of the Darmstadt studio—Ernst Ludwig Haus—built by Olbrich and named in honor of the grand duke of Hesse, the colony's patron and, coincidentally, the brother of Empress Alexandra of Russia.

The first analysis of theoretical writings of the modern style appeared in the December 1902 issues of *Zodchii*, which contained a three-part essay by Evgenii Baumgarten on the recently published third edition of Otto Wagner's *Moderne Architektur*. Baumgarten, a graduate, in 1891, of the Academy of Arts, was both attracted to and alarmed by Wagner's insistence that the new architecture was essential to the formation of a modern environment. After summarizing the prefaces to the three editions of this seminal work—in which Wagner states his intentions, attacks those who oppose them, and proclaims his opponents' defeat—Baumgarten discussed the book's contents, occasionally expressing his dismay, for example, at the elevated role Wagner envisioned for the modern architect, which Baumgarten rejected as a "dithyramb" that posed "a danger for young minds already disposed to delusions of grandeur."[13]

Although Baumgarten acknowledged the exemplary importance of Wagner's architectural work, he objected strenuously to Wagner's famous dictum "Etwas unpraktisches kann nicht schön sein" (Nothing impractical can be beautiful).[14] Refusing to allow for rhetorical exaggeration, Baumgarten interpreted the statement literally:

> Although Professor Wagner's instructions are practical, we are compelled to take a negative view of this theoretical argument. In leaning toward the utilitarian, he falls into an obvious absurdity. Proposing that the contemporary architect "come to terms" with the statement "nothing that is not practical can be beautiful," he lowers the architectural art, praised with such emotion, to the level of an applied craft. With such criteria for appraising the beautiful, there is nothing left to say about original artistic creativity.[15]

Wagner attempted to reconcile opposites: elitism and democracy, the supremacy of art and practicality, the utilitarian and the ideal. And the critic's outrage at what he construed as an assault on aesthetics was perhaps greater because Baumgarten himself had tentatively advocated a union of technology and style in the new architecture (as in his article "The Aesthetics of Iron Construction"). Baumgarten thought that architects yielded all too readily to the practical and relied on derivative, hackneyed solutions—views in fact irrelevant to Wagner's book:

> Under the guise of responsiveness to contemporary needs, the theory of Professor Wagner proposes aesthetic suicide. Of course it is necessary to build houses solidly, cheaply, quickly, and conveniently, but the beauty of a house has no relation to the technique of construction. The human soul requires architectonic beauty.... For that reason any inspired artist values the legacy of past geniuses. By a successive link with them he finds the firm ground for independent activity that Professor Wagner vainly seeks in bourgeois contemporaneity, floating "without rudder or sails" along the troubled waters of everyday practical life.[16]

Baumgarten's obsessive use of the term "bourgeois" in his polemic with Wagner's treatise indicates certain ingrained cultural attitudes of Russian intellectuals, but his remarks miss the purpose of Wagner's writings. In fact, no Russian architect of the prerevolutionary period reached the radical break with tradition that Wagner advocated and Adolf Loos achieved in Vienna. The concern with beauty, as defined by Baumgarten, continued to occupy Petersburg and Moscow architects in both commercial structures and apartment houses, virtually all of which were intended for the "bourgeoisie."

By 1903 the *style moderne* had achieved precedence in the Russian architectural press, although buildings designed in the academic, eclectic manner still appeared in *Zodchii.* The new style received further support at Russia's first major contemporary design exhibit in Moscow, from December 1902 through January 1903. The show, which had a strong Austrian component, included rooms and furnishings by Charles Mackintosh, Olbrich, William Walcot, Fedor Shekhtel, and Ivan Fomin. Although critical responses were not uniformly favorable, no other event associated with the new style received so much attention in the press. The extensive coverage in such journals as Diagilev's *Mir iskusstva* revealed that the *style moderne* had established itself in the most influential cultural circles.[17] There were, however, a few skeptical comments, such as a critique that appeared in *Iskusstvo stroitel'noe i dekorativnoe* under the title "Ist zu wienerisch."[18] Baumgarten's report[19] on the exhibit in *Zodchii* was complimentary and noted Otto Wagner's influence on the furniture design.[20]

Despite its interest in almost all varieties of modern design in Europe, Russian critical commentary throughout 1903 continued to focus on the Vienna Secession and its factions. In the September and October issues of *Zodchii*, Aleksandr Dmitriev[21] surveyed recent Secession exhibits at Darmstadt, Düsseldorf, Turin, and Paris. Among its frequent summaries of items from journals such as *Der Architekt* and *The Studio*, *Zodchii* featured commentary by Aleksandr Dmitriev on Otto Wagner's entrance portal for the telegraph offices of *Die Zeit* (1902). Reacting against the curvilinear excesses of the new style, Dmitriev praised the "classical resolution" with which Wagner defined the rectangle of glass and aluminum over an iron framework.[22] Such comments suggest a turn to classical values in Petersburg architecture at the end of the decade.

The most intelligent critical appraisal of the new style—and of the role played by German architects such as Otto Wagner in defining it—was a short book by Vladimir Apyshkov entitled *The Rational in the Latest Architecture* (1905). A 1901 graduate of the Institute of Civil Engineers, Apyshkov wrote the book as a concise, lucid textbook for his students at the institute. Like many Russian observers of the new style, he acknowledged England's importance in healing the nineteenth-century rift between structure and appearance, architecture and decoration. In his view, however, John Ruskin was less important to modern architecture than Otto Wagner.[23] He not only interpreted Wagner's work precisely—including the dictum on beauty and practicality that so alarmed Baumgarten—but also linked Wagner's work with a new rationalism in design.

According to Apyshkov, architecture at its best has always arisen from a balanced system that relates function, form, climate, and material—as in ancient Greece. The ultimate goal of architecture is still to transform "the useful into the beautiful"; to this end he rejected what he interpreted as the radical functionalism of another German theoretician, Leopold Bauer.[24] Apyshkov's defense of aesthetics is distinguished by its recognition of Otto Wagner's leadership in the aesthetics of modern design. In his own interpretation of Wagner, Apyshkov noted: "The true task of an artist in a given situation is to introduce the decorative element without contradicting utilitarian demands."[25]

In terms of architectural practice, the appearance of the "new style"—or *style moderne*, as it was frequently called in Russian—in the major cities of Russia at the turn of the century has many obvious influences but no clear point of origin.[26] As the critical debates analyzed above make clear, there was a well-informed awareness of developments in Western and Central Europe that would stimulate a major aesthetic and technological redirection of Russian architecture. The rise of the *style moderne* in Moscow and St. Petersburg revealed considerable differences between these two urban centers, and the style in general assumed a protean character that reflected its varied sources of inspiration, from Vienna to Glasgow.

The essential precondition for the new style was the continued expansion not only of the population of large cities such as Moscow and St. Petersburg, but also the rise of a class of private patrons whose wealth derived from capitalist, entrepreneurial activity. The decades preceding World War I formed an unprecedented period of private initiative and investment in the construction of Russian cities. The results were varied in both style and taste as architects strove to accommodate the deluge of work involved in the creation of an urban infrastructure under social and economic conditions that were still far from modern.[27]

Although St. Petersburg applied the new aesthetic on a massive scale, with similarities to apartment house construction in Berlin and Vienna, Moscow seems to have created the most distinctive examples of the style in the work of Fedor Shekhtel, Lev Kekushev, William Walcot, and others. Of these, the greatest was Fedor Shekhtel (Schechtel), born in 1859 of Volga German parentage in the city of Saratov. Shekhtel's education took place in Moscow, and it was there that he did his greatest work, including the building that can be seen as the "signature" of the *style moderne* in Russian architecture—the Stepan Riabushinskii house (see Fig. 11.1).[28]

Almost a century after its construction, one is still confronted with the remarkable novelty of the Riabushinskii house in its Moscow setting, for no other private dwelling of the *style moderne* displayed the new aesthetic with such panache and with such thorough knowledge of contemporary design. Through Russian architectural journals, Shekhtel was well informed of the work of European architects, from Hector Guimard to Joseph Olbrich. Indeed, Olbrich's detached private dwellings—in contrast to the art nouveau townhouses of Paris and Brussels—provide the closest comparison with Shekhtel's design for Riabushinskii. Of special relevance in this regard is the house that Olbrich built for the sculptor Ludwig Habich at the Matildenhöhe community, with the same block arrangement of mass and a flat, overhanging roof.[29] Yet it must be emphasized that the Riabushinskii house, begun in 1900 and completed around 1902, is contemporary with—or predates—the house designs of Olbrich at Matildenhöhe.

Shekhtel also incorporated modernized Russian motifs in the design of the Riabushinskii house: the heavy entrance porch is reminiscent of

FIGURE 11.1 Riabushinskii House. Photograph by William Brumfield.

the *kryl'tso* frequently attached to the entrance of seventeenth-century Muscovite churches; and the two small windows on the upper right corner of the front façade, whose size and shape contrast with the art nouveau contours of the main windows, suggest the narrow paired windows of a medieval Muscovite *terem.* The ebullient use of the applied arts for the exterior decoration—the mosaic frieze and the wrought-iron work, for example—refer not only to the crafts revival in Russia, but also to the polychromatic patterns of medieval Russian architecture. At the Riabushinskii house, Shekhtel had merged the old, Muscovite concept of the *teremok* and the *palaty* (chambers) with the most contemporary interpretation of the *osobniak*, or detached private dwelling.

Following the principle of textural juxtaposition, Shekhtel created two large porches, one placed within the left corner of the house, and the other extending from the main entrance. In each case the concrete assumes a molded shape that is the antithesis of the rectilinear outlines of the basic structure—another feature that has its parallels in Olbrich's work, notably at Matildenhöhe. This contrast is particularly evident in the arched openings of the side porch, adjacent to the dining room. The front porch also exploits the plasticity of the material in a rusticated treatment that suggests the entrance to a cave or grotto. The comparison is not accidental, for the aquatic theme, noted in such elements as the fish-scale pattern in wrought iron, continues on the great mahogany door, which is decorated with its own aquatic motif: a brass lotus pattern in the block style of the Secession.

Another of Shekhtel's works that bears close comparison with the work of Otto Wagner and Josef Hoffmann is the house he built in 1901 for Aleksandra Derozhinskaia, wife of a wealthy Moscow industrialist (see Fig. 11.2). This design rejects elaborate decorative effects in favor of a monumental definition of mass and space in an international modern style. The main window, extending to the cornice line of the two-storied structure, serves as a form of triumphal arch in its projection from the façade. Above this dominant element stand two flanking brick-faced cylinders linked by concrete and suggesting the brutalism of the "machine age." The style of Wagner and Hoffmann is also evident in the work of the Moscow architect Ilarion Ivanov-Schitz, particularly in his public and commercial buildings with classical detailing.[30]

In comparison with Moscow there were fewer private mansions in St. Petersburg, which had a denser pattern of urban construction. Nonetheless, an affinity for the Central European variant of the new style in domestic architecture is particularly evident in the mansion of the prima ballerina Matilda Kshesinskaia (see Fig. 11.3), built by Alexander von Hohen, who also worked for the imperial family. The structure suggests a determined reinterpretation of Viennese architecture as defined by Otto Wagner and the Secession. Kshesinskaia maintained close relations with the imperial court, and her position in the cultural world provided additional incentive for entertainment on a lavish scale. The architect thus combined a family dwelling with a grand social space—all decorated in an eclectic manner.

It is on the exterior that von Hohen's concept of the new style emerges in a coherent, if austere, design. From the rough stone surface of the base, the walls ascend in a progressive refinement of material: a rusticated granite facing reaches midway to the first floor, before yielding to the precise surface of pale yellow pressed brick for the upper part of the walls. Upon this durable façade Hohen applies decorative work that, by Moscow standards, is muted in color and dominated by heavily articulated iron work. On the street façade a dark ceramic tile frieze, decorated with attached iron garlands and segmented by large wrought-iron consoles, extends beneath the flat, projecting cornice. No section of the mansion is without some form of ornamental iron work in the linear patterns of the Vienna Secession, from the cupola railing above the

FIGURE 11.2 Derozhinskaia Mansion. Photograph by William Brumfield.

FIGURE 11.3 Kshesinskaia Mansion. Photograph by William Brumfield.

corner of the central structure to the curved glass and metal bay of the conservatory attached to the main hall.

After 1910, the increasing acceptance of reduced, "stripped" classicism among Russian modernist architects appears to have derived not so much from the historicist neo-classical revival in Moscow and St. Petersburg (e.g., the work of Ivan Fomin), as from the monumental design by Peter Behrens for the new German Embassy in St. Petersburg (1911–12) (see Fig. 11.4). Behrens received less attention in the Russian press than his Viennese contemporaries, whose designs were better suited to commercial architectural tastes in Russia; yet this leading representative of the *Werkbund* was widely known to Petersburg architects through German architectural journals, which in any event provided much of the copy for the Russian professional press.[31]

Although Soviet scholarship ignored the German embassy building—which subsequently became Intourist headquarters in Leningrad, and is now used by Dresdner Bank—-it was one of the most impressive structures built in St. Petersburg just before the war. Indeed, the silence on it—as well as the negative comments of some Russians at the time of the building's completion—is undoubtedly related to its uncompromising projection of the power of the German Reich; yet it was highly

FIGURE 11.4 German Embassy, St. Petersburg. Photograph by William Brumfield.

praised by the anonymous author of an article, accompanied by three pages of photographs, in the 1913 issue of *Moskovskii arkhitkturnyi mir.*

The article noted the success of Behren's approach to a difficult trapezoidal site by joining two independent structures surfaced in rough-hewn red granite. To the main part, facing St. Isaac's Square and containing the embassy's state and reception rooms, Behrens attached a second rectangular structure—extending along the side street and containing consular offices and staff quarters—at an angle to the main façade. This ingenious arrangement interlocks the two parts, which are nonetheless separate and have separate identities. (The angle of placement screens the front edge of the side wing from the cathedral square.)

What is particularly interesting about this commentary on Behren's design—in a Moscow journal—is the praise of the austere, noble simplicity of the façades:

> On St. Isaac's Square a façade rises with fourteen gigantic columns of three stories, with beautifully spaced, very modest windows. The three windows over the main portal are distinguished by balconies. Nothing breaks the simplicity and severity of the architectural lines, the proud rhythm of the stonework. The integrity and unity of the architectural impression is nowhere disturbed by any ornamentation.
>
> The entire system of attached columns and windows on the main façade is enclosed is a well-proportioned frame: a small base below, two wide pilasters on the sides, and a high cornice above, crowned with a stepped roof with an impressive [statuary] group of horses and handlers. Now one comprehends the idea of the artist, who deliberately refrained from continuing the columns along the entire façade [including the corner of the side wing]; it is also clear that the criticism of the exterior architecture for lacking unity is unjustified.[32]

The anonymous critic associated integrity, unity, and simplicity of structure with an absence of architectural ornamentation. Neither the *style moderne* nor the neoclassical revival had approached form and function so rigorously; Behrens had abandoned the historicism decried by Otto Wagner without vitiating the symbolic authority of monumental architecture. Unfortunately, the chauvinist sentiments evident in earlier commentary on this building would burst forth in August 1914, when the embassy was sacked by a mob and its equestrian statuary was toppled into the street below.

Despite the anti-German articles that appeared in architectural journals as well as elsewhere in the Russian press during World War I, the 1915 volume of *Zodchii* included two major articles that surveyed advanced functional design, with emphasis on German architecture. A two-part article by the engineer A. Stepanov on industrial architecture gave the first detailed survey of the German *Werkbund* school of architecture in a Russian journal, and made particular note (with photographs) of Peter Behrens's work for AEG.[33]

Shortly thereafter, A. M. Ginzburg published an article entitled "The Role of Germany in the Russian Construction Industry," in which he provided an itemization of the areas in Russian construction most affected by innovations from Germany.[34] Ginzburg's motives are not those of a disinterested observer: from his conclusion it is clear that he considered this "dependence" a dangerous situation, stemming from inadequacies in the Russian economic and technological spheres, and one that threatened to turn Russia—even without the war—into an economic colony of Germany.

After the 1917 revolution and ensuing civil war, architectural links between Germany and the new Soviet state assumed unprecedented closeness, as avant-garde artists and architects in both countries sought common cause. Not only did members of the Soviet avant-garde such as El Lissitsky cooperate with the Bauhaus, but a number of prominent German architects sought work in the Soviet Union during the industrialization campaigns of the late 1920s and 1930s.[35] Yet that is another topic, one that deserves its own analysis. Although the above survey has been necessarily selective, it should be evident that the increasing importance of German and Austrian architects to Russian architecture since the beginning of the nineteenth century has led to a cultural interaction that can be compared to the similar connections between these cultures in music, painting, and literature.

Notes

1. The connection with Barbarossa was made by Vasilii N. Tatishchev (1686–1750) in *Istoriia Rossiiskaia s samykh drevneishikh vremen* (St. Petersburg, 1774), vol. 3, p. 127. Tatishchev was among the earliest Russian historians in the modern sense, but his medieval manuscript source for this claim is not extant. Nonetheless, the idea receives support G. K. Vagner, *Skul'ptura drevnei Rusi XII v.: Vladimir–Bogoliubovo* (Moscow, 1969), pp. 84–85. A more recent, and sympathetic, examination of this possibility is contained in N. L Nickel, "Bezugsmotive der sächsischen romanischen Bauornamentik zu den Schmuckmotiven der Vladimir-Suzdaler Architecktur," in *Dmitrievskii sobor vo Vladimire*, ed. E. A. Smirnova (Moscow, 1997), pp. 81–92.

2. For an analysis of the role of foreign architects in the development of Russian architecture, see William Craft Brumfield, *A History of Russian Architecture* (New York and Cambridge, 1993).
3. The basic monograph on Veldten's career is Militsa Korshunova, *Iurii Fel'ten* (Leningrad, 1988). In English, see Brumfield, *History*, pp. 266–71.
4. The genesis of the New Hermitage project is described in V. M. Glinka et al., *Ermitazh. Istoriia stroitel'stva i arkhitektury zdanii* (Leningrad, 1989), pp. 404–7.
5. Schinkel's Capella is described and illustrated in A. N. Petrov et al., *Pamiatniki arkhitektury prigorodov Leningrada* (Leningrad, 1983), pp. 450–51.
6. On Russian perceptions of the German educational system in architecture, see William C. Brumfield, *The Origins of Modernism in Russian Architecture* (Berkeley, 1991), pp. 2, 298 n9.
7. See William Craft Brumfield, "The French Connection: Victor Hugo, Nikolai Benois and the Medieval Revival in 19th-Century Russian Architecture," *The Harriman Review* 8 (1994) 4:1–13. The relation between text and architecture in the modern era is also discussed in Philippe Hamon, *Expositions: Littérature et architecture au XIXe siècle* (Paris, 1989), pp. 49–50.
8. G. Ravich, "Besedy stroitel'ia," *Stroitel'* 11–12 (1899), pp. 411–46.
9. Pavel Makarov, "Novyi stil' i dekadenstvo," *Zodchii* 6 (1902), p. 66.
10. Makarov, "Novyi stil' i dekadenstvo," *Zodchii* 9 (1902), pp. 101–5.
11. Pavel Makarov, "Arkhitektor i dekorator Iosif Gofman," *Zodchii* 23 (1902), pp. 263–64.
12. Pavel Makarov, "Darmstadtskaia vystavka 1901 goda," *Zodchii* 33 (1902), pp. 31–74; and *Zodchi* 34, pp. 379–82.
13. Evgenii Baumgarten, "Sovremennaia arkhitektura," *Zodchii* 48 (1902), p. 545.
14. Wagner's phrase appears in *Die moderne Architektur*, 3rd ed. (Vienna, 1902), p. 70.
15. Baumgarten, "Sovremennaia arkhitektura," *Zodchii* 48 (1902), p. 558.
16. Ibid., p. 571.
17. Sergei Diagilev, "Sovremennoe iskusstvo," *Mir iskusstva* 3 (1903), pp. 22–24.
18. A. Mikhailov, "Ist zu wienerisch," *Iskusstvo stroitel'noe i dekorativnoe* 3 (1903), p. 20.
19. Evgenii Baumgarten, "Sovremennoe iskusstvo," *Zodchii* 16 (1903), pp. 211–13.
20. Further discussion of the show is contained in Aleksandr Dmitriev, "Vpechatleniia ot vystavki arkhitektury i khudozhestvennoi promyshlennosti novogo stilia v Moskve," *Zodchii* 9 (1903), pp. 113–15. Dmitriev noted: "The participation of Olbrich is explained, in all probability, primarily by the sponsorship of the exhibit by Grand Duchess Elizaveta Fedorovna [wife of the governor-general of Moscow], a native of Darmstadt."
21. Aleksandr Dmitriev, "Sovremennoe dekorativnoe iskusstvo," *Zodchii* 39 (1903), pp. 543–57.
22. Aleksandr Dmitriev, "Aliuminievyi portal redaktsii 'Die Zeit,'" *Zodchii* 29 (1903), p. 353.
23. Vladimir Apyshkov, *Ratsional'noe v noveishei arkhitekture* (St. Petersburg, 1905), pp. 39–42.
24. Ibid., p. 56. In mentioning Bauer, Apyshkov refers to Bauer's praise of the Eiffel Tower and the bicycle as examples of the union of material and form, and

thus more satisfying aesthetically than the Basilica of St. Peter. Apyshkov claims that Bauer's views derive from Gottfried Semper, *Der stil in den technischen und tektonischen Künster, oder praktische Aesthetik* (Munich, 1860–63).

25. Apyshkov, *Ratsional'noe,* p. 58.
26. Among the Russian publications that examine the work of *style moderne* architects, two of the most authoritative are: E. A. Borisova and T. P. Kazhdan, *Russkaia arkhitektura kontsa XIX-nachala XX veka* (Moscow, 1971); and E. I. Kirichenko, *Russkaia arkhitektura 1830–1910-kh godov* (Moscow, 1978). The most comprehensive work in English is Brumfield, *The Origins of Modernism.*
27. For a broad analysis of these changes, see James L. West and Iurii A. Petrov, eds., *Merchant Moscow: Images of Russia's Vanished Bourgeoisie* (Princeton, 1998).
28. A detailed analysis of Shekhtel's work, including the Riabushinskii house, is contained in Brumfield, *The Origins of Modernism,* 120–73.
29. On the work at Matildenhöhe, see Ian Latham, *Joseph Maria Olbrich* (New York, 1980), p. 149.
30. For examples of the work of Ivanov-Schitz, see Brumfield, *Origins of Modernism,* pp. 274, 278, 281.
31. The design of Behren's house at Matildenhöhe, for example, was featured in Pavel Makarov's report on the 1901 Darmstadt exhibition, *Zodchii* 33 (1902), pp. 372–73.
32. Early responses to the German Embassy building include this unsigned article in *Moskovskii arkhitekturnyi mir* (1913), pp. 103–7; and Marian Lialevich, "Dom germanskogo posol'stva," *Peterburgskaia gazeta* 19 (1913). A critique of the embassy—with favorable commentary on Behrens's rough-hewn interpretation of neoclassicism—appeared under the initial "G.," "Dom germanskogo posol'stva," *Arkhitekturno-khudozhestvennyi ezhenedel'nik* 2 (1914), pp. 5–7. After the beginning of the war, another article in the same journal showed a predictable reaction against the embassy building and its symbolic projection of German might, in Martell, "Razgadannyi rebus," *Arkhitekturno-khudozhestvennyi ezhenedel'nik* 26 (1914), pp. 253–55.
33. A. Stepanov, "O promyshlennom zodchestve (Zametki i mysli)," *Zodchii* 12 (1915), pp. 119–23; *Zodchii* 13 (1915), pp. 131–37.
34. A. M. Ginzburg, "Rol' Germanii v russkoi stroitel'noi promyshlennosti," *Zodchii* 16 (1915), pp. 155–59.
35. Discussions of the participation of German architects in Soviet architecture include: Anatole Kopp, "Foreign architects in the Soviet Union during the first two Five-Year Plans," in *Reshaping Russian Architecture: Western Technology, Utopian Dreams,* ed. William C. Brumfield (New York and Cambridge, 1990), pp. 176–214; and Milka Bliznakov, "Soviet Housing during the Experimental Years, 1918–1933," in *Russian Housing in the Modern Age: Design and Social History,* ed. William C. Brumfield and Blair A. Ruble (New York and Cambridge, 1993), pp. 108, 136–38.

Chapter 12

SCIENCE, PHILOSOPHY, MUSIC

Chekhov's Three Germans

Anna Muza

A human condition famously portrayed in Russian literature is defined by the overwhelming need to direct one's life to a larger social, historical, or metaphysical truth. A work that epitomizes this quest is Chekhov's *The Three Sisters*, in which spiritual anguish and the search for a higher inspiration are common to all characters with one distinct exception. It is not Natasha, however, untroubled by any spiritual concerns, but Baron Tusenbach, a key figure in the play's explicit and implicit philosophical debate, who experiences the anxiety of an outsider among the initiated. In the middle of a characteristic general discussion of man's relation to and responsibility for life as it will be in "two hundred, three hundred years from now," the Baron interrupts his own hymn to work: "No doubt you're thinking: the German is getting sentimental. But I'm a Russian, word of honor. Why, I don't even speak German. My father was Russian Orthodox ..." (Act I).[1] Nothing even remotely suggests, at that moment or later on, that Tusenbach's background may be on anyone's mind but his own; yet it is not incidental that the Baron remembers and denies his Germanness when the conversation concerns the purpose of life and the ennobling effect of work.

Slightly out of place in the Prozorov drawing room, Tusenbach's remark belongs in a larger context of Russian literature where reflections on the human raison d'être are repeatedly framed by the opposition of Russian and German approaches to life. Characters of German descent duly re-emerge in Chekhov's writings that are dedicated to teleological questions, to the purposeful and aimless, predetermined and incidental in human existence: *The Duel, The Three Sisters, The Cherry Orchard.* However, Chekhov's treatment of the German theme in this traditional context is antitraditionalist. He reconsiders and parodies a set of cultural and literary commonplaces, and redefines the connotations and literary function of Germanness.

By Chekhov's time, the discourse of German character in Russian literature had reached the status of a topos. Pushkin, Gogol, Goncharov, Tolstoy, to mention only the major writers, had associated the German view of life with the cult of purpose: but a purpose so petty and so utterly egotistical as to become a parody of a purpose. Regardless of their status or welfare, Germans in Russian letters cherished the golden rule of their forefather in *The Queen of Spades*, who would not "sacrifice the necessary in the hope of acquiring the superfluous." Germanness meant pragmatism, predilection for structures and schedules, doggedness, narrow-mindedness, worshipping of rules, and insensitivity to exceptions. In *War and Peace*, the Rostovs' eldest daughter, the rational and unperceptive Vera, alien to her family, would marry a German—and find a perfect match in the ridiculously pedantic and conceited Berg. Goncharov's *Oblomov* introduced a direct opposition of the German and Russian modes of being within the same circle of intertwined categories: sense of purpose, careful organization of life, ability to act. Stolz, one of the more likable Germans in Russian literature, owed his humanity to his fine Russian mother whose "songs and whisper" prevented him from becoming a *bürger*. His German element expressed itself in his excellent common sense, dedication, and purposefulness—but also in his "fear of imagination," hostility to "painful and tormenting" reflections, and a lack of passion and audacity—particularly pitiful in Russian eyes. Stolz's formula of life resembled an embellished translation of Hermann's motto: "... a steady and slowly burning fire is better than vehement flames, however poetic their blaze."[2]

Poetry, whether poetry at large or German poetry in particular, has been divorced from the German character. It appears that, while pondering the ultimate questions of life's meaning and purpose, Russian writers have been particularly irritated by the alleged German capacity to reduce these questions to manageable domestic proportions. Goethe, Schiller, Hoffmann, or Heine, worshipped by generations of Russians, had no bearing on the thoughts and aspirations of the vast majority of Germans portrayed in Russian literature. When Gogol in "Nevsky Prospect" introduced two Petersburg tradesmen, "Schiller, not the Schiller who wrote 'Wilhelm Tell' and the 'History of the Thirty Years' War,' but the famous Schiller, the ironworker of Meshchanskaya Street" and "Hoffmann, not the writer Hoffmann, but Hoffmann the shoemaker from Officerskaya Street," his travesty touched upon the presumed gap between German letters and German men. The Schiller who wrote Wilhelm Tell was, according to Vissarion Belinsky, a spokesman for mankind, a herald of humanity; the ironworker Schiller who, "when he was only twenty … had planned out his life, […] had set himself to accumulate a capital of fifty thousand in ten years" and "permitted himself to kiss his wife not more than twice in twenty four hours …" was, according to Gogol, "a complete German in the full sense of the word."[3]

In 1889 the twenty-nine-year-old Chekhov reread *Oblomov* and wondered "why up to now [he had] considered Goncharov a first-rate writer." His disappointment concerned the portrayal of both major characters: Oblomov now seemed "blown-up" and unworthy of the "status of a social type," Stolz appeared "half-concocted and three quarters stilted."[4] Chekhov's reappraisal of his "demigod" resounded two years later in *The Duel*, a novel-long study of two contrasting personalities: the weak and self-indulging Laevsky and the stern and resolute scientist von Koren. In Chekhov's version of the relationship between a Russian of no will and a German of strong will, the sentimental friendship of Goncharov's heroes was replaced by intense animosity. Laevsky's and von Koren's antagonism is part of the story's elaborate framework of more or less obvious literary references and motifs (a tale of rivalry and adultery taking place in the Caucasus and culminating in a duel).[5] Laevsky wishes to exonerate his good-for-nothingness by placing himself among the 'superfluous men.' Von Koren, whose self-admiration needs no crutches, personifies the Russian lexicon of Germanness. Not only does he meticulously plan his own life, he wishes to organize and classify all human species and lectures peremptorily on hard work, determination, and progress. His very pursuit of zoology is a perfect vehicle for the alleged German preoccupation with classification and order, mocked by Tolstoy in *War and Peace*: "Die erste Kolonne marschirt, die zweite Kolonne marschirt…."[6] His unqualified cult of exact knowledge and lack of humanity illustrate the old verdict formulated by Belinsky thus:

> Everybody knows that there is no nation in the world more learned than the Germans … Yet the Germans have a defect, too, which is that they stay *Burshen* till the age of thirty, and for the remaining and larger part of their lives continue as *Philistern*, and consequently have no time to become human beings.[7]

The expected 'human deficiency' of the German character is taken by Chekhov to an almost grotesque extreme in the scientist's proposition to shoot socially useless human organisms or at best sentence them to hard labor. Realizing that this project is not a joke, von Koren's friend Samoilenko, a doctor, cries out in despair: "You are a very learned man of the greatest intellect, Fatherland's pride, but you have been spoilt [corrupted] by the Germans. Yes, by the Germans! The Germans!"[8] The doctor thus praises von Koren as a fellow Russian, "Fatherland's pride," and attributes his delusion to an evil foreign influence. A vastly ignorant man, he must believe that von Koren's "learned" heroes, Darwin and Spencer, are German. Yet by his awkward phrasing the kind doctor inadvertently suggests that the influence in question is not exactly *foreign* to von Koren and that if he has been 'corrupted' by the Germans then indeed from the outset. The narrator adds the following postscript to the doctor's invective:

> Since leaving Dorpat, where he had studied medicine, Samoilenko had rarely seen Germans and had not read a single German book, but, in his opinion, everything harmful in politics and science stemmed from the Germans. He himself could not say where he had gotten this opinion, but he maintained it firmly.[9]

The repetition of "opinion" in this context ironically accentuates the mythic quality of Samoilenko's irrational judgment. Laevsky himself, in a moment of uncontrollable rage, "as if in a fog," suddenly calls his enemy "a German Yid," thus 'doubling' the accusation. The axiom of inherent German "corruption" is equally taken for granted by a semiliterate character in an early Chekhov story, who remarks that someone "lives with a German instead of a wife";[10] and by the refined Baron Tusenbach, embarrassed by his own name: "Upon my word, I'm Russian …"

Chekhov's mistrust of stock formulas of human identity informs the striking and much criticized transformation and rapprochement of the two enemies in the story's finale. Importantly, contemporaries found *The Duel's* resolution artificial: Chekhov's characters must have been largely perceived as familiar literary 'types,' whose consistency (i.e., predictability) is disrupted by an inappropriate (i.e., unprecedented) narrative twist. The comment of Dmitry Merezhkovsky, by no means an average reader, is most characteristic: "The type of Laevsky is lovely, fully alive, if it were not for the last chapter which remains forced."[11]

However, one of Chekhov's intentions is precisely to break through the shell of 'literariness' surrounding the two antagonists. A careful reader may notice that various literary patterns are subverted in the story prior to the denouement. Thus, the emphatic lack of connection between

Laevsky's unromantic romance and his duel with von Koren is meaningful and slightly parodistic against the narrative pattern established by his oft evoked predecessors, from Onegin to Bazarov. As far as von Koren is concerned, his very participation in a life-threatening ritual is not compatible with his proto-character. The recurrence of the duel in the fate of Chekhov's Germans is hardly incidental: ten years later in *The Three Sisters,* it will end the life of Tusenbach, killed by the bloodthirsty captain Solenyi. On the whole, the duel has no appeal for Chekhov, always suspicious of grand ostentatious gestures: both in the story and in the play, the duels are provoked by men who see their lives through the prism of literature and conceive of themselves in literary terms. At the same time, the act of putting one's life at stake on the outcome of chance is the exact opposite of the alleged German way of planning and measuring, and the readiness to risk everything for the sake of immaterial values signals a deviation from the image of the German incapable of any extravagant *élans.*

Von Koren's involvement in the duel is marked by contradictions. It is ironic that the zoologist's beloved project of legitimizing the execution of socially useless human organisms should materialize in the form of a procedure not only illegitimate but also one in which a useless organism may prevail. After Laevsky shoots in the air, however, the contest does for a moment resemble an execution, as if ultimately shaped by von Koren's will: "'I'll kill him now,' thought von Koren, aiming at Laevsky's forehead and fingering the trigger." He fires, contrary to his own reasonable conclusion of yesterday that "to be arrested for Laevsky … isn't worth the candle," but misses—not only his target but also his chance to participate in the Darwinian elimination of the weak. Neither outraged nor embarrassed by the outcome, von Koren is but profoundly "exhausted," "worn out." Such wavering between reason and emotion, theory and experience, and a lack of understanding of his own motives unsettle the scientist's fixed identity and presage his changed image in the end.

The von Koren of the story's finale is a sad and pensive man who comes *prostit'sia,* to bid farewell to and ask forgiveness of, his former enemy. This change is a subdued reflection of Laevsky's profound transformation, the very suddenness and unpredictability of which undermine von Koren's 'scientific' diagnosis: "I see now, to my great joy, that I was mistaken about you.…" Similarly undermined are the literary personae of the German scientist and the Russian superfluous man. Von Korean's final melancholy realization, "No one knows the real truth," annihilates not only his own earlier positivism but also the opposition underlying the entire narrative and, as sensed by the disappointed reader, largely the narrative itself.

Von Koren's farewell line, "No one knows the real truth," resonates in Laevsky's mind as he observes von Koren's departure. The scientist leaves on a rainy night in a small boat, trying to reach his ship moored farther away from the shore and indiscernible in the dark. In all its poignant detailed reality, the last scene, characteristic of Chekhovian departures with their awkwardness and commotion, is, of course, a materialized metaphor of the uncertainty of human destination. The shift in the discourse accompanying von Koren, from the scientific and rationalist to the lyrical and poetic, indicates yet another change implicit in *The Duel*—the literary metamorphosis of the German 'type.'

In a sense, von Koren's end contains the beginning of Baron Tusenbach—an amiable and somewhat dreamy young man whose sonorous name, Tusenbach-Krone-Altschauer, seems to be the only vestige of his origin. His good nature, however, does not save him from having an enemy, and the confrontation of Tusenbach and Solenyi evokes *The Duel* in an inverted way: in *The Three Sisters* (1901), it is a stubborn and crude Russian who persecutes and wishes to destroy a gentle and sensitive German. This reversal, which almost suggests a Chekhov self-parody, also affects the juxtaposition of Russian and German philosophies. Solenyi's outlook is but a travesty of an enlarging vision: the captain invents himself as a sinister romantic figure and wishes to enhance the importance of his life by taking someone else's. Tusenbach, on the contrary, yearns to share in the human effort of making life "beautiful and marvelous," has "a passionate craving for life, for struggle, for work...." Although Solenyi professedly scorns Tusenbach as a rival in love, what truly irritates him about the Baron is the latter's idealism and "philosophizing": "Tsip, tsip, tsip ... Don't feed the Baron any mush, just give him a chance to talk philosophy" (Act I).

Alongside Laevsky, Solenyi is one of Chekhov's characters who, in Gary Saul Morson's words, "try to endow their lives with a spurious meaningfulness by imitating literary characters and scenes": a kind of behavior which for Chekhov "was a particularly loathsome form of lying."[12] Tusenbach, on the contrary, is a genuine idealist, a philosopher and a poet. He believes in beauty, self-sacrifice, and gradual refinement of the human soul; work for him is a sacred faith, and love, pure and elevated worship. He perceives life as a purposeful yet mysterious process whose immanent law is beyond human understanding: "Migratory birds ... fly and will go on flying, no matter how many philosophers may turn up among them." This view of the inherent meaningfulness of all existence, not necessarily rational or intelligible, brings about one of the best-known lines in Chekhov's drama, a tiny 'poem in prose': "Meaning … Look there, it is snowing … What is the meaning of that?" (Act II).

Ever insisting on his Russianness, Tusenbach assures Irina that there is "nothing German" in him except for the "stubbornness" with which he follows her. The Baron misreads his German element in the conventional Russian way. Tusenbach's roots are not in German stubbornness and mercantilism but rather in the German philosophical tradition, contemplation, music, and intense emotionality. Robert Mann has suggested a 'Faustian' dimension in the Chekhov character so passionately focused on the "Faustian problem of work and happiness."[13] "Work and happiness" equally preoccupy the noble heroes of Schiller's dramas, tremendously popular in Russia throughout the nineteenth century. For example, Ferdinand von Walter in *Kabale und Liebe* intends to sacrifice his privileged existence for a humble life of honest labor with his beloved, and Baron Tusenbach abandons his military career and prepares to start a new life of honest labor with *his* beloved: both die violent deaths instead of being reborn. It seems unlikely that in Tusenbach Chekhov refers to a specific character in Goethe or Schiller; rather, he evokes a distinctive set of images and motifs, defining German literature for the Russian reader. Tusenbach is a German related to the Schiller who wrote *Wilhelm Tell*, not the famous tinsmith of Meshchanskaya Street in Gogol's "Nevsky Prospect." The Baron's sentimental longing for work parodies both the cliché of German industriousness and productivity and the aspirations of young idealists. Only a man who grew up with "servants taking off his boots" can envy workmen who "must sleep soundly." In a peculiar ('Faustian') blindness he fails to give thought to the effect real work has had on Irina: "My brain is drying up, I've grown thin, old, ugly…."

A Russian German, Tusenbach also resembles a Germanized Russian, Vladimir Lensky—a poet "with a Göttingen soul" whose ideas and spiritual pursuits were shaped in "the misty Germany":

> The purpose of our life to him
> was an alluring riddle;
> he racked his brains over it
> and suspected marvels.
> He believed that a kindred soul
> to him must be united …[14]

Chekhov's view of Tusenbach presents a mixture of empathy and irony, similar to Pushkin's attitude to his "young lover." The two characters suffer untimely death in a duel at a time of enthusiasm and happiness. The two authors suggest that this end, cruel and absurd as it is, not only casts doubt on the victims' idealistic view of life but also forestalls some unalluring changes or bitter disappointments. Pushkin laments Lensky as a poet yet adds that with age he might have lost "the fervor of the soul" and turned into a "happy and cornute" old man who had "learned life in its reality/at forty, had the gout/drunk, eaten...." His fiancée's conspicuously short mourning makes one also wonder about the kindred soul. The overall structure of Chekhov's dramatic work prompts similar suppositions about Tusenbach's future. Three married couples that we observe in the play leave little hope that Tusenbach may have a happier or better life with Irina, who marries him out of despair and who has already been profoundly disappointed in work, which the Baron so eagerly anticipates.

In *The Duel*, Chekhov used the conventional opposition of German and Russian sensibilities to reveal its literary limitations and psychological inadequacy. In *The Three Sisters*, he parodied this opposition and slyly infused his unwilling-to-be-German hero with the spirit of German philosophy and literature. In his last work, *The Cherry Orchard* (1904), Chekhov carried his polemical intention still further and liberated the German character from any antithetical structures, offering a novel variation on the German theme.

The German governess Charlotta, one of the most peculiar characters in Chekhov's world, pointed to a dimension so unusual and unexpected that it went unrecognized. Although Charlotta is more 'authentically' German than her predecessors and, unlike Baron Tusenbach, actually does speak German, contemporaries failed to see the relevance of her national identity. Thus, Stanislavsky believed that Charlotta's immediate prototype was an eccentric *English* governess whose company Chekhov enjoyed in the summer of 1902 when the first throbs of *Cherry Orchard* stirred in his mind. Indeed, several things about Charlotta, such as her dog, hunting gun, contempt for men, and slender build, befit a daughter of Albion more than that of Allemand. Stanislavsky shared his insight with Chekhov and could not understand why the writer got perplexed and insisted rather vehemently that Charlotta "was German, had to be German."[15]

Why German? The presence of a foreign governess in a noble Russian family would be commonplace, yet one function that Charlotta does not perform and that no one expects her to perform is precisely that of a governess. It is indeed striking that not once in the play does Charlotta speak to Anya, the young lady entrusted in her care. Charlotta's domain is not the rules of German grammar or proper demeanor but buffoonery and magic tricks. These tricks, with their bewildering losses, finds, and substitutions sustain and mock major themes of *The Cherry Orchard*—fate, chance, human ability to control life. Charlotta's foreignness has often been seen, broadly, as a synonym of her 'difference,' her *emploi* of the wise fool. Savely Senderovich has interpreted the German governess as a "private joke" hinting at Chekhov's wife the actress Olga Knipper and the "magic power of transformation possessed by a dramatic actress."[16] Indeed, the role was originally meant for Knipper, but Chekhov failed to persuade her that the "darling" (*milaia*) Charlotta was not inferior to the apparently more central and 'serious' part of Mme. Ranevskaya.[17]

Whatever personal details Charlotta may have absorbed, this character ultimately belongs in a larger literary family. In Act III Charlotta demonstrates her power over things and people in a tour-de-force performance at Ranevskaya's ill-timed party, while the fate of the Cherry Orchard and its inhabitants is being sealed. One of the fascinated guests gets truly enchanted with the conjurer. To his declaration of love, however, Charlotta only shrugs her shoulders: "Are you capable of love? *Guter Mensch, aber schlechter Musikant.*"

With the latter reference, the German governess, who has complained of not having as much as a "genuine passport," provides her genuine identification. I believe that Charlotta quotes from E. T. A. Hoffmann's *Kater Murr* (1820), in which Iohannes Kreisler, one of Hoffmann's favorite recurring characters, discusses the human race in terms of this dichotomy: "… like the judge of the world, I quickly divided people into two different groups, one of which consisted of the good people who are bad musicians, or rather none at all, the other one of the real musicians. No one was to be damned, but all were to be blessed in different ways."[18] In Hoffmann, Kreisler is a romantic artist, a Musician par excellence: homeless, penniless, kinless, he persists in his vagabond existence for the sake of his art, whose demands place him outside of the mundane goals and considerations of the "good people." The passportless and homeless Charlotta, who grew up traveling "from fair to fair" performing in the circus and who does not know "where [she] comes from or who [she] is," also is one of the Musicians. Her eccentric dog who "eats nuts, too" may be an offspring of Kreisler's companion Kater Murr.

In the play whose central conflict is defined by the choice between the useful and useless, rational and sentimental, pragmatic and beautiful, Charlotta represents the superfluous, the artistic, the bizarre—aspects of human experience that, like a conjurer's tricks, cannot be reduced to a logical explanation. Appearing as a ventriloquist, Charlotta emits "a mysterious woman's voice that apparently comes from under the floor: Oh yes, the weather is magnificent, dear lady." The bodiless sound rising from the underneath is a comical counterpart of the famous sound of the broken string "as if coming from the sky." Both refer to smaller or larger mysteries hidden from human sight; both underscore, mockingly or dramatically, the limits of human judgment tied to the ostensible surface of things. Strangely enough, the man who falls under Charlotta's spell at the party, the naive Simeonov-Pishchik, is indeed "blessed" at the end of the play—rescued, unlike the owners of the Cherry Orchard, from financial ruin by a miraculous circumstance.

A more obvious and less humorous, yet important, conflation of Germanness and musicality can be found in an earlier work of Russian literature, Turgenev's *The Nest of the Gentry* (1858).[19] In the novel, the expatriate music teacher Lamm is a suffering solitary figure, enclosed in the world of music and elevated emotion, a genuine albeit unknown artist, contemptuous of good men "who don't understand." Interestingly, in Turgenev's play *A Month in the Country* (1855), commonly seen as a precursor of Chekhov's drama, the German tutor Schaaf fully conforms to the *emploi* of the "comic foreigner" in popular theatre: he speaks heavily accented grotesque Russian, bubbles with self-importance, chases the maids, and so on. Admonishing the Moscow Art Theatre that Charlotta should speak pure unaccented Russian, Chekhov was well aware of the need to protect this "important part" from the trappings of habitual shallow comedy.[20] Unhappy but not pathetic, shrewd but not didactic, funny but not ridiculous, Charlotta comprises both Turgenevan 'high' and 'low' versions of Germanness, mixed here to an entirely new effect.

In his well-known criticism of the MAT's rendering of *The Cherry Orchard*, Vsevolod Meyerhold discussed the play in musical terms and compared Act III to a movement of the symphony, where the "basic elegiac melody" finds a counterpoint in the "monotonous cacophony of the distant band and the dance of the living corpses (the philistines)": "Not for nothing does Chekhov make Charlotta dance amongst the philistines in a costume familiar in the puppet theatre—a black tail-coat and check trousers."[21] Meyerhold did not discuss Charlotta's German connection, yet an admirer of Hoffmann, a musician, a conjurer in his own right (and, incidentally, of German extraction), he was perfectly attuned to the themes expressed by a fellow artist in the guise of the German governess.

Chekhov's redefinition of the German character is a part of his dialogue with Russian literary tradition—a dialogue whose scope and significance have gained critical attention only in a relatively recent time. This particular endeavor includes, but is not limited to, mockery and reversal of the established convention. The writer's ultimate purpose is not to replace a 'prosaic German' with a 'poetic German' but to modify the function of the German character in the narrative or dramatic structure. The German as an antithesis, an estranging device which helps to articulate "Russianness," is for Chekhov an artificial construct, obscuring rather than elucidating "the real truth." The antithetical model is carried to its extreme and exhausted in *The Duel.* In *The Three Sisters,* Chekhov does not place Tusenbach's German origin in the usual opposition to the "Russian character" of Solenyi but rather lets it have a significance of its own. The Baron, whose triple name and title are as excessive and inappropriate in the Russian provinces as the sisters' fluency in foreign tongues, brings with him a philosophical legacy and universal vision that extend the symbolic space of the play far beyond the proverbial Moscow. The German governess in *The Cherry Orchard* is both related to and detached from all other characters. Her loneliness, the loneliness of a foreigner in a strange land, epitomizes the condition more or less familiar to everyone else. Her homelessness, the homelessness of a companion who has no life of her own, reflects the literal and figurative state of the entire group. Yet her art, her connection with the mysterious and fateful, make her superior to ordinary mortals whom she shuffles in her tricks like cards. In Charlotta, the Chekhovian transformation of the meaning and purpose of the German character reaches its ultimate form.

Notes

1. *Anton Chekhov's Plays,* trans. and ed. E. K. Bristow (New York and London, 1977). All quotations from Chekhov plays are from this edition, occasionally with slight modifications.
2. Ivan Goncharov, *Oblomov* (Moscow, 1947), p. 145.
3. *Tales from Gogol,* trans. R. Portnova (London, 1945), p. 151.
4. *Letters of Anton Chekhov,* sel. and ed. A. Yarmolinsky (New York, 1973), pp. 113–14.
5. The literary, cultural, and biblical allusions in the story are discussed, in particular, in Andrew R. Durkin, "Allusion and Dialogue in *The Duel,*" in *Reading Chekhov's Text,* ed. R. L. Jackson (Evanston, 1993), pp. 169–78, and Willa

Chamberlain Axelrod, "The Biblical and Theological Context of Moral Reform in *The Duel*," *Russian Literature* XXXV–II (1994): 129–52.

6. Commenting ironically on steadfast meticulousness of German strategic planning, always upset and ignored in the chaos of actual battles, Tolstoy evokes the Kolonne prescription twice: in 1, 3, XII and 4, 2, IV.
7. Vissarion Belinsky, "Peterburg i Moskva," in *Sobranie sochinenij v trex tomakh*, vol. 2 (Moscow, 1948), p. 782.
8. Anton Chekhov, *The Duel*, in *Seven Short Novels*, trans. B. Makanowitzky (New York, 1963), p. 31.
9. Ibid., pp. 31–32.
10. The story is *Loshadinaia familia*, and the Russian original pales in translation into a genderless language: "Zhiviot ne s zhenoj, a s nemkoj ..."
11. A survey of contemporary responses to *The Duel* can be found in Aleksandr Chudakov, *Poetika Chekhova* (Moscow, 1971), pp. 234–37. Merezhkovsky's letter to Chekhov is quoted on p. 234.
12. Gary Saul Morson, "*Uncle Vanya* as Prosaic Metadrama," in *Reading Chekhov's Text*, pp. 214–15.
13. Robert Mann, "The Breaking Clock: Chekhov and Goethe," *Germano-Slavica* 2 (1988): 84.
14. *Evgenii Onegin*, 2: VII–VIII, quoted in Vladimir Nabokov's translation.
15. Konstantin Stanislavsky, *Moia zhizn' v iskusstve* (Moscow, 1962), pp. 325–26.
16. Savely Senderovich, "*The Cherry Orchard*: Chekhov's Last Testament," *Russian Literature* XXXV–II (1994): 236–39.
17. On the whole, Chekhov's correspondence with Olga Knipper and other members of the Moscow Art Theatre during auditioning and casting in October to November 1903 reveals the author's remarkable concern about Charlotta's amusing but technically minor "character part." See Anton Chekhov, *Polnoe sobranie sochinenij i pisem*, vol. XX (Moscow, 1951).
18. E. T. A. Hoffmann, *The Life and Opinions of Kater Murr*, in *Selected Writings*, ed. and trans. L. Kent and E. Knight (Chicago, 1969), p. 133.
19. I am indebted to Professor Omry Ronen for pointing out this parallel.
20. *Letters of Anton Chekhov*, p. 172.
21. Vsevolod Meyerhold, "The Naturalistic Theater and the Theater of Mood," in *Meyerhold on Theatre*, trans. and ed. with a critical commentary E. Braun (New York, 1969), p. 28.

Chapter 13

NIETZSCHE'S HIDDEN VOICE IN SOCIALIST REALISM

Bernice Glatzer Rosenthal

Socialist Realism, the official aesthetic of the Soviet Union in Stalin's time and for decades after, was adopted by the first Congress of the Union of Soviet Writers in September 1934. The Union's Charter specified that Socialist Realism "requires of the artist a true, historically concrete depiction of reality in its revolutionary development ... the artistic depiction of reality must be combined with the task of the ideological transformation and education of the workers in the spirit of Socialism." Although the writers were assured of "exceptional prospects for manifesting creative initiative, of a choice of diverse forms, styles, and genres,"[1] Socialist Realism became a straitjacket. The official aesthetic applied to all the arts, not just to literature; other Unions adopted it subsequently.

Socialist Realism was a political aesthetic from the time of its inception. It was formulated by Maksim Gorky, the famous writer (not a member of the Communist Party); Anatoly Lunacharsky, formerly Commissar of Enlightenment (Education); Andrei Zhdanov, supervisor of the Party's "ideological front"; Karl Radek, a high official in the Communist International; and others between 1932 and 1934, fleshed out between 1934 and 1939, and codified after World War II.

Soviet propagandists reviled Nietzsche as "the philosopher of fascism," but his hidden voice guided the formulation of Socialist Realism. Gorky's and Lunacharsky's debt to Nietzsche has been well documented.[2] Through them and others—Marxists and non-Marxists—Nietzschean ideas were embedded in Russian culture before the Bolshevik Revolution and, unacknowledged, continued to permeate Soviet culture.[3] Nietzsche's trenchant slogans, alluring images, and provocative ideas stuck with his readers long after their initial impression. His memorable phrases and emotionally charged images became detached from their context and took on a life of their own. The most obvious appropriation of Nietzsche in Stalin's time, the Soviet Superman, has been treated elsewhere.[4] This chapter will examine the politicization of Nietzsche's idea of art as a lie and the quasi-Apollonian language that Socialist Realism mandated.

Art as a Lie

In *The Birth of Tragedy* Nietzsche describes art as an illusion that veils the horror of the abyss. In *On the Genealogy of Morals*, he writes: "The will to truth requires a critique … the value of truth must for once be experimentally *called into question*," and he contrasts the scientist's "will to truth" with "art, in which precisely the *lie* is sanctified and the *will to deception* has a good conscience."[5] Nietzsche did not invent the idea of art as a lie—propagandists were using art to misrepresent reality centuries earlier—but he expressed the idea with great forcefulness and gave it a philosophical rationale. The distinction between illusion and lie is blurry, but illusion is the gentler term. It does not imply deliberate deception and lacks the moral opprobrium of the word "lie." Moreover, there are harmless white lies and malevolent black lies. Russians distinguish between *vran'ë*, loose talk or bluster, and *lozh'*, a hurtful lie. In any language there are lies of omission and lies of commission, inventing "facts."

Nietzsche maintained that the ruler defines what is "true" and that reality is a matter of perspective. The "real world" is false, cruel, contradictory, seductive, and without meaning. Human beings need lies to conquer this reality; they "have need of lies in order to live."[6] And yet, Nietzsche attacked liars and hypocrites and praised "truthfulness as the highest virtue."[7] The formulators of Socialist Realism appropriated Nietzsche's ideas selectively; they required lies to retain power and to accomplish their goals.

Socialist Realism was the Stalinist version of "myth-creation" (*mifotvorchestvo*), a concept promulgated by the Symbolist poet and theorist Viacheslav Ivanov during the Revolution of 1905. He advocated a resurrected Theater of Dionysus in which the Symbolist poet, acting as a high priest, would lead the chorus (the people), in the creation of a new myth that would restore cultural coherence and end social conflict.

Unlike Nietzsche, Ivanov believed that myths express eternal truths and that new myths are reformulations of old ones. From the new myth would stem a new cult, a new culture, and a new society, in that order. Other Symbolists held similar views.

Gorky and Lunacharsky proposed a Marxist counterpart to "myth-creation," a surrogate religion called god-building, that would inspire people to fight and die for socialism. Gorky dreamed of a Russian Superman who would lead the masses in a struggle for liberation. Lunacharsky envisioned a perfected superhumanity and believed that art would transform consciousness. The Nietzschean Marxist Aleksandr Bogdanov wrote a utopian novel, *Red Star* (1908), about a flourishing communist society on the red planet Mars to induce the proletariat to continue its struggle. Other versions of "myth-creation" were championed by the Futurists and the Acmeists after 1909 and by the Soviet avant-garde in the 1920s.

The first Five-Year Plan (1928–33) featured crash industrialization, collectivization of agriculture, and a cultural revolution intended to eliminate "bourgeois" elements from Soviet life. The masses had been led to believe that after a few years of stupendous effort, they would be living in paradise. Instead the standard of living plummeted. Peasants killed their cattle rather than turn them over to the collective farms, and planted less. Bread was rationed. Ordinary consumer goods were available only to the privileged (who were not supposed to exist in a socialist society), and millions of peasants were deliberately starved to death in the "terror-famine" of 1932–33.[8] In the cities, worker resentment flared into demonstrations and strikes. Even stalwart Bolsheviks thought that Stalin had to go.[9] Critics of Stalin's policies were repressed, but force had its limits. What was needed, Party leaders realized, were inspiring visions that would controvert experienced reality and keep hope alive.

Socialist Realism was designed to create these visions, to create a beautiful illusion and pass it off as truth. Many Soviet artists and writers truly believed the myth of the radiant communist future. Some became disillusioned later on; others kept the faith to the end of their days. But the formulators of the official aesthetic knew exactly what they were doing; they were appropriating Nietzsche's concept of art as a lie to sustain Bolshevik power. Socialist Realism was designed to bolster morale, prevent slackening of effort, and cover up mistakes. The writer's task was to depict reality, not as it was at the time, but as it *would be* under socialism, and then treat the future *as if* it were the present.

At the first Congress of Soviet Writers, Zhdanov said that Socialist Realism was "openly tendentious" and called it a "romanticism of a new type, a revolutionary romanticism," as distinct from old-style romanticism, which depicted a non-existent utopia and non-existent heroes. "*Soviet literature must be able to show our heroes, must be able to catch a glimpse of our tomorrow. This will not be a utopia, because our tomorrow is being prepared today by our systematic and conscious work.*"[10] Karl Radek described Socialist Realism as "not only knowing reality as it is, but knowing whither it is moving."[11] The future-orientation of Socialist Realism (emphasizing the future at the expense of the present generation) stems from Nietzsche and Marx, and distinguishes it from critical realism and from Norman Rockwell's sentimental depiction of everyday life. The kitsch that became a hallmark of Socialist Realism emerged in the mid-1930s and became pronounced after the war.

Gorky dominated the Congress. Critical realism, he said, was "bourgeois" and therefore obsolete and irrelevant in Soviet society where the negative aspects of reality were rapidly disappearing and fantasy was becoming reality, but it (critical realism) was appropriate for the capitalist world. Lunacharsky died shortly before the Congress met, but in his pre-Congress report on Socialist Realism (1933) he explained that the Soviet Union was building a splendid palace for the workers that was still incomplete. The "bourgeois realist" sees that the palace has no roof and gloats "Here's your socialism!" "From the point of view of socialist realism, this is unreality, a lie, a substitution of carrion for life." The Socialist Realist sees that there will be a roof and describes the building as if it had one, because truth is not static; it "flies … and we must see it like that." The "bourgeois realist" is "voluntarily or involuntarily a counterrevolutionary, a wrecker." If "truth" is objectionable from the point of view of Socialist Realism, it is a lie.[12]

Gorky had long questioned the value of truth and praised the beautiful lies that give humankind success. His 1902 play *The Lower Depths* includes a Nietzschean dialogue on truth. The character Bubnov says "give 'em the whole truth, just as it is." To which Kleshch replies: "What truth? Where's the truth? No work, no strength, not even a place to live. The only thing left is to die like a dog! This is the truth … What do I want the truth for? I want to breathe more freely—that's all I ask. What have I done wrong? Why should I have been given the truth?" To which Luka responds: "You've been saying we need the truth. But it isn't always truth that is good for what ails a man—you can't always cure the soul with truth." To prove his point, Luka tells about a simple man who believed in the existence of a "true and just land" until a learned man showed him that it does not exist. Unable to bear this bitter truth, the simple man "went home—and hung himself."[13] In the same act, Luka advises Natasha to marry Peppel, also known as Vassya, saying: "He's all right, he's a good fellow. Only you have to remind him as often as you can that he's a good fellow—so that he doesn't forget it. He'll believe you. Just keep telling him—you're a good man Vassya, remember that! And besides, my dear, where else can you go?" (51). Throughout the play, Luka purveys comforting lies. The character Satin, arguably Gorky's voice, says "Lies are the religion of slaves and bosses. Truth is the god of the free man" (64).

Socialist Realism represents the victory of Luka over Satin, of comforting illusions over bitter truths. Gorky continuously told the people that the future is here, that they are heroic, energetic, creative, hard-working, self-disciplined, and rational, because telling them that might make it come true, and besides "where else could they go?" To return to capitalism was unthinkable, but if the people knew that the socialist paradise would not be reached in their lifetime, they might lose confidence in the Party and succumb to despair. Then who would build socialism?

Gorky wanted a "created truth" that would serve as an energizing myth. In 1929, he broke with an old friend because she told the truth about Soviet Russia. In a letter to her, he wrote

> What is important to me is the rapid all-round development of the human personality, the birth of a new man of culture, of workmen in a sugar refinery who read Shelley in the original ... such men do not need a petty accursed truth in the midst of which they are struggling. They need the truth they create for themselves.... I am now one-sided.[14]

The scale and boldness of the first Five-Year Plan reactivated Gorky's hope of creating a Russian Superman. The full human cost was not yet apparent and confidence was still high. Gorky regarded hardship and suffering as temporary obstacles to be overcome by reason and will. The horrors of dekulakization (the brutal dispossession and persecution of the so-called "rich" peasants who were not rich at all) did not disturb him, for he regarded the kulaks as "vermin" and "yet-unthrottled snakes."

In 1932, when the defects of the Plan were all too evident, Gorky claimed:

> In the Soviet Union a new man is developing ... He has faith in the organizing force of reason ... He feels himself the creator of a new world and although he lives in difficult conditions, he knows that he is creating other conditions. That is his goal and the deed of his rational will, therefore, he has no basis for pessimism. He is young not only biologically but historically. He is strong ... [he] recognizes his own path, his own meaning in history, and he is constructing his [own] culture.[15]

Nietzsche's voice is audible in Gorky's vision of the new man and in his resolute optimism. Also discernible are the Nietzschean polarities of sickliness/health, decadence/vitality, and old age/youth. Youth, a theme of Nietzscheans since the 1890s, was used to contrast Soviet society with expiring capitalism. That capitalism was in its death throes was taken for granted because of the Great Depression.

Gorky's positive thinking mingled Zarathustra's "yes to life" with the Bolsheviks' faith in their historical mission and with early twentieth-century theories of thought transfer and hypnotic suggestion.[16] In 1913 Gorky attacked the Moscow Art Theater for staging Dostoevsky's novels because the tormented, neurotic characters were bad role models for workers. Gorky disapproved of Aleksandr Afinogenov's play *Lozh'* (The Lie, 1933), which depicted Party leaders who lie as a matter of policy. One character says that the whole country lies and cheats, for it is being cheated itself. A long monologue by the protagonist bemoans the collision between word and deed in contemporary Soviet society. Afinogenev sent the manuscript to Gorky, who replied that such a play would be extremely useful if it were performed in a closed theater, before a thousand educated Leninists who were unshakeably persuaded of the validity of the general line, but it could never be shown to an audience in which there were "not socialists but the [bastards] of the past, in exile from the shambles of their bourgeois life."[17] Nietzsche associated pessimism with nihilism, but he also talked about a pessimism of strength. Gorky ignored the latter and misleadingly associated pessimism with skepticism (as well as with nihilism and with weakness). He wanted people to emulate the determined, self-sacrificing heroes of folk myth. The hero of his short story "The Old Woman Izergil" (1895) tears out his flaming heart and uses it as a beacon to guide his tribe to safety.

At the Congress of Soviet Writers, Gorky lambasted the years 1907 to 1917 as the "most shameful and shameless decade in the history of the Russian intelligentsia," pointed to the harmful influence of Dostoevsky and Nietzsche, and declared that Nietzsche's ideas "form the basis of the fanatical creed and practice of fascism."[18] Reiterating his advocacy of a new literature of positive heroes based on folk myth, which always exudes a positive optimistic spirit, Gorky proclaimed that "pessimism is alien to folklore despite the fact that [its] creators lived a hard life."[19] Tragedy, suffering, and neurosis have no place in Soviet literature. He called myth a form of "creative invention" that changes the world in a practical way and said that the dying bourgeois world has lost its capacity for creative invention; hence its writers could not depict positive heroes.

Gorky told the assembled writers that they had a leadership role in guiding the creative energies of the entire country. Fulfilling this role required them to depict worker-heroes, to inculcate a socialist morality, to curb artistic individualism, to improve the quality of their writing, and to foster collective responsibility. In addition, leadership entailed periodically purging the "toxic weed" or "microbe" of philistinism (*meshchanstvo*) from Soviet literature and eliminating parasitic characters and superfluous people.

In contemporaneous articles Gorky glorified slave labor by political prisoners and praised the secret police. He co-edited and contributed an article to *Belomor*—a collective work that extolled the building of the Baltic-White Sea canal by forced labor—in which he praised the "heroic labor" of the women convicts on the Moscow-Volga Canal (who were dying like flies) as a demonstration of the "truth of socialism."[20] It was, at least of Soviet socialism, but Gorky was not being ironic. In another article, he predicted that "in fifty years, when things will be a little calmer and the first half of this century will seem like a splendid tragedy, a proletarian epic, it is probable that art as much as history will be able to do justice to the wonderful cultural work of the rank-and-file Chekists [secret police] in the camps."[21]

In painting, the Nietzschean idea of art as a lie is exemplified by Sergei Gerasimov's "Holiday on the Collective Farm" (1935). A vision of plenty in a land where peasants barely survived, this work is typical of the inverse relationship between image and truth that came to characterize Socialist Realism: the darker the reality, the more light in the picture. If set outdoors, there was bright sunshine; if indoors, all sorts of "electric suns." The more the purge destabilized ordinary life, the more the arts depicted a stable and ordinary world. The more overcrowding in communal apartments, the more space in the photo; the more sacrifice demanded of the individual, the more talk of individuality; the greater the terror, the more people had to smile (pessimism was equated with disloyalty); the more tragedy in real life, the more comedy in film; the more fear and distrust pervaded life, the more emphasis on intimacy and love. Intimacy was a bad word in the 1920s; its return in the 1930s belied the lack of private living space.

The Quasi-Apollonian Language of Socialist Realism

The language mandated by Socialist Realism was a response to three decades of debate about language and myth, which Nietzsche connected in *Birth of Tragedy.* A new myth entailed a new language, or, as the Symbolists put it, a "new word," by which they meant a new truth—new values by which to live. They regarded "the word" as a palpable entity with theurgical properties. A "new word" would generate a new myth and, eventually, a new reality. The Symbolist concept of life-creation (*zhiznetvorchestvo*), grew out of their notion of myth-creation and pertained to the transfiguration of everyday life (*byt*) by means of art, especially poetry. Their fixation on "the word" influenced some Marxists and was perpetuated by the Futurists. Symbolist poets tried to replicate the "spirit of music" and to convey a mood rather than an idea. Futurist poets highlighted the spoken word and the cacophony of modern life. Rejecting Symbolist mystification, they thought of the word as a "thing" (*veshch'*), and theorized about language as power and about constructing a new language out of new signs. Their attention-getting antics (they titled their first manifesto "A Slap in the Face of Public Taste") led their opponents to call them literary hooligans. Lenin considered the Futurists bourgeois nihilists. Although he recognized the importance of language for agitation and propaganda, Lenin did not theorize about it.

The Futurist poets Velimir Khlebnikov and Aleksei Kruchenykh tried to create a trans-rational language, *zaum',* literally beyond the intellect or the mind.[22] Their conceptions of *zaum'* differed, but each was indebted to Nietzsche. *Zaum'* was intended to express the full range of human emotions, to restore the spontaneity of experience (the latter is more Bergsonian than Nietzschean), and to destroy the hold of reason on human thought. Nietzsche believed that grammatical structures and categories determine our thought and prevent us from going outside them. "'Reason' in language: oh what a deceitful old woman! I fear we have not got rid of god, because we still believe in grammar."[23] And, "this ancient mythology established the belief in cause and effect after it had found a firm form in the functions of language and grammar."[24] "Linguistically we do not know how to rid ourselves of them."[25] To Kruchenykh, smashing the old laws of grammar and syntax amounted to smashing the old "tables of values," as Zarathustra recommended.[26] Kruchenykh and Khlebnikov rejected "frozen" (Apollonian) language, words whose meaning is forever fixed, as unsuited to a state of inspiration. Compare Nietzsche:

> Apollo's deception: the eternity of beautiful form; the aristocratic legislation: "thus shall it be forever."!
>
> Dionysus: sensuality and cruelty. Transcience could be interpreted as enjoyment of productive and destructive force, as continual creation.[27]

Zaum' was a Dionysian language. It is important, not for the relatively few poems its advocates produced, but for bringing the language issue to the vital center of literary debate, where it remained for over two decades. The question of the pictorial language—abstract or representational—was almost as contentious. The analogical paintings produced by some Futurists were pictorial counterparts to *zaum.'*

After the Bolshevik Revolution, the controversy over language became politicized. Rival literary and artistic schools claimed to be the sole authentic voice of the proletariat, hence entitled to government support—not only funding but also supplies of paper (which was rationed) and access to a printing press (private presses were nationalized). The Futurists demanded a dictatorship over the arts, run by themselves. So did zealots of the proletarian culture movement, Proletkult, founded by Bogdanov, Lunacharsky, and others in October 1917, which numbered half a million members at its peak in 1921. Proletkult was a major conduit for Nietzschean ideas; Symbolists and Futurists taught in its schools and studios.

Between 1921 and 1927, small-scale private enterprise was permitted, but the market for serious literature and art was minuscule, so the struggle for government subsidies and monopoly status became ever more vicious. It was assumed that sooner or later one socialist art form and language would prevail. Rival schools wanted to make sure it was theirs. The rivals can be roughly divided into avant-garde and realists. The latter included militant "proletarian" writers, who were not of proletarian origin. The avant-garde argued the psychological and power-political importance of language, ideas that derived from Nietzsche and his popularizers. Marx and Engels said very little about language.

Until the end of 1927, Party leaders were preoccupied with rebuilding the devastated economy and the struggle to succeed Lenin. Stalin's defeat of Trotsky set the stage for the first Five-Year Plan and the Cultural Revolution. The Russian Association of Proletarian Writers (RAPP) became the de facto dictator of literature until April 1932, when the Party ordered RAPP and all other literary and artistic groups to disband, signaling the end of the Cultural Revolution. The same decree announced that writers who wanted to participate in socialist construction would be integrated in a Union of Writers with a communist faction therein, and that an organizing committee would plan the first Congress. Between 1932 and 1934, qualities associated with Dionysus—spontaneity, iconoclasm, sheer energy—were rejected in favor of the Apollonian qualities of rationality, order, and purposefulness. The language question became a vehicle for issues such as authorial autonomy that could not be discussed openly. The language that Socialist Realism mandated was the opposite of *zaum',* the other side of the Nietzschean coin so to speak.

The myth of the radiant future had to be couched in words and forms that unsophisticated workers and peasants could understand. This meant clear, simple language, easy-to-follow plots in literature, and representational rather than abstract forms in the pictorial arts. Nietzsche conceived of the Apollonian as a dream state, an illusion. Apollonian language does not necessarily correspond to reality. The language of Socialist Realism was clear, but its purpose was to construct an illusion, indeed, a lie. We must distinguish between language used to mystify or deceive and the trenchant literary style that Nietzsche praised in *Birth of Tragedy,* "the language of Sophocles' heroes ... Apollonian precision and lucidity."[28] Nietzsche admired strict Roman style "not mere emotional garrulousness,"[29] noted the severe constraints that classic French dramatists imposed on themselves, and seemed to endorse "profound obedience to law in language."[30] The lingering effect of Nietzsche's influence led the formulators of Socialist Realism to a quasi-Apollonian language that combined clarity, precision, and obfuscation (or mystification), and trained the masses in the mentality and behavior that the Party desired.

When Gorky returned to the Soviet Union to stay in 1931, his first articles were about language. Fidelity to literary norms, models, and canons, he insists, is part of the writer's educational mission. On the first page of his essay *"O sotsialisticheskom realizme"* (On Socialist Realism, 1933), he states that "the technique of literary work leads—first of all—to the study of language, the basic material of every book.... [The] authentic beauty of the language, acting as a force, is created by precision, clarity, sonorous words, which form the pictures, the characters, the idea of the book."[31] The writer must combine his "materials" in a masterful way, selecting the clearest, most exact, and strongest words for maximum impact. The language must be purged of all ambiguities and unnecessary phrases. Gorky attacks Andrei Bely (the Symbolist writer) and Khlebnikov for their "verbal chaos" and criticizes Fedor Panferov's novel *Bruski* (Ingots, 1928–37) for its grammatical errors and use of dialect. Panferov's language expresses "peasant power" and is therefore anti-Soviet, Gorky charges, explicitly linking correct language with correct political views. In the same politicizing spirit, Gorky associates eloquence with individualism, because it displays the author's talent or erudition. Again and again, he declares that writers must study how to write clearly, correctly, and simply. Complexity "darkens the meaning." "Truth demands simplicity; the lie, complexity." Just as there is only one truth, there is only one correct speech. George Orwell observed that "when there is a gap between one's real and one's declared aims, one turns as it were instinctively to long words and exhausted idioms, like a cuttlefish squirting out ink."[32] He forgot that lies can also be expressed clearly and simply.

In 1909, Gorky regarded language and myth as products of the people's creativity. In the early 1930s, he wanted language and myth to be controlled "from above." Ivanov envisioned "myth-creation" as a collective process and wanted to abolish the barriers that separate actors and spectators. Socialist Realism restored the barriers and curbed popular spontaneity, a process that began in 1920–21. *Zaum'* was intended to express personal experience and convey the nonrational nature of reality. Socialist Realism depicted a constructed experience, mediated through the eyes of writers and artists, and described in a "frozen" (Apollonian) language that was used to create an illusion.

Inscribing "new values on new tablets"[33] entailed establishing new linguistic norms or re-establishing old ones in a new context. If the "death of God" invalidated old words, old truths, and old values, the "new word" entailed a new truth and new values. For Nietzsche, language is the embodiment of a particular moral system; it controls what can and cannot be said and done. Socialist Realism reconstituted on a new basis the horizon that had been broken by Futurism, and pulled together a new world, partly by means of language. Nietzsche implied that since no objective reality exists, language is the world, a concept taken up by French post-structuralists. "What things *are called* is incomparably more important than what they are...." It is "enough to create new names and estimations and probabilities in order to create in the long run new 'things.'"[34] Among the "new things" that Soviet writers were supposed to help create were a new Soviet man, a new culture, and a new society. Nietzsche noted the power of the rulers to "name" the enemy, a power that Stalin frequently exercised.

In Stalin's time, emphasis was on the literary text (Stalin was often portrayed reading or writing). According to Theodor Adorno, "The eye is always the organ of effort, work, concentration: it approaches something specific in an unambiguous way. The ear, in contrast, is unconcentrated and passive. Unlike the eye, it does not have to be opened."[35] Adorno was discussing Wagner, but his observation helps explain the Soviet preference for reading over listening and for sight over sound. A picture can seem objective. Sound (and not just the spoken word) leaves more scope for subjectivity, which the Symbolists and the Futurists valued and the "rational" and "scientific" Marxists rejected. The Futurist emphasis on the spoken word was perpetuated, however, in omnipresent loudspeakers and radios blaring the latest production statistics to give the impression of success and spur the workers to even greater efforts. The purge trials of 1936 and 1937 were broadcast to the entire nation.

In addition to clarity and simplicity, Socialist Realism mandated correct grammar, logical connectedness, words with a fixed meaning, and the elimination of dialect and slang. The Nazis had a similar system of regimenting the language. The requirement of a clear and simple language shrunk the permissible vocabulary and precluded the expression of complex or subtle thoughts. Nietzsche called philologists the destroyers of every faith that rests on books and noted the incapacity of theologians for philology. The language of Socialist Realism was designed to prevent any subversion of the Soviet civil religion (the new myth). The official aesthetic implied an authoritative voice ("thou shalt"), a passive reader, and an impersonal truth, delivered in a way that no misunderstanding was possible and no questions were raised. Writers were to use denotative rather than connotative words, to avoid ambiguity, and (unlike Nietzsche), to eschew words with multiple connotations. The standardization of the language was also intended to uplift the masses. The writer was to act as a Henry Higgins, teaching the masses to speak and write correctly, and to behave in a "cultured" manner. Apollo was the god of individuation, but the standardization of language and plot precluded distinctive authorial styles. Apollo was also the god of beauty, but the language of Socialist Realism was not eloquent, though it sometimes aspired to elegance.

As the high priests of the new myth, writers were to use "realistic speech" and shun naturalism and formalism—naturalism because it was "backward and reactionary," and formalism because it reflected the "inner torment and confusion" of "hyper-individualist authors." At the Congress of Soviet Writers, Radek called James Joyce's *Ulysses* "a heap of dung crawling with worms" seen through a microscope and magnified.[36] Several speakers used the rhetoric of equality and class struggle to justify standardizing the language. After the French Revolution, one speaker claimed, the language of the nobility became the language of the people. Nietzsche praised the courtly language of Voltaire; the Soviets wanted the former "slaves" to speak like "masters." There was general agreement that dialect, slang, and grammatical error were vestiges of class or regional inequality that would soon disappear. Another speaker asserted that priests use an alien language to mislead the people. Yet another speaker emphasized Lenin's belief that anarchistic ideas were reflected in anarchistic speech. Later in the decade, John Dos Passos was attacked for his stream-of-consciousness writing.

The emphasis on discipline, mastery, consciousness, and control in language was related to the Party's desire to inculcate these virtues in the masses. Nietzsche's work contains many allusions to the interrelation of language and consciousness, and to the need for discipline, self-control, and restraint. He explicitly relates linguistic style to morality.[37] Stalin's style was very ordered and logical, with each point clearly laid out and labeled 1, 2, 3, or a, b, c. His speeches were used as models to teach children writing.

Attacks on "formalism" escalated after the Congress. The term alluded, very vaguely, to modernist experimentation in literature and the arts, but it really had no fixed meaning. Particularly unwelcome were techniques that delegitimated, demystified, estranged, bared the device, undermined the old canon, and unmasked illusions, for they were indeed subversive. Writers and artists were to shun modernist fragmentation; their goal was to present a coherent world-picture, a clear ideological perspective, a new totality.

Belief in the "magic of words" (the title of Bely's influential 1909 essay), was another reason for the tight control of language. The occult-derived aspects of Socialist Realism have been treated elsewhere,[38] so here I will simply note that Bely, Ivanov, Gorky, Lunacharsky, and Khlebnikov were deeply interested in the occult, as well as in Nietzsche. Khlebnikov believed the kabbalistic notion that changing one or two letters of a word, exploiter to explorer, for example, would be replicated in life; an exploiter would actually become an explorer. If so, typing errors could do real harm! *Pravda* went through several proofreadings before publication. As the official voice of the Soviet Union, it had to be a model of correct spelling and grammar. The language of Socialist Realist literature and art was not new, strictly speaking, but it certainly was distinctive. The official aesthetic rejected modernism and canonized the Russian classics, but its purpose was to create the sense of a new reality—hence the frequently expressed desire for a Red Tolstoy and a Socialist Faust.

In the pictorial arts, representational images replaced modernist abstractions. Stylized, square-built workers and peasants dominated the pictorial arts of the first Five-Year Plan. Subsequent portrayals of workers and peasants were more realistic, but the subjects were still impersonal mythic types. Photographs showed fields filled with tractors as if they were the norm. Photomontages reassembled fragments of reality in a coherent new vision that was compelling because it looked real. The technique was the perfect vehicle for "documenting" Stalin's leadership in the Bolshevik Revolution and socialist construction, excising Trotsky and other purged Bolsheviks, and magnifying Stalin to superhuman stature. The propaganda journal *USSR in Construction* consisted almost entirely of photographs and photomontages. Although it was published in five languages (Russian, English, German, French, and Spanish), the text was minimal. Gorky sat on the editorial board. The issue devoted to the Baltic-White Sea canal opens with a photograph of Stalin's head in the lower left quarter superimposed over a shoreless sea, and the last photograph is a silhouette of Stalin and his entourage superimposed on the prow of a barge entering a lock—even though the canal was too shallow for ocean-going ships and Stalin never visited it. Nor did Stalin visit the front during World War II. Nevertheless photomontages and paintings put him there—yet another example of art as a lie.

* * * *

This essay does not exhaust the question of the impact of Nietzsche on Socialist Realism, nor does it argue that Nietzsche was its only source. My point is that Nietzsche's idea of art as a lie, his connection of language and myth, and his insistence on the psychological and power-political importance of language guided the formulators of Socialist Realism, despite the fact that they never acknowledged his voice.

Notes

1. *Pervyi vsesoiznyi s'ezd sovetskikh pisatelei* (Moscow, 1934), pp. 712–14.
2. George L. Kline, *Religious and Anti-Religious Thought in Russia* (Chicago, 1968), pp. 103–26; Mary Louise Loe, "Gorky and Nietzsche: The Quest for a Russian Superman," in *Nietzsche in Russia*, ed. Bernice Glatzer Rosenthal (Princeton, 1986), pp. 251–74; Arch Tait, "Lunacharsky: A 'Nietzschean Marxist'?" *Nietzsche in Russia*, pp. 275–92; Edith Clowes, *The Revolution of Moral Consciousness: Nietzsche in Russian Literature 1890–1914* (DeKalb, IL, 1988), pp. 175–224; Hans Günther, *Der Sozialistische Übermensch* (Stuttgart, 1993), passim.
3. Bernice Glatzer Rosenthal, ed., *Nietzsche and Soviet Culture: Ally and Adversary* (Cambridge, 1994).
4. Günther, *Der Sozialistische Übermensch*; Margarita Tupitsyn, "Superman Imagery in Soviet Photography and Photomontage," in *Nietzsche and Soviet Culture,* pp. 287–310.
5. Friedrich Nietzsche, *On the Genealogy of Morals*, trans. Walter Kaufmann (New York, 1967), p. 152. N.B.: Nietzsche citations are to page number, not section.
6. Nietzsche, *The Will to Power*, trans. Walter Kaufmann and R. J. Hollingdale (New York, 1968), p. 451.
7. Nietzsche, *Ecce Homo,* trans. Walter Kaufmann (New York, 1967), p. 328.
8. Details in Robert Conquest, *Harvest of Sorrow* (New York, 1986).

9. Stephen Cohen, *Bukharin and the Bolshevik Revolution* (New York, 1975), pp. 343–44; Lars Lih, *Stalin's Letters to Molotov* (New Haven, 1995), p. 225.
10. *Problems of Soviet Literature*, ed. H. Scott (London, 1935), pp. 21–22. Italics in original.
11. Ibid., p. 157.
12. Anatolii Lunacharskii, *Sobranie sochinenii,* 8 vols. (Moscow, 1967), 8:497.
13. *The Lower Depths and Other Plays by Maxim Gorky*, trans. Alexander Bakshy (New Haven, 1945), pp. 47–49.
14. BertramWolfe, *The Bridge and the Abyss: The Troubled Friendship of Maxim Gorky and V. I. Lenin* (Stanford, 1967), p. 60.
15. *Maksim Gorkii: Sobranie sochinenii,* 30 vols. (Moscow 1953), 26:289.
16. Mikhail Agursky, "An Occult Source of Socialist Realism: Gorky and Theories of Thought Transference," in *The Occult in Russian and Soviet Culture*, ed. Bernice Glatzer Rosenthal (Ithaca, 1997), pp. 247–72.
17. Anthony Kemp-Welch, *Stalin and the Literary Intelligentsia* (New York, 1991), p. 248.
18. *Problems of Soviet Literature*, pp. 45, 48–49.
19. Ibid., p. 36.
20. *Maksim Gorkii*, 27:127.
21. Ibid., 27:509.
22. Bernice Glatzer Rosenthal, "A New Word for a New Myth: Nietzsche and Russian Futurism," in *The European Foundations of Russian Modernism*, ed. Peter Barta (Lewiston, NY, 1991), pp. 219–49.
23. Nietzsche, *Twilight of the Idols*, trans. R. J. Hollingdale (New York, 1968), p. 38.
24. *Will to Power*, p. 336.
25. Ibid., p. 296
26. Nietzsche, *Thus Spoke Zarathustra*, trans. R. J. Hollingdale (Harmondsworth, 1961), p. 51.
27. *Will to Power*, p. 539.
28. Nietzsche, *The Birth of Tragedy*, trans. Walter Kaufmann (New York, 1967), p. 67.
29. *Twilight of the Idols*, p. 105.
30. *Will to Power*, p. 441.
31. *Maksim Gorkii*, 27:5.
32. George Orwell, "Politics and the English Language," in *A Collection of Essays by George Orwell*, ed. Sonia Brown Orwell (Garden City, NY, 1954), p. 173.
33. *Zarathustra*, p. 52.
34. Nietzsche, *The Gay Science*, trans. Walter Kaufmann (New York, 1974), pp. 121–22.
35. Theodor Adorno, *In Search of Wagner* (Manchester, 1981), pp. 99–100.
36. *Problems of Soviet Literature*, pp. 53–55.
37. Nietzsche, *Beyond Good and Evil*, trans. Walter Kaufmann (New York, 1966), p. 100.
38. Agursky, "An Occult Source"; Irina Gutkin "The Magic of Words: Symbolism, Futurism, Socialist Realism," in *The Occult in Russian and Soviet Culture*, pp. 225–46; Bernice Glatzer Rosenthal, "Political Implications of the Early Twentieth Century Occult Revival," ibid., pp. 397–413.

Chapter 14

TRANSNATIONAL AUTHORSHIP ON THE GERMAN-SLAVIC BORDER

The Examples of Nietzsche and Nabokov

John Burt Foster, Jr.

I.

In October 1888, feeling deeply alienated from Germany, Friedrich Nietzsche was living in Italy; yet despite his great admiration for that country, he was also toying wishfully with the idea that the last syllable of his surname betokened Slavic ancestry. When he learned that the Danish critic Georg Brandes would lecture on his work in St. Petersburg, Nietzsche was delighted. Especially flattering was Brandes's request that he send one of his books to a certain Princess Tenishev.[1] For a comparatist like myself, it is tempting to speculate that this lady belonged to the same family as the founder of the famous Tenishev School, which Vladimir Nabokov (as well as Osip Mandelstam) would attend some two decades later. By this time, in the years just before World War I, Nietzsche had become famous in Russia as "the German Dostoevsky," and was even being interpreted as a major religious thinker.[2] Nabokov, meanwhile, as a descendant of Grauns and von Korffs, actually did have some German ancestors.[3] During most of the 1920s and 1930s he lived in exile in Berlin, among the large Russian colony that, as Alexander Dolinin discusses in his essay in this volume, had gathered there after the 1917 revolution. Here Nabokov began his career as an émigré novelist; though he often played down the influence of this environment upon him, he has in fact given conflicting accounts of his command of German. It is also worth noting that German was the first foreign language into which his novels of the 1920s were translated.

This quick sketch of Germano-Slavic cultural contact and interchange might once have served to introduce a standard binary comparison of Nietzsche and Nabokov. One possible topic would have been Nabokov's sharply opposed evaluations, many years later, of two major works of prose fiction, both begun in 1911 and both strongly influenced by Nietzsche: Andrey Bely's *Petersburg*, in Nabokov's view one of the "greatest masterpieces of twentieth-century prose," as contrasted with Thomas Mann's "asinine" *Death in Venice*.[4] But recently, as a result of spirited debates about the methods and assumptions of comparative literature as a field, comparisons of this kind have drawn sharp criticism.[5] They are said to depend on a limited and outdated sense of cross-cultural exchange, one that is narrowly "international" rather than boldly "transnational." Let me briefly explain this shift in focus.

The older term "international" encapsulated three salient and, in the end, contradictory attitudes within comparative study. On the one hand, there was the defining interest in tracing literary relations across national and linguistic boundaries, behind which there was also a mood of skepticism and even defiance about the cultural barriers assumed by late nineteenth-century European hypernationalism. Concerning this nationalism, we might recall that W. E. B. Du Bois, who lived in Europe at that time, went so far as to posit separate Latin, Anglo-Saxon, Teutonic, and Slavic races, on the model of black-white relations during the segregation era in the United States.[6]

And yet, despite the comparatist's deep suspicion of these barriers, there was also an ultimate readiness to endorse the nation-state as the basic unit for cultural analysis. This inconsistency explains the Spanish-American comparatist Claudio Guillén's position in *The Challenge of Comparative Literature*, originally published in Barcelona in 1985 after a career spent mainly in the United States. This searching synopsis of the field—based on lectures in Spain and China, two countries where cross-cultural inquiry has faced major obstacles in the twentieth century, perhaps ones as great as those in Germany and Russia in the 1930s and 1940s—rejects the international approach. In studying the fascinating richness of literature, Guillén insists that "the point of departure is not to be found in national literatures, nor in the interrelationships between them."[7] Instead, the basic impetus for comparatism lies in a "yearning," as Guillén puts it, to move beyond the categories of the various national literary histories. However, rather than "transnational," with its flavor of actively cutting across borders, Guillén (or rather his translator) prefers a loftier, more serenely cosmopolitan word—"supranational."

But elsewhere, often outside official comparative literature organizations and journals, transnationalism has become the term of choice for more actively polemical approaches to issues of literature, culture, and the nation. One example would be *Diaspora: A Journal of Transnational Studies*, which invites essays on "intranational and transnational phenomena now challenging the nation-state, including nomadic ideas, works of art, and mass media productions that traverse frontiers and borderlands."[8] There is also the Society for Transnational Cultural Studies and its journal *Public Culture*, based at the University of Chicago. The journal is anthropological rather than literary, but along with stressing the porousness and provisionality of national boundaries in cultural analysis, it has a global vision that emphasizes postcolonialism, the diasporas of non-European populations, and the multi-ethnic cultures of the United States.

This recent shift in vision and terminology in comparative studies will provide the impetus for my remarks on Nietzsche and Nabokov. In its spirit I will begin by questioning our very habit of calling the first of these writers a German philosopher, and the other a Russian, or even a Russian-American, novelist. By setting the record straight about these matter-of-fact labels, we shall get some first insights into the nature of Nietzsche's and Nabokov's transnationalism.

Nietzsche, to be sure, did write philosophy in German. But whether his ideas and style gave a special aggressive edge to German nationalism has been hotly debated for most of this century. In my view, the attempts by both German chauvinists and anti-German polemicists to nationalize a radically transnational writer are badly misplaced. Nietzsche wrote scathingly about "the new idol" of late nineteenth-century nationalism,[9] and his prose abounds with terms such as "über," "um," and "jenseits," all of which signal the need to move above and beyond received conceptions. This is in fact one implication of the prefix "trans" in such English constructions as "transnational." Beyond this point about Nietzsche's thought, let us also remember that he lived outside Germany most of his adult life, and (as already noted) that he liked to imagine other identities for himself, including a Slavic one. Oddly, though he had to renounce his Prussian citizenship in order to join the faculty at the University of Basel in 1869, no one to my knowledge has ever called Nietzsche a Swiss philosopher. Yet this formula, by acknowledging his ties with a multilingual country at the junction of several European traditions, more accurately reflects Nietzsche's cultural multiplicity. Even "Swiss" is misleading, however, for Nietzsche apparently forgot to complete the formalities of getting his new citizenship—and thus was technically stateless for the rest of his life.

There is less room for debate about Nabokov's transnational identity, since no journalists or historians have tried to brand him as a radical nationalist. Still, it does bear repeating that although his family eventually fled the Russian revolution, they did not do so as loyalists of the old regime; his father was a leader in the liberal February revolution against the Romanovs, and was later killed by Russian monarchists in a botched assassination aimed at his friend Paul Miliukov. As for Nabokov himself, though he wrote acclaimed fiction in both Russian and English, this striking bilingualism fails to capture the cultural multiplicity of his career. In the late 1930s, for example, he wrote several works in French, while in the 1960s and 1970s he was an expatriate American in French-speaking Switzerland. It is even more telling that from age twenty until past age forty—in the very period when he became a writer and wrote his first major works—Nabokov was a stateless person and on three occasions a refugee.[10] Indeed, as the holder of a League of Nations passport, he never knew how he would be treated by border personnel as he moved around Western Europe.[11]

In short, as a result of multiple border-crossings, changes in citizenship, and problematic relationships with national identity itself, both Nietzsche's and Nabokov's biographies resist clear-cut national labels. Let us now turn to two more specifically literary expressions of this transnationalism, both of them connected, as the title of this essay suggests, with the status of Nietzsche's and Nabokov's authorship. In particular, I am interested in such basic issues as, first, the place where they wanted to locate themselves as writers and, second, the language that they chose to put on the page. What, in other words, is the nature of the transnational space within which they functioned as authors, leaving aside the cultural and institutional frameworks within which their works were in fact published, read, and discussed?

II.

One key factor involves the literary and cultural genealogies that both writers constructed for themselves, which in crucial ways look back to Romanticism broadly defined. The early nineteenth century, of course, saw the upsurge of nationalism, which led many European language communities to posit strong links between native literary figures and their claims to unique national identity. In the wake of these assertions of cultural particularity, both Nietzsche and Nabokov chose readily available figures as special heroes—none other than Goethe and Pushkin, both of whom had been canonized as supreme geniuses in the German and Russian literary systems, even as guarantors of their nations' claims to civilized status.[12] But in interpreting these cultural heroes, Nietzsche and Nabokov were careful to separate the idea of literary genius, which they continued to venerate, from the idea of national essence, which they pointedly criticized.[13]

Thus, in his culminating tribute to Goethe in *Twilight of the Idols*, one of his last books, Nietzsche praises Goethe's supreme creative gifts. Even more importantly, he identifies the wholeness, freedom, and affirmative spirit of Goethe's creativity with his own master-myth of Dionysus, derived from his sharply revisionist thesis about classic Greek culture and religion that dates back to *The Birth of Tragedy*.[14] At the same time, though Nietzsche does not mention Goethe's ideas about *Weltliteratur*, which would later inspire several generations of comparatists,[15] he does take pains to denationalize Goethe's genius, calling it "not a German event, but a European one." Indeed, he lavishly praises one deed of Goethe's that had been anathema to German hypernationalists—his willingness to meet with Napoleon.[16] Removing Goethe from the emerging canon of German-national literature, Nietzsche shifts him to a broader European context whose own basic meaning would require further discussion, given Nietzsche's hotly debated rereading of its Greek roots.

If anything, Nabokov's admiration for Pushkin was even more striking than Nietzsche's for Goethe. During the 1950s, at the height of his powers as a novelist, he devoted years to a four-volume annotated translation of Pushkin's *Eugene Onegin* (1964). Yet even as he paid remarkable homage to a literary masterpiece, Nabokov disputed *Onegin*'s status as a Russian national classic in at least two ways. First, he added a long essay detailing Pushkin's African ancestry, by way of an Ethiopian great-grandfather; second, he interspersed his commentary with many pointed examples of French influence on Pushkin's Russian phraseology. Contrary to romantic nationalism, therefore, Nabokov envisioned a Pushkin whose multiple ethnic roots and hybridized literary language suggest an incipient transnationalism, an identity which at once foreshadows and inspires Nabokov, born exactly a century later.

Beyond reinserting these domestic icons into more diverse cultural settings, however, Nietzsche and Nabokov both wrote prose whose very texture sought to cross national boundaries. In fact, because the transnationalism of both writers went beyond the dichotomies of simple biculturalism to affirm the many-sidedness of cultural multiplicity, during their careers they crossed borders into several different languages and cultures. As a result, though the German-Slavic border played an important role both in Nietzsche's imagined cultural identity and spreading reputation and in Nabokov's family background and direct experience as an exile, it was not, I shall argue, the single most important factor in their transnational authorship. Nietzsche's fantasies about St. Petersburg and Nabokov's life in Berlin simply did not have the same decisive influence on their sense of cultural borders as, for example, his years in Riga had on Herder.

Multiplicity rather than biculturalism is directly addressed in crucial passages by both writers. In *Thus Spoke Zarathustra*, which Nietzsche regarded as his best book and in which he began to elaborate his doctrine of the will to power, this idea first appears in a chapter called "The Thousand and One Goals." In it, Zarathustra tries to sum up his observations of "many lands and many peoples," beginning with the Greeks, the Persians, the Jews, and the Germans. But in the end he does justice to this diversity only by speaking of a hyperbolical "one thousand goals."[17] These goals are then explained by the "one goal" of the will to power, which in this context becomes the exuberantly multiple activity of cultural creation itself.

Similarly, in a publisher's prospectus that Nabokov drew up for his autobiography *Speak, Memory*,[18] he stresses his adventurous course through "many different lands and people and modes of living," a formula which by accident or design neatly chimes with Zarathustra's language in "The Thousand and One Goals." As *Speak, Memory* took shape, Nabokov realized that one reason for this unusual cultural mobility had been his radically pluralistic education. Like other Russians of their class, the Nabokovs had created a multilingual home for their children. Even more striking, the book's middle chapter, on Nabokov's boyhood teachers, starts out as a cultural panorama: after a Russian village schoolmaster we are shown the son of an Orthodox deacon, who is followed by a Ukrainian, a Lett, and a Pole, who make way for the more complex case of Lenski, "a Lutheran of Jewish extraction." The repeated recourse throughout his education to "a representative of another class or race" took place, Nabokov feels, because his father wanted "to expose us to all the winds that swept over the Russian Empire."[19]

This awareness of "all the winds" or "a thousand goals" is basic to the transnationalism of both writers, and takes us decisively beyond the German-Russian binary. Not surprisingly, therefore, despite Nabokov's partly Germanic roots or Nietzsche's self-identification as a Slav, the paramount linguistic expression of their transnationalism lies elsewhere—in the strategic use of French in Nietzsche's late philosophy and in Nabokov's fully evolved sense of Russo-English bilingualism. Both examples are strong ones, not just because of their intrinsic importance for the writers but also because of their historical context, since Nietzsche and Nabokov wrote at times of severe Franco-German and Soviet-American political tension.

III.

By Nietzsche's strategic use of French I mean something more precise than his well-known interest in French literature and culture. I am referring to passages in which the French words have a transnational charge, both because Nietzsche's thought crystallizes at that point to propose a far-reaching thesis and because the expression draws attention to itself as non-German. The rationale for this practice becomes clear in *Twilight of the Idols*, where Nietzsche states, "I am often asked why, after all, I write in German."[20] In this perspective, his French is a stylistic device for avoiding German at least at moments of crowning insight. A good brief example appears in his autobiography *Ecce Homo*, which is sprinkled with French words. At one point, while attacking his German contemporaries for not understanding him, Nietzsche accuses them of overlooking fine distinctions, then exclaims that he is just such a distinction: "wehe mir! ich bin eine *nuance*" (Alas, I am a nuance).[21] In this witty bilingual aside, Nietzsche's transnational urge to assimilate a foreign culture combines with his preference for French at moments of culminating insight.

Several key concepts in Nietzsche's last writings emerge in the same manner, especially "ressentiment" in the *Genealogy of Morals* and "décadent" in *The Case of Wagner*, both of which I have analyzed in more detail elsewhere.[22] With "décadent," Nietzsche uses a French word to demystify Wagner's claims to purely German cultural significance. For example, once the critic deflates Wagner's grandiose Nordic myths, he turns out to be pursuing the same problems as "those which now preoccupy the little *décadents* of Paris." In slyly inserting the polemical French term, Nietzsche the culture critic speaks from a standpoint similar to that of his contemporaries, the late nineteenth-century comparatists who wanted to reveal the more-than-national meanings for national cultural phenomena.

With "ressentiment," we should recall the dramatic first appearance of this term midway through Part I of the *Genealogy of Morals*. Nietzsche's opening remarks have ground to a halt, an imagined "free spirit" gives a provocative speech, and now Nietzsche needs to clarify the psychological implications of his main argument about the differing moral sensibilities embedded in the contrast "gut und böse" versus "gut und schlecht,"[23] or good and evil versus good and bad. Used in this German linguistic context, the French word "ressentiment" concretizes Nietzsche's often expressed admiration for the psychological finesse of the French. Soon, however, the passage moves into an area of language creation that is even more transnational. For when Nietzsche defines "ressentiment" as a reactive moral stance, he seems to impose a private etymology on the word. Just as a "reaction" responds to a prior "action," so "ressentiment" follows a more forceful "sentiment," a logic also found in *Ecce Homo* when Nietzsche translates "ressentiment" as "Nachgefühl,"[24] or "after-feeling." Thus, even as he gallicizes his basic thesis, Nietzsche's own expressive needs cause him to reverse the French verb "ressentir," where the prefix is an intensifier, not a sign of secondary status.[25]

An even richer instance of Nietzschean French occupies a transnational space that begins to resemble Nabokov's intricate Anglo-Russian bilingualism. In Part III of the *Genealogy*, Nietzsche weighs Kant's idea of disinterested aesthetic judgment against Stendhal's famous definition of beauty, later incorporated into Adorno's aesthetics—"une promesse de bonheur" (a promise of happiness).[26] The passage shows that Nietzsche preferred the forthright concreteness of Stendhal's hedonism to Kant's philosophical abstractions. In linguistic terms, however, the confrontation seems less clear-cut: the key terms fluctuate between German and French, so that Kant's "ohne Interesse" can appear as "désintéressement," and Stendhal's phrase becomes "das Schöne *verspricht* Glück." Rather than simple Francophilia, such interplay suggests a continuous testing of both languages. That Nietzsche calls the French writer not by his real name of Henri Beyle, as he had done the year before in *Beyond Good and Evil*,[27] but by his Germanic pseudonym gives the passage an added transnational twist. As Nietzsche's prose crosses and recrosses the Franco-German linguistic frontier, moreover, it defies one of the most sharply contested political boundaries of his day.

A transnationalism, which in Nietzsche takes the form of vivid but localized stylistic gestures, becomes in Nabokov a systematic practice of bilingual writing. Not that Nabokov avoids linguistic border-crossings on the order of Nietzsche's. Especially noteworthy for the theme of this volume is the Germano-Slavic language he invents for the imaginary East European dictatorship in *Bend Sinister* (1947).[28] But the most ambitious and best-known example is Russo-American; it appears in his Gogol book of 1944, written five or six years after he started working extensively in English. Here Nabokov anglicizes the Russian word "poshlost'," which loosely means self-satisfied mediocrity, by turning it into "poshlust" and giving it English-language implications of its own, focused mainly on American consumerism.[29]

As for Nabokov's achievement as a fully bilingual writer, I have already mentioned his success as a novelist with both Russian émigré and American audiences. But this statement oversimplifies the complex interplay between the two languages that occurred over his entire life as a writer. Only four especially telling episodes can be noted here. Nabokov's native tongue was Russian, but because he grew up in an Anglophile home, he learned the Roman alphabet before the Cyrillic one. Later, in the reverse linguistic exchange, he chose to become a Russian writer while studying in England in the early twenties. Then, in a complex maneuver lasting slightly more than a year, Nabokov shifted over to English. From late 1937 to early 1939, he completed his most ambitious Russian novel, *Dar* (The Gift), personally translated another Russian novel into English (*Laughter in the Dark*), and started working on his first novel in English (*The Real Life of Sebastian Knight*)—yet at intervals he also wrote two plays and several accomplished short stories in Russian! Finally, after his success with *Lolita* in the late 1950s, Nabokov translated that novel into Russian, while at the same time closely supervising the reverse activity—the rendition of most of his Russian works into English.

This crisscross pattern of Anglo-Russian interchange may seem dizzying, but the composition of Nabokov's autobiography goes even further in projecting an authorship that oscillates between two linguistic realms. Leaving aside the part of *Speak, Memory* that was written in French in 1936, and also the passages that were fictionalized in Nabokov's Russian novels of the 1920s and 1930s,[30] we have an autobiography that was written in English during the 1940s and published in 1951. Nabokov then translated the book into Russian in 1953, somewhat enlarging it in the process. In 1965 he further enlarged the English edition to produce a final English version (published in 1966) that also uses some passages unique to the Russian edition. Nabokov's summary of this textual history reveals his sharp sense of transnational intricacy in his writing: "This re-Englishing of a Russian re-version of what had been an English re-telling of Russian memories in the first place."[31] After conceding the difficulty of what he calls a "diabolical task," he voices pride in what he views as an unprecedented experience of cultural mobility: "such multiple metamorphosis, familiar to butterflies, had not been tried by any human before."[32] This is a strong statement, for Nabokov was an entomologist of some note who had been fascinated with metamorphosis since boyhood.

Nonetheless, Nabokov's spirited assertion of his cross-cultural prowess needs to be qualified by an underlying political reality. Whatever his success with moving back and forth between Russian and English in the realm of language, during his entire adult life he could never do likewise with the corresponding national borders. The existence of this barrier already informs the ingenious open ending of his early novel *Glory* (1930), in which the young hero, entranced by the challenge of just walking over the border into the Soviet Union, eventually does so, only to vanish completely; and it returns in his late novel *Ada* (1969), which is set in a parallel world where North American and Russian geography can intermingle freely. Nabokov died in 1977, never anticipating that within a decade his novels would be published in his native land, free of Soviet censorship. As with Nietzsche's French, therefore, bilingual exchanges in Nabokov's authorship represent a transnational practice in which language builds connections that seek to defy the intractability of disputed or heavily policed borders.

IV.

Let me conclude by relating Nietzsche's and Nabokov's experiences of transnationalism to our current sense of wider cultural horizons, as shown by the vogue, in both journalism and scholarship, for such terms as "globalization," "multiculturalism," or the "postcolonial." Now that we can consider the broader transnational, often intercontinental, vistas of writers like Maxine Hong Kingston and Wole Soyinka; of comparatists like Edward Said and Gayatri Spivak; and of places like the Caribbean, Singapore, or other sites illumined by Deleuze and Guattari's concept of a "minor literature,"[33] neither Nietzsche's nor Nabokov's transnationalism might seem quite so provocative. Moving beyond European hypernationalism within a European setting, as they did, is not the same thing as undertaking similar projects elsewhere, where intercultural transactions must step outside an exclusively European framework.

Still, there are two qualifications to this sense of limits. As shown, both Nietzsche and Nabokov did have an incipient awareness of global perspectives, the first most strikingly as the thinker of "one thousand goals," who tried to allow for a true multiplicity of cultures; the second in a more specifically topical way, as when he publicized Pushkin's African background in the race-sensitive United States. Even more importantly, if the evolving practice of comparative literature expresses a yearning to surpass the cultural hegemony of national boundaries, both Nietzsche and Nabokov are powerful exponents of that desire. No matter how carefully we pinpoint the dimensions of their transnationalism for purposes of analysis, in its effect on readers no such impulse can be confined to the conditions in which it arose.

To enlarge on this second point, Nabokov's books are now marketed in the United States with Soyinka's and Kingston's in a series of so-called "international" authors—though by now I hope it is evident that the term is a misnomer. Or, to take a more thoughtful example, a new global anthology edited by the well-known comparatist Mary Ann Caws samples world literature on regional principles, then turns to what it calls "Writing Across Boundaries."[35] Nabokov appears in this unit as one early example of a trend that includes Czeslaw Milosz, Aimé Césaire, Derek Walcott, and Salman Rushdie. Similarly, but more specifically, Paul Gilroy's recent book on the cultural space that he calls the black Atlantic foregrounds the term "transnational"; it opens with two sets of paired epigraphs which link Nietzsche with Frederick Douglass and the Caribbean intellectual Edouard Glissant.[35] These Nietzsche citations, far from being existentialist or deconstructive set-pieces, have been picked to resonate with Gilroy's analysis of black expressive culture as it moved back and forth among Africa, the Caribbean, the United States, and Europe. Thus, though Nietzsche never thought in terms of a "black Atlantic," his insights as a culture critic who recognized the need for wider-ranging analytic frameworks still have validity for this more broadly intercontinental form of transnationalism.

Despite the arrival of writers with fuller global horizons, therefore, they do have clear affinities with Nabokov's and Nietzsche's more localized defiance of cultural nationalism and isolation. To be sure, neither the Franco-German nor the Soviet-American divide means nearly so much now as when both writers were active, not to mention their less decisive participation in the ups and downs of Germano-Slavic interchange. But in a world in which major barriers still persist even as its parts grow ever closer, both Nietzsche's and Nabokov's authorial border-crossings—as well as those of their models Goethe, Pushkin, and Stendhal—should still be instructive, perhaps even inspiring, for their many transnational successors.

Notes

1. See, for example, Nietzsche's "Letter to Peter Gast," written in Turin on 14 October 1888, in Friedrich Nietzsche, *Werke in drei Bänden*, ed. Karl Schlechta, vol. 3 (Munich, 1966) p. 1321. Other references to Princess Tenishev appear in letters of 20 October, 13 November, 9 December, and 21 December of the same year (Nietzsche, *Werke*, vol. 3, pp. 1325, 1331, 1339, 1344).
2. Mihajlo Mihajlov, "Nietzsche and Neo-Idealism," in *Nietzsche in Russia*, ed. Bernice Glatzer Rosenthal (Princeton, 1986), pp. 137–41.
3. Vladimir Nabokov, *Speak, Memory: An Autobiography Revisited* (New York, 1967), pp. 54–56.
4. Vladimir Nabokov, *Strong Opinions* (New York, 1973), p. 57. For a discussion of *Death in Venice* as a work that is vitally concerned with German-Slavic cultural borders and cultural interchange in the period just before World War I, see John Burt Foster Jr., "Why is Tadzio Polish? *Kultur* and Cultural Multiplicity in *Death in Venice*," in *Thomas Mann's* Death in Venice*: A Case Study in Contemporary Criticism*, ed. Naomi Ritter (Boston, 1998), pp. 192–210.
5. Two key examples of these debates are Susan Bassnett, *Comparative Literature: A Critical Introduction* (Oxford, 1993) and Charles Bernheimer, ed., *Comparative Literature in the Age of Multiculturalism* (Baltimore, 1995).
6. W. E. B. Du Bois, "The Conservation of Races," in *W. E. B. Du Bois Speaks: Speeches and Addresses 1890–1919*, ed. Philip S. Foner (New York, 1970), pp. 76–77.
7. Claudio Guillén, *The Challenge of Comparative Literature*, trans. Cola Franzen (Cambridge, MA, 1993), p. 3. This book is a translation of *Entre lo uno et lo diverso: Introducción a la literatura comparada* (Barcelona, 1985).
8. *Diaspora: A Journal of Transnational Studies* 4 (2) (Fall 1995): 241.
9. Nietzsche, *Werke*, vol. 2, pp. 313–15.
10. In addition to leaving Russia in 1919, Nabokov (who had a Jewish wife) had to flee Nazi Germany in 1937 and then France on the eve of Nazi occupation in 1940.
11. Brian Boyd, *Vladimir Nabokov: The Russian Years* (Princeton, 1990), pp. 437–38n.
12. In comparing the German and Russian situations, however, it should be noted that Pushkin and Goethe were not exact parallels in assuming the role of representative national writers. Though Pushkin was seen as a supreme genius by the Russians themselves, he was not nearly so well known abroad. Goethe, on the other hand, often appeared outside Germany to be the outstanding German writer, but many Germans continued to have doubts about him. For the nature of Goethe's reputation when Nietzsche came of age, see Herbert Lindenberger, "The Normality of Canon Change," in *The History in Literature: On Value, Genre, Institutions* (New York, 1990), pp. 134–36, which summarizes Peter Hohendahl, *Building a National Literature: The Case of Germany 1830–1870* (Ithaca, 1989). For Pushkin's canonical status in the early twentieth century among both Soviet and émigré Russians, see Irina Paperno, "Nietzscheanism and the Return of Pushkin in Twentieth-Century Russian Culture (1899–1937)," in *Nietzsche and Soviet Culture: Ally and Adversary*, ed. Bernice Glatzer Rosenthal (Cambridge, 1994), pp. 211–32.
13. For more on Nietzsche's denationalization of Romantic ideas of the genius, see Geoffrey H. Hartman, *The Fateful Question of Culture* (New York, 1997),

pp. 5–7, who notes, however, the potential for antidemocratic tendencies within the cult of genius.

14. Nietzsche, *Werke*, vol. 2, pp. 1024, 1031. All the translations of Nietzsche into English in this essay follow Walter Kaufmann's versions in *The Portable Nietzsche* (New York, 1954) and *Basic Writings of Nietzsche* (New York, 1968).
15. Guillén, *Challenge*, pp. 37–45.
16. Nietzsche, *Werke*, vol. 2, pp. 1024, 1025.
17. Ibid., pp. 322–23.
18. Vladimir Nabokov, "Letter to Kenneth D. McCormick," 22 September 1946, in *Vladimir Nabokov: Selected Letters, 1940–1977*, ed. Dmitri Nabokov and Matthew Bruccoli (New York, 1989), pp. 69–70.
19. Nabokov, *Speak, Memory*, p. 153.
20. Nietzsche, *Werke*, vol. 2, p. 1026.
21. Ibid., p. 1150.
22. For a fuller conceptual analysis of these terms, see John Burt Foster Jr., *Heirs to Dionysus: A Nietzschean Current in Literary Modernism* (Princeton, 1981), pp. 68–75, 99–109.
23. Nietzsche, *Werke*, vol. 2, p. 782.
24. Ibid., p. 1078.
25. Paul Robert, ed., *Dictionnaire alphabétique et analogique de la langue française*, 2nd ed., Alain Rey, vol. 8 (Paris, 1985), p. 319.
26. Nietzsche, *Werke*, vol. 2, pp. 845–46.
27. Ibid., p. 722.
28. Vladimir Nabokov, *Bend Sinister* (1947; rpt. New York, 1974). In his introduction to a later edition of *Bend Sinister*, Nabokov would explain that "The language of the country … is a mongrel blend of Slavic and Germanic with a strong strain of ancient Kuranian" (x). The country's popular periodicals include both a *Simplizissimus* and a *Strekoza* (15), and people live in *dachas* (83) while eating *knakerbrod* (88). But the Slavic element usually dominates, as seen when a character recites the first three lines of Hamlet's most famous soliloquy: "Ubit' il' ne ubit' Vot est' oprosen. / Vto bude edler: v rasume tzerpieren / Ogneprashchi i strely zlovo roka—" (118).
29. Vladimir Nabokov, *Nikolai Gogol* (New York, 1944), pp. 63–74. For more on poshlust and its function in various cross-cultural contexts, see John Burt Foster Jr., "Poshlust, Culture Criticism, Adorno, and Malraux," in *Nabokov and his Fiction: New Perspectives*, ed. Julian W. Connolly (Cambridge, 1999), pp. 216–35.
30. For a discussion of these developments, see John Burt Foster Jr., *Nabokov's Art of Memory and European Modernism* (Princeton, 1993), pp. 52–69, 110–29.
31. Nabokov, *Speak, Memory*, p. 12.
32. Ibid., p. 13.
33. Gilles Deleuze and Félix Guattari, *Kafka: Toward a Minor Literature*, trans. Dana Polan, foreword Réda Bensmaïa (Minneapolis, 1986).
34. Mary Ann Caws and Christopher Prendergast, eds., *The Harper Collins World Reader: The Modern World* (New York, 1994), pp. 2656–707.
35. Paul Gilroy, *The Black Atlantic: Modernity and Double Consciousness* (Cambridge, MA, 1993), pp. xiii, 1.

Chapter 15

"The Stepmother of Russian Cities"

Berlin of the 1920s through the Eyes of Russian Writers

Alexander Dolinin

Nobody knows for sure how many Russians lived or at least passed their time in Berlin at the beginning of the 1920s. After coming back from a trip to Germany at the very end of 1922, Vladimir Mayakovsky[1] gave the figure of more than four hundred thousand; Victor Shklovsky, in the second edition of his *Sentimental Journey* published in 1924, wrote about the "three hundred thousand Russians of various nationalities" who "wander aimlessly in the cracks of a perishing city."[2] Historians cautiously claim that at the peak of the emigration well over one hundred thousand Russians lived in southwest Berlin,[3] in and around Charlottenburg, renamed Charlottengrad by local comedians.[4] Whatever the numbers, the Russian presence in Berlin was highly visible and audible, at least for the invaders themselves. When Martin, the protagonist of Vladimir Nabokov's *Glory*, arrives in Berlin from Switzerland, he is surprised to find "the free-mannered, loud-voiced Russia that chattered everywhere, in the trams, in the shops, on street corners, on the balconies of apartment houses."[5] In a sense, it seemed as if Russians had colonized an alien city, forcing their customs, cultural codes, and tastes upon the submissive natives. As another Nabokov, Nicolas, a renowned composer, remarked in his amusing memoirs:

> The Russian refugees in Berlin were by no means a humble, frightened lot that fretted and trembled in a hostile environment. They seemed to have taken over Berlin and transformed it into a Russian camp.
>
> There were Russian newspapers, Russian theaters, Russian schools and churches, Russian cabarets and libraries, Russian literary clubs and publishing houses, Russian foreign-exchange speculators, Russian bookstores, Russian art galleries, grocery stores, confiseries and antique or bargain stores that sold false and real Fabergé jewelry and a lot of genuine and fake icons.
>
> The center part of West Berlin, from the Wittenberg Platz past the Gedächtnis Kirche and down the Kurfürstendamm, seemed to have surrendered to the Russian takeover. Every other person in the street, in cafés and stores, spoke Russian or a heavy, consonant-laden German. All around, on the walls and on the advertising pillars, one saw announcements, sometimes in Russian, not only of Russian opera performances and concerts given by Russian artists, but also of Russian political meetings, officers' clubs, welfare committees. Dowager ballerinas opened ballet schools, writers held forums and public readings of prose and poetry.[6]

The frenzied literary and intellectual life of "Russian Berlin" has been well documented and studied. There are diaries, letters, essays, travelogues, and memoirs left by its inhabitants and visitors—Aleksandr Bakhrakh, Nina Berberova, Ilia Ehrenburg, Iosif Gessen, Roman Gul', Evgenii Lundberg, Nicolas and Vladimir Nabokov, Victor Shklovsky, Gleb Struve, to name only a few. There are numerous works that chronicle the Berlin days and works of the most prominent authors, both émigrés and Soviet travelers who in the 1920s visited Germany often.[7] However, most of the studies so far have treated "Russian Berlin" as an entity that existed separately from "German Berlin" and have not taken into account cultural interactions between the two. Within the limitations of this chapter, I address this omission and discuss the perception of German Berlin—in the words of Khodasevich, "the stepmother of Russian cities"[8]—by Russian writers and poets in the 1920s, focusing on topoi rather than individual visions.

The 1920s are usually regarded as the "golden age" or "the magnificent decade" in the history of Berlin.[9] After the civil unrest, political turmoil, and economic catastrophe at the beginning of the post-war period, the city entered what Gottfrid Benn later called "its most wonderful years, its Parisian years, full of talent and art."[10] For some contemporaries, it even outstripped Paris in the race for the title of the Cultural Capital of Europe. It was "as if all the eminent artistic forces were shining forth once more, imparting to the last festive symposium of the minds a many-hued brilliance before the night of barbarism closed in," nostalgically writes in his memoirs the famous conductor Bruno Walter.[11] He and other witnesses speak about the accomplishments of Berlin innovative theaters ("the best in all Europe," by the evaluation of novelist and playwright Vicky Baum[12]); about the exciting experiments in the visual arts, cinema, architecture, and music; about an unprecedented cultural explosion. Side by side with the avant-garde—constructivism, "New Objectivity," dadaism, atonal music, and so on—there flourished a vigorous mass culture that offered a rich variety of entertainment: jazz, dancing halls, cabarets and nude shows, sports—especially boxing and bicycle races—balls, amusement parks, movie palaces, and erotic journals of all kinds. The all-pervading hedonistic thirst for liberation and experimentation spurred technological progress (Berlin, for example, was the first European city that introduced traffic lights), transformed fashions and lifestyles, and sanctioned unconventional codes of behavior. Public displays of nudity and shameless adoration of the naked body became the epitome of the New Berlin. The sexual permissiveness in the city was unrivaled anywhere in the world. The protagonist of Erich Kästner's novel *Fabian: The Story of a Moralist* (1931), who sets out to explore Berlin manners, confronts various kinds of licentiousness everywhere: nude bars, a club where one can immediately pick up a partner for a one-night stand, adultery in the open, hungry girls in bathing suits offering themselves for food and money in dancing halls, a lesbian restaurant "La cousine," a male brothel. Christopher Isherwood, a homosexual English intellectual who came to Berlin at the end of the 1920s to enjoy sexual freedom and to taste "the seething brew of history in the making," wrote in *The World in the Evening*:

> In Berlin, it wasn't enough merely to want sex; you were expected to specialize, to ask for a teen-age virgin, a seventy-year-old woman, a girl with a whip and high boots, a transvestite, a policeman, a pageboy or a dog. There were various kinds of brothels and bars to supply your needs. And, in case you couldn't make up your mind what you wanted, there was a Museum of Sexual Science where you could study photographs of hermaphrodites, sadists' torture instruments, fantasy drawings by nymphomaniacs, female underwear worn by officers beneath their uniforms, and many other marvels.[13]

The weird mixture of high and low cultures, divisive radical politics, sexual permissiveness, and cynicism imbued Weimar Berlin with a fascinating aura of modernism and even futurism. As Stephen Spender, another young Englishman who came to Germany to taste the "seething brew," wrote in his autobiography *World Within World*:

> Modernism in this Germany was ... a popular mass-movement. Roofless houses, expressionist paintings, atonal music, bars for homosexuals, nudism, sun-bathing, camping, all were accepted, and became like bright, gaudy, superficial colors in which the whole country was painted.[14]

Walter Ruttmann's full-length documentary *Berlin: Symphony of the Big City*,[15] popular novels of the period (for example, Döblin's famous *Berlin Alexanderplatz: The Story of Franz Biberkopf*), essays of writers and journalists (Walter Benjamin and Siegfried Kracauer among them), all celebrated, or sometimes condemned, the capital of German modernism as the metropolis of a new type—an ever changing city that incessantly effaced its past and reinvented its present. Fragmented into scores of disconnected particles, it substituted an infinite variety of avatars, configurations, and patterns for any recognizable, relatively unvarying "face." To quote a well-known essay by Kracauer, "Kurfürstendamm: A Street Without Memory," living in Berlin was "not like a line but a series of points; it [was] new every day like the newspapers that are thrown away when they have become old."[16] Irrespective of their views of Berlin's pursuit of innovations, all the "insiders" agreed that it produced "a strange electricity,"[17] emanated energy, vitality, zest, and hence lured and enchanted people. As the playwright Zuckmayer remarked, Berlin "was worth more than a mass." He recalls that in the 1920s one spoke of the city as one speaks of a highly desirable woman—"arrogant, snobbish, uncultivated, common," but at the same time the center of everyone's fantasies and the goal of everyone's desires: "everyone wanted her, she enticed everyone."[18]

It is this erotic infatuation with the charms of the New Berlin that most Russian descriptions of the city conspicuously lack. In fact, the only Russian authors who seemed to pay attention—albeit superficially—to the modernism of Berlin were advocates of the Communist regime, either visitors from Soviet Russia or émigrés thinking of going back home. Thus Ilia Ehrenburg in his travelogues of 1928–30 portrays Berlin as the ultimate epitome of the modern age, "an apostle of Americanization," the most comfortable and advanced urban space where one immediately experiences "our times" through dealing with "cigarette packs, Pottsdamerplatz traffic, the latest model of a gas stove, three tariffs for cabs and thirty tariffs for women."[19]

Ehrenburg's observations, as always, were derivative: he followed the commonplaces of German journalism as well as Roman Gul's autobiographical book *Zhizn' na fuksa* eulogizing new Berlin—"the embryonic soul of post-war Germany"—for its stunning innovations: "Movie-theaters of UFA with a magnificent palace *UFA am Zoo*, radio-towers, gas holders, Unkers and Simmens factories, electrical trains, subway, bars, dancing halls—this is what Berlin is famous for.... Germany is being Americanized, and Berlin takes the lead."[20] Sergei Tretiakov, an ardent champion of both communism and modernism who visited Berlin in 1930, also could not help admiring new inventions and gadgets such as "paper cups for milk,... light shows of signs on Berlin façades, traffic signals, and stainless steel."[21]

Those Russian exiles in Berlin who, to put it mildly, did not approve of Bolshevism usually looked upon the upsurge of changes without any sympathy, for they suspected a kinship between modernism and communist ideology. They had good reasons for their suspicions because German modernism was closely associated with leftist politics. "Nearly all the German intellectuals whom we knew accepted and practiced a kind of orthodoxy of the Left," testifies Stephen Spender. "This attitude influenced the theater, the novel, the cinema, and even music and painting."[22] Thus to express admiration for trendy novelties often meant to take a certain political stand. In Nabokov's early story "The Reunion," a well-placed Soviet engineer sent to Berlin on business comes to see his brother Lev, a poor émigré, and starts to discuss the exciting changes: "'They've outdone America,' he said, 'Just look at the traffic. The city has changed enormously.'" And again, after a brief strained conversation he repeats the same banalities *à la* Ehrenburg, or Gul': "The Americanization.... The traffic. The remarkable buildings." As for Lev, he remains indifferent to these clichés as *his* Berlin has nothing to do with them. It is a gloomy, dark, unfriendly place: "Street lamp. Puddle. Dark post office building. Old beggar woman standing as usual by the stamp machine. She extended a hand with two matchboxes. The beam of the street lamp touched her sunken cheek; a bright drop quivered under her nostril."[23] For the paradigmatical exile, the attractions of the Americanized, futuristic Berlin are just hollow illusions; he exists beyond their spell, in the other world of his permanent isolation, and from his position as an outsider observes the underside of an alien city, ignoring its brilliant exterior.

Images of the city created by Russian authors in exile, such as Nabokov's depiction of a wet street and dark post office in "The Reunion," usually contrast and defy the canonical myth of dazzling, sparkling, flashing Berlin.[24] Instead of the multicolored, kaleidoscopic urban festival, they portray a dull, monotonous, and monochrome cityscape. For example, in Khodasevich's *European Night*, Berlin is the cumbersome (cf. "Po skvazhinam gromozdkogo Berlina"), rotten ("I tam, skol'zia v nochnuiu gnilost'"), dirty ("Vot tak i shlepai po griazi"), demonically dark ("Doma—kak demony, / Mezhdu domami—mrak; / Sherengi demonov, / I mezhdu nikh skvozniak") space associated with illness, sterility, and violent death.[25] Echoing his bitterness, Nina Berberova would later curse Berlin of the 1920s in her autobiography: "[S]ickly Germany, sickly money, the sickly trees of Tiergarten, where we sometimes strolled in the morning … Sickly trees, sickly streetwalkers on the corner of Motzstrasse."[26] Andrei Bely's curious book of essays *A Chamber in the Kingdom of Shadows* (Odna iz obitelei tzarstva tenei), written after he returned to Moscow from Germany, is a diatribe against the corrupt, putrid, insane Berlin that hides the "wildest chaos of degeneration and death" behind the façade of formal bourgeois decency. Throughout the book Bely plays upon repetition of sounds in the first syllable of "Berlin," "burgher," "bourgeois," and Russian adjectives *seryi* (gray) and *buryi* (dull brown, drab), identifying the "drab city" with gray and brown hues to convey its cheerless atmosphere of boredom and stagnation. Suffice it to cite a poetically organized, cadenced description of Berlin in which sound combinations "BUR" and "SER" are repeated 16 (!) times:

> On [Berlin] … neStERpimo zhareet uzhasnoiu, BURoiu kopot'iu letom, i SERo-BURovataia mgla povisaet nad nim oseniami i zimami … shlepaiut pod nogami takie zhe BURye, mokrye ot dozhdia trotuary … i vse eto—v BURoi, toskliveishei dymke, i BURye, skuchnye, pRESnye BIURgery speshno begut v BURovatykh pal'to vdol' tekh ulits, vdol' SkvERov, vdol' ploshchadi i provalivaiutsia v dyru, ziiaiushchuiu poSERedine, chtoby vyskochit' gde-nibud' … iz tochno takoi zhe dyry … i nestis' vdol' takogo zhe BURovatogo, prenelepogo riada domov v BURovatoi tomitel'noi mgle, pod BUReiushchim nebom, nad BURym asfal'tom.[27]

> It is unbearably heated by the terrible, brownish soot in summer, and the gray-brownish haze hangs over it in fall and winter; and the same brownish sidewalks wet from rain plash under your feet … and everything is covered by brownish, dullest haze, and brownish, boring, stale burghers hurriedly run in their brownish coats along the streets and gardens and cross a square and fall down a hole gaping in the middle, only to reappear out of another hole looking exactly alike and run along the same brownish, stupid rows of houses in the brownish oppressive fog, under the brownish sky, over the brownish asphalt.

Where the gaze of enchanted Berliners and sympathetic foreigners finds variety, diversity, and difference, Andrei Bely discovers only the terrifying sameness, the nightmarish pageant of copies, doubles, repetitions, clones:

> All the houses are alike. One can not tell a street from a street, a house from a house. Though the buildings themselves are rather monumental, splendid, imposing, in the eyes of the viewer their splendor and grandeur converge into drab, gray-brownish, monotonous boredom of organized madness in which it is impossible to understand a street, a house, a man living within.[28]

That image of Berlin as the city of lifeless monotony was to be repeated again and again in the writings of Russian malcontents. "Berlin ... is hard to describe," complains Victor Shklovsky in his *Zoo, or Letters Not about Love,* because for him everything in the city looks identical:

> The houses are as alike as suitcases.... There are many streetcars, but there is no point in riding around town, because one part of town is just like another. Palaces are from some store that sells ready-made palaces. Monuments as alike as place settings.... One *Kirche* in Berlin looks so much like another *Kirche* that we distinguish them only by the streets on which they stand.[29]

Though in his 1923 essay Ehrenburg declared that he loved Berlin, his vision of the city seems to echo that of Shklovsky:

> Berlin is gloomy, monotonous and lacks *couleur locale*.... It is hard to tell apart its long, straight streets—all of them are just exact copies of one another. You can walk for an hour, for two hours and everything around you will always look the same.... The city is an exhibition, a huge dummy, a blueprint seen in a dream.[30]

In his book *The Soul of St. Petersburg,* Nikolai Antsiferov remarks that "a city opens its face only to those who have, at least temporarily, been citizens of this city, been involved in its existence and become a part of the complex whole."[31] The refusal (or inability) of Russians in Berlin to discriminate, to see differences, "to understand a street, a house, a man living within" actually betrays their uninvolvement, their refusal (or, again, inability) to merge into "the complex whole" of the city. German Berlin seen through Russian eyes always remains alien, hostile, incomprehensible, and impenetrable; the paradigmatical Russian point of view is that of Khodasevich's Berlin poems in which the lyrical persona is cruelly locked out[32] or separated from the city by a thick glass wall.[33] Shklovsky's simile likening Berlin houses to suitcases is also symptomatic: the inner life of an alien city, or, in Antsiferov's terms, its soul, is hidden from an outside observer as things in a locked receptacle without a key. For Yury Dzhanumov, a minor émigré poet, Berlin is "a repulsive, cumbersome, huge warehouse";[34] for Mikhail Gorlin, another minor poet, the houses of the city are like impenetrable strong-boxes.[35]

Attempts to unlock the city and to establish some human contact with its soul are not frequent and, as a rule, turn out to be unrewarding. Whereas almost any non-Russian story of a newcomer's arrival in Berlin involves the protagonist's initiation into the extravagant sexual scene of the city, Russian authors either ignored risqué themes or bluntly denounced Western licentiousness. True, some of them—for example, Shklovsky in *Zoo*, Gul' in *Zhizn' na fuksa*, and Aleksei Tolstoy in his story "Black Friday"—do portray proverbial Berlin prostitutes, but they look at them, as it were, from a safe distance, precluding any desire. The women are invariably so old, ugly, and repulsive that they forbid, as Aleksei Tolstoy puts it, even simple curiosity.[36] The closest contact with the taunting sexuality of Berlin that the Russian writers were ready to describe is a night visit to some illegal private dive (*Nachtlokal*) where guests are entertained with striptease and nude dancing. Seen through Russian eyes, though, the strippers of *Nachtlocal* lack even the slightest vestige of sex appeal. Thus characters of Aleksei Tolstoy's "Black Friday" feel nothing but utmost revulsion watching a dance of ugly naked women.[37] Both Shklovsky and Ehrenburg stress the utter boredom and banality of illicit voyeuristic experience. In their portrayal, a *Nachtlocal* is just a respectable bourgeois flat used at night for a small family business—the host serves drinks, his daughters do clumsy, amateurish nude dancing and then, after the show is over, discuss Dostoevsky with Russian visitors or hand out the coats: "Everything was arranged very solidly and probably kept in the family. As for debauchery, there was … none."[38] Only in Andrei Bely's mythogenic discourse does a visit to *Nachtlocal* acquire a deeper meaning but it is entirely negative. He depicts an ordinary drunken pimp in his black bowler hat canvassing for customers as the mysterious creature with a dog-like head (*pes'egolovyi chelovek*) who on ancient Egyptian frescoes always escorts the dead into the kingdom of shadows, thus transforming a banal night-life scene into a symbol of spiritual death.[39]

Undermining the image of Berlin as a paradise of life-giving Eros, some Russian authors, in contrast to their Western counterparts, found the essence of the city not in the "primitive vitality" of orgies and nude dances but in a sad spectacle of masturbation. While cafés are filled with music, notes Shklovsky, "in the dark public toilets of Berlin, men indulge in mutual onanism. They are suffering from a devalued currency and hunger; their country is perishing."[40] In a poem "Pod zemlei" (Underground), which David Bethea calls "his most terrifying and powerful expression of the condition of exile,"[41] Khodasevich parallels an underground public toilet with Hades, and the pathetic pleasures of a mad masturbator "pressed to the wall by his wild dreams" with the "rage and grief of the poet" whose "walking stick incessantly taps on the alien granite."[42] Outcasts barred from entering the body and soul of the city, the Russian writers perceived Berlin as an inhuman, inanimate space—the embodiment of sterility; hence the recurrent images of stone, stone walls, and pavements in the Berlin poems of Khodasevich or the obsession with motifs of iron and railways elsewhere.

While German writers chose the city's most bustling, most humanly charged sites as the metonymic representations of Berlin—the Alexanderplatz, the Kurfürstendamm, and the Tiergarten,[43] their Russian counterparts found the central emblem of the alien city in Gleisdreieck, a huge multilevel station that, besides regular commuter trains, also served underground ones, which surfaced through a special arch built in a nearby apartment house. Shklovsky called Gleisdreieck "the iron heart of Germany" and described it as a demonic maze of iron constructions: "All around, along the roofs of the long yellow buildings, run tracks; tracks run along the ground and along high iron platforms, where they intersect other iron platforms as they rise to platforms still higher. Thousands of fires, lanterns, spires, iron balls on three legs and semaphores—semaphores everywhere."[44] Ilia Ehrenburg also meditated on the "iron idyll" and "iron nets" of the station.[45] Boris Pasternak, who stayed in Berlin for almost a year, frequented Gleisdreieck, showed it to visiting Mayakovsky,[46] and wrote a poem that depicts a majestic sunset observed from the station:

> Chem v zhizni probavliaetsia chudak,
> Chto kazhdyi den' za nebol'shuiu platu
> Sdaet nad revom propasti cherdak
> Iz Potsdama speshashchemu zakatu?
>
> On vystavliaet rozu s rezedoi
> V klubiashchuiusia ná verstu korzinu,
> Gde semafory sporiat krasotoi
> So snezhnoi dal'iu, pakhnushchei benzinom.

V rukakh u krysh, u trub, u nedotrog
Ne sumerki,—karandashi dlia grima.
Tuda iz mraka vyrvavshis', metro
Komkom grimas letit na kryl'iakh dyma.[47]

What does an eccentric subsist on in life / who every day, just for a song, sublets / the loft above the roaring abyss / to the sunset rushing in from Potsdam? // He puts out roses and reseda into a basket steaming for miles / where semaphores vie in beauty / with the snowy vista redolent with gasoline. // The roofs, the chimneys, the touch-me-nots hold in their hands / not the twilight but mascara / and the underground train, bursting out of darkness, flies there / on the wings of smoke as a lump of grimaces.

At first glance the poem seems to be a straightforward impressionistic picture of beautiful sunsets over Berlin observed from the Gleisdreieck platform. However, what looks like a playful juxtaposition of urbanistic details and natural phenomena contains a number of stock images that in the discourse of the 1920s acquired fixed symbolic meanings pertaining to the doom of Western civilization: the rubescent sunset foretelling disasters, the roaring abyss, the smell of gasoline, snow covering the earth, twilight, and darkness. In the final stanza of the poem, motifs of make-believe, masking, and pain (cf. "mascara"/masks and both connotations of the word "grimaces"; cf. also German "Grimm"—fury, rage) explode a peaceful cityscape, and the movement of the underground train (*metro*) rushing from darkness into the fires of hell becomes a metaphor of Berlin as a doomed, dying *metro*polis—the "faceless Babylon" as Pasternak called it in a letter to S. Bobrov.[48] Though Pasternak never brings the theme of doom into the open, the imagery of the poem and the sound structure of its last stanza—which evokes homophonic and anagrammatic echoes of such words as *trup* (a dead body) in *trub*, *umer* (died) in *sumerki*, *gorod* (city) in *nedotrog*, and *smert'* (death) in *vyrvavshis' metro*—indicate that he read the city in eschatological terms close to those of Oswald Spengler's famous *Decline of the West* or, in the Russian translation, *Zakat Evropy* (The Sunset of Europe).

Many Russian writers found an influential model for conceptualization of their Berlin experience in Oswald Spengler's concept of the modern city as a demonic megalopolis—or, better, a necropolis—where Western civilization meets its imminent end. According to Spengler, Berlin in particular was "the autumnal city" inhabited by "the spiritually dead" and marked by unfruitfulness and sterility. Glaring symptoms of irreversible decay and degradation were everywhere: among other things, he pointed to the drying up of life-giving carnality, the childlessness and "race-suicide" (meaning homosexuality and other unconventional forms of sexuality), and such concessions to barbarism in arts and lifestyles as "Expressionism, boxing contests, nigger dancing ... and racing." The world-city—"this stony mass"—takes an "essentially metaphysical turn towards death"; "the spirit-pervaded stone of Gothic buildings, after a millennium of style-evolution, has become the soulless material of this daemonic stone-desert."[49]

In fact, Russian writers' condemnation of Berlin, as well as the images and metaphors they used in their critique, in many cases echoes Spengler's historicist rhetoric. When Andrei Bely, for example, castigated Berlin as "the bourgeois Sodom," Tartarus, Hades, or the Egyptian Kingdom of Shadows and lamented the demise of the old European culture conquered by the animalistic, "barbarous Dionysus" in the image of a pagan African deity—"the 'Negro' of Berlin, the 'Negro' of the new Europe but, in fact, the image of its death, its doom,"[50] he reinforced the Spenglerian predictions of the imminent "decline of the West." Bely's demonic, insane, doomed Berlin, Khodasevich's stony hell, Shklovsky's "iron maze"—all these and similar concepts served as filters that separated, from immediate experience, only those images of Berlin that would not contradict the general historicist idea of its inevitable apocalyptic fall. An embodiment of this position is the Russian modernist poet (probably modeled upon Ilia Ehrenburg) of Khodasevich's aborted poem "In a Café," who watches shabby dancers and gloats over the unseemly, shameless holes in their leotards, which he interprets as a symbol of the downfall of "the cursed European world."[51]

However, there were several émigré writers and poets (mostly of the younger generation) who worked out and successfully used a different strategy in coping with the alien city. Instead of denouncing the "bourgeois Sodom" and looking for signs of its impending historical doom, they would try to put down or debunk Berlin by questioning and transcending its actuality. For them, the city and its inhabitants become almost immaterial, spectral, turned into a quasi dream, while their inner oneiric worlds supersede the phenomenal. Walking down a Berlin street, the exilic dreamer is actually not there but in the realm of memory and imagination, and, to quote Nabokov's *The Gift*, ignores "the pale ghosts of innumerable foreigners flickering among those natives like a familiar and barely noticeable hallucination."[52] External and internal realities change their places: memories of the past, visions, dreams, recollected Russian literary texts are represented as something more immediate than the immediacies of routine existence in Berlin. Vladimir Korvin-Piotrovsky, for example, begins a poem with well-known Berlin place-names only to dismiss them as his fantasy or dream and to plunge into the "more real" world of Pushkin's *Feast During the Plague:*

> Charlottenburg, Kurfürstendamm,—ne veriu—
> Ia vydumal, prosnus' i ne poimu—
> Spoi pesenku, zadumchivaia Meri,
> Kak pela Dzhenni drugu svoemu ...[53]

> Charlottenburg, Kurfürstendamm,—I don't believe it—I have invented them; I'll wake up, and I won't understand any of it—Sing me a song, o melancholy Mary, as Jenny sang one for her friend ...

In many instances a detail of life in Berlin serves as a trigger sending the exilic mind off to its oneiric vision or, conversely, bringing it back from reveries. Thus, Nabokov's characters in *Mary* and *The Gift* roam the streets of Berlin now reliving their past and oblivious of their surroundings, now alert to it again. The memories of a storm in Moscow obliterate the immediate Berlin present in Pasternak's "*Babochka-buria*" (A Storm Butterfly).[54] Sergei Gorny's story "*Brat'ia*" (Brothers) climaxes in an epiphanic moment on Friedrichstrasse when "the street and store windows and black heads of the crowd swim away ... and disappear."[55] In this way, the writer constructs his personal "oneiric house" within the city and hence gains an opportunity to domesticate Berlin on his own terms, making it a minor, subordinate part of the overarching dream.

Of the three strategies we have delineated so far—fixation upon modernist novelties, historicist demonization, and oneirization—it is the third that proved to be most productive artistically. At any rate, it facilitated the transfiguration of Berlin into a unique poetic vision achieved in Vladimir Nabokov's Russian writings. Looking at Berlin from the position of an uninvolved outsider, Nabokov goes far beyond mere questioning and undermining its reality. Not only does he break the totality of the city into a myriad of discrete fragments, but he reconfigures them so as to create the defamiliarized, mysterious Berlin of his own. In his seminal story "A Guide to Berlin," Nabokov focuses on various "insignificant trifles" of everyday life in the city—black pipes in the snow, the streetcar, workmen tearing up the pavement, a baker on his tricycle, animals in the Berlin Zoo, a pub—deriding the objections of a typical Russian émigré who does not care about the vignettes sketched by the storyteller: "'It's of no interest,' my friend affirms with a mournful yawn. 'What do trams and tortoises matter?

And anyway the whole thing is simply a bore. A boring, foreign city, and expensive to live in, too.'"[56] For Nabokov, on the contrary, the only details that matter are those devoid of any social or historical meaning; he sets out "to portray ordinary objects as they will be reflected in the kindly mirrors of future time; to find in the objects around us the fragrant tenderness that only posterity will discern and appreciate in the far-off times when every trifle of our plain everyday life will become exquisite and festive in its own right."[57]

This personal, "exquisite and festive" Berlin of Nabokov lies beyond the scope of our general survey and requires special study.[58] It is clear, however, that in spite of all its originality it did not emerge in a vacuum but owes much to the various Russian readings of Berlin discussed above. Fashioning his vision of the city, Nabokov carried on a dialogue with the images and concepts shaped by his predecessors; he responded to Khodasevich and Bely, Shklovsky and Gul'. He learned from them, alluded to them, laughed at them, argued with them. In the final analysis, Nabokov's Berlin is both a subversion and an acme of the overall tradition that implanted the "stepmother of Russian cities" in the consciousness of Russian culture.

Notes

1. V. V. Mayakovsky, *Polnoe sobranie sochinenii v trinadtsati tomakh,* vol. 12 (Moscow, 1959), p. 462.
2. V. Shklovsky, *Zoo, or Letters Not about Love,* trans. and ed. Richard Sheldon (Ithaca and London, 1971), p. 135.
3. R. C. Williams, *Culture in Exile: Russian Emigrés in Germany, 1881–1941* (Ithaca and London, 1972), p. 114.
4. Russifying Berlin place-names seems to have been a popular amusement of the day in the Russian community and the German press. A Russian journalist in 1923, besides alluding to the quite common CharlottenGRAD, jokes about plans to rename the Kurfürstendamm "Prospekt Kurfiursta" (see S. Segal, "Russkaia zhizn' v Berline," *Russkii illiustrirovannyi mir* 1 [1923]: 12). Mayakovsky mentions some street in Berlin that has been renamed Nevsky Prospekt in jest (Mayakovsky, *Polnoe sobranie,* vol. 12, p. 462). The most Russianized places in Berlin were located near the Kaiser-Wilhelm-Gedächtniskirche (the Memorial Church), around the eastern end of the Kurfürstendamm and the Tauentzienstrasse, to which Andrei Bely gave the name of a lively Moscow street, Kuznetsky Most (Andrei Bely, *"Odna iz obitelei tsarstva tenei"* [Moscow, 1924], p. 29). Iosif Gessen recalls a German newspaper cartoon entitled "A

Portrait of the Future," which depicted numerous Russian stores and offices on the "Kurfiurstendammskii prospekt" and "Tauentsinskaia ulitsa" with occasional notices in the shop-windows: "German spoken here" (I. V. Gessen, *Gody izgnaniia. Zhiznennyi otchet* [Paris, 1979], p. 139).

5. Vladimir Nabokov, *Glory* (New York and Toronto, 1971), p. 35.
6. Nicolas Nabokov, *Bagazh: Memoirs of a Russian Cosmopolitan* (New York, 1975), p. 119.
7. On the Russian Berlin, see, among others, L. Fleishman et al., *Russkii Berlin 1921–1923. Po materialam arkhiva B. I. Nikolaevskogo v Guverovskom institute* (Paris, 1983); F. Mierau, ed., *Russen in Berlin: Literatur, Malerei, Theater, Film, 1918–1933* (Leipzig, 1990); Th. R. Beyer, "Andrej Bely: The Berlin Years 1921–1923," *Zeitschrift für Slavische Philologie* 50 (1) (1990): 90–142; K. Schlögel, ed., *Russische Emigration in Deutschland 1918 bis 1941. Leben im Europäischen Bürgerkrieg* (Berlin, 1995).
8. An expression from his Berlin poem "Vse kamennoe. V kamennyi prolet ..." (1923); see V. Khodasevich, *Sobranie sochinenii.*, vol. 1 (Ann Arbor, 1983), p. 153.
9. For basic information regarding Berlin of the 1920s, see P. Gay, *Weimar Culture* (New York, Hagerstown, San Francisco, London, 1970), pp. 128–44; W. Laqueur, *Weimar: A Cultural History 1918–1933* (London, 1974), pp. 224–53; A. De Jonge, *The Weimar Chronicle: Prelude to Hitler* (New York, 1978), pp. 125–46; Ch. W. Haxthausen and H. Suhr, eds., *Berlin: Culture and Metropolis* (Minneapolis and Oxford, 1990), pp. 95–165; Th. W. Kniesche and S. Brockmann, eds., *Dancing on the Volcano: Essays on the Culture of the Weimar Republic* (Columbia, SC, 1994), pp. 71–180. Otto Friedrich's popular *Before the Deluge: A Portrait of Berlin in the 1920s* (New York, 1995, 2nd ed.) written in a journalistic vein, can serve as a general introduction to the period and historical sources.
10. Quoted in Laqueur, *Weimar: A Cultural History*, p. 277.
11. B. Walter, *Theme and Variations: An Autobiography* (New York, 1941), p. 234.
12. V. Baum, *It Was All Quite Different* (New York, 1964), p. 274.
13. Ch. Isherwood, *The World in the Evening* (New York, 1954), p. 75.
14. S. Spender, *World Within World: The Autobiography* (New York, 1994), p. 108.
15. On the film, see S. Hake, "Urban Spectacle in Walter Ruttmann's *Berlin: Symphony of the Big City*," in Kniesche and Brockmann, eds., *Dancing on the Volcano*, pp. 127–42.
16. S. Kracauer, *Schriften* 5 (3) (Frankfurt, 1990), S. 170.
17. Baum, *It Was All Quite Different*, p. 276.
18. Quoted in Gay, *Weimar Culture*, p. 132.
19. I. Ehrenburg, *Sobranie sochinenii v vos'mi tomakh*, vol. 4 (Moscow, 1990), p. 30.
20. R. Gul', *Zhizn' na fuksa* (Moscow and Leningrad, 1927), pp. 133–34.
21. S. Tret'iakov, *Den' Shi-Khua. Liudi odnogo kostra. Strana-perekrestok* (Moscow, 1991), p. 309. In her emotionally charged travelogue, Vera Inber likewise focuses on spectacular novelties of the Berlin modernism. See V. Inber, *Kniga chetvertaia* (Kharkov, 1929), pp. 21–24 and 28–32. Exasperated by such reports, Aleksei Tolstoy in a private letter pleaded with a friend to inform him of the real life in Berlin as he was sick and tired of stories about traffic lights on Pottsdamerplatz (see Fleishman et al., *Russkii Berlin 1921–1923*, p. 121).

22. Spender, *World Within World*, p. 131.
23. *The Stories of Vladimir Nabokov* (New York, 1995), pp. 300–307.
24. Cf., for instance, a passage in Josephine Baker's memoir: "Berlin. I found it dazzling. The city had a jewel-like sparkle, especially at night, that didn't exist in Paris" (J. Miller and T. Smith, eds., *Berlin* [San Francisco, 1996], p. 169).
25. See his poems "Berlinskoe," "S berlinskoi ulitsy," "Net, ne naidu segodnia pishchi ia," "Vse kamennoe. V kamennyi prolet ..." (Khodasevich, *Sobranie sochinenii.*, vol. 1, pp. 147–51).
26. N. Berberova, *The Italics Are Mine* (New York, 1992), pp. 165–66.
27. Andrei Bely, "*Odna iz obitelei tsarstva tenei,*" p. 7.
28. Ibid., p. 31.
29. Shklovsky, *Zoo, or Letters Not about Love*, p. 67.
30. Ehrenburg, *Sobranie sochinenii v vos'mi tomakh*, vol. 4, p. 8.
31. N. Antsiferov, *Dusha Peterburga* (Leningrad, 1990), p. 37.
32. See his "Vse kamennoe. V kamennyi prolet ..." in which "a key rattles against the stone, a door bolt clatters; now you just keep tramping the stony streets until five o'clock" (Khodasevich, *Sobranie sochinenii*, vol. 1, p. 153). Cf. also the theme of being locked out in the first and last chapters of Nabokov's *The Gift*.
33. See in particular his "Berlinskoe" (Khodasevich, *Sobranie sochinenii*, vol. 1, p. 147–48).
34. G. V. Adamovich and M. L. Kantor, eds., *Iakor': Antologiia zarubezhnoi poezii* (Berlin, 1931), p. 192.
35. M. Gorlin, "Nochami," *Sovremennye zapiski* 51 (1934): 209.
36. A. N. Tolstoy, *Sobranie sochinenii v desiati tomakh*, vol. 4 (Moscow, 1958), p. 362.
37. Ibid., pp. 380–81.
38. Shklovsky, *Zoo, or Letters Not about Love*, p. 5; cf. Ehrenburg, *Sobranie sochinenii v vos'mi tomakh*, vol. 4, p. 13.
39. Andrei Bely, *Odna iz obitelei tsarstva tenei*," pp. 59–60.
40. Shklovsky, *Zoo, or Letters Not about Love*, p. 40.
41. D. Bethea, *Khodasevich: His Life and Art* (Princeton, New Jersey, 1983), p. 292.
42. Khodasevich, *Sobranie sochinenii*, vol. 1, pp. 152–53.
43. See E. Schütz, "Beyond Glittering Reflections of Asphalt: Changing Images of Berlin in Weimar Literary Journalism," in *Dancing on the Volcano*, pp. 119–21.
44. Shklovsky, *Zoo, or Letters Not about Love*, p. 68.
45. Ehrenburg, *Sobranie sochinenii v vos'mi tomakh*, vol. 4, pp. 8–9.
46. See E. Pasternak, *Boris Pasternak: Materialy dlia biografii* (Moscow, 1989), pp. 373–74.
47. B. Pasternak, *Sobranie sochinenii v piati tomakh*, vol. 1 (Moscow, 1989), p. 535.
48. E. Pasternak, *Boris Pasternak*, p. 384.
49. O. Spengler, *The Decline of the West*, vol. 2 (New York, 1970), pp. 99–103.
50. Andrei Bely, "*Odna iz obitelei tsarstva tenei,*" p. 60.
51. Khodasevich, *Sobranie sochinenii*, vol. 1, p. 239.
52. V. Nabokov, *The Gift* (New York, 1991), p. 166. Cf. also a statement in Nabokov's autobiography: "As I look back at those years of exile, I see myself, and thousands of other Russians, leading an odd but by no means unpleasant existence, in material indigence and intellectual luxury, among perfectly unimportant strangers, spectral Germans and Frenchmen in whose more or less illu-

sory cities we, émigrés, happened to dwell. These aborigines were to the mind's eye as flat and transparent as figures cut out of cellophane, and although we used their gadgets, applauded their clowns, picked their roadside plums and apples, no real communication, of the rich human sort so widespread in our own midst, existed between us and them" (V. Nabokov, *Speak, Memory: An Autobiography Revisited* [New York, 1966], p. 276).

53. V. Piotrovsky, "Stikhi k Pushkinu," *Rul'*, 5 February 1928.
54. On this poem, see V. V. Ivanov, "Stikhotvorenie Pasternaka 'Babochka-buria'," in *Tekst i zdanie* (Wroclaw et al., 1983), pp. 303–12.
55. S. Gorny, *Tol'ko o veshchakh* (Berlin, 1937), p. 238.
56. *The Stories of Vladimir Nabokov*, p. 159.
57. Ibid., p. 157. For a general discussion of the story, see D. Barton Johnson, "A Guide to Nabokov's 'A Guide to Berlin'," *Slavic and East European Journal* 23 (1979), pp. 353–61. Omry Ronen was the first to suggest that "A Guide to Berlin" should be read as Nabokov's response to Shklovsky's descriptions of the city in *Zoo, or Letters Not about Love.* (See O. Ronen, "Zaum' za predelami avangarda," *Literaturnoe obozrenie* 12 [December 1991], pp. 36–43). He has convincingly developed the thesis in his "Puti Shklovskogo v 'Putevoditele po Berlinu'," *Zvezda* 4 (April 1999), pp. 164–72.
58. For basic approaches to the problem, see: A. Engel-Braunschmidt, "Die Suggestion der Berliner Realität bei Vladimir Nabokov," in Schlögel, ed., *Russische Emigration in Deutschland 1918 bis 1941*, pp. 317–78. I discuss some of its aspects in A. Dolinin, "Clio Laughs Last: Nabokov's Answer to Historicism," in J. W. Connolly, ed., *Nabokov and His Fiction: New Perspectives* (Cambridge, 1999), pp. 197–215.

Chapter 16

NABOKOV AND GOETHE

Omry Ronen

The year 1999 saw three anniversaries that traversed conventional boundaries between national cultures. August 28 marked two hundred and fifty years since the birth of the author of *West-Eastern Divan*, the first master-builder of a poetic bridge between East and West, the first German writer to be widely recognized and admired west of the Rhine, and the creator of the concept of *Weltliteratur* as a system of reciprocal interliterary reflections and refractions.

On June 6 Russia and the international community of Slavic studies celebrated the bicentennial of the birth of Pushkin, whose terse poetry, drama, and prose had condensed, miniaturized, and encapsulated, as it were, the great occidental accomplishment of the ancients, and of Dante, Ariosto, Shakespeare, and Goethe, as well as some traditions of the Orient, to implant it in the youngest national verbal art of Europe, the literature written in the modern Russian language.

The historical role of the third jubilarian, Nabokov, born on 10 April 1899, is in a certain way unique even in comparison with the other two. A multilingual artist, with matching achievements in Russian and in English, Nabokov did indeed attain an interliterary and international significance that Dostoevsky had claimed for Pushkin when, in his critically naive, "politico-patriotic" (Nabokov's term) speech, he insisted on what he presumed to be Pushkin's exclusive and typically Russian gift of all-embracing, all-human, universal sensitivity, sensibility, receptivity, and understanding of other nations, which should serve Russia's mission of "the all-human reunion of all the tribes of the great Aryan family." Aside from the preternatural claims for Russia, stated with apodictic fervor, but made somewhat comical by Dostoevsky's calling up to the Aryan colors the manifestly "non-Aryan" greatest Russian poet (in Pushkin's own words, "an unsightly descendant of Negroes"), it is clear from the purely literary-historical standpoint that, unlike the exotically perceived moral prophecy and evocative power of Tolstoy and Dostoevsky, Pushkin's extraordinary art, limpid and enigmatic at once in the original Russian, had eluded translators and, for this reason, had not become part of the occidental body of great writing—not merely acknowledged, but actually read as living poetry—before Nabokov transposed his compact and idiosyncratic yet universally appealing and aesthetically effective model of Russia's "Gold Fund" of literature, with Pushkin as its cardinal value, to the Western, and above all American, reader.

Mary McCarthy touched in passing the essence of Nabokov's unprecedented contribution to universal culture when she described *Pale Fire* as a novel of the future, a supranational novel.[1] An all-embracing artist who created a rich variety of characters and landscapes, as well as images of languages, images of national literatures and literary works in various languages, and literary styles and individual stylistic manners in his fiction, poetry, drama, and criticism, Nabokov is more than any other writer of this century relevant to the idea and the reality of world literature.

Because the fascinating theme of the exchange of ideas between Goethe and Pushkin, with Pushkin's "Scene out of *Faust*" identified as the inspiration for the scene of the arrival of the bark in *Faust* II.5,[2] and the extensive Pushkinian subtext in the life work of Nabokov have been explored quite minutely,[3] I should like, in the spirit of the past triple anniversary, to address the least studied and the most controversial side in the Goethe-Pushkin-Nabokov triangle: Nabokov's lifelong involvement with Goethe.

A profound attraction tinged with a streak of an equally deep revulsion, it should best be examined in the twofold context of Nabokov's earnest search for congenial figures and features in German literature and culture, on the one hand, and certain persistent, intellectually and emotionally indispensable motifs of his art, on the other.

For Nabokov, a descendant of Teutonic crusaders on his paternal grandmother's side, a longtime refugee in Berlin, and then a refugee *from* Berlin, Germany was a theme of great and violent intensity; its main components can be identified especially clearly in the story "Cloud, Castle, Lake" (1937). The question that Nabokov's favorite personages put to contemporary Germany is formulated in *The Gift* by Fedor Godunov-Cherdyntsev when he recollects the kindly and idealistic German tutors of Russian children in the nineteenth century ("Where are they? Where are their delicate skeletons?"), or wonders whether there are still any poets in Germany. In this respect, there is a symptomatic similarity between Nabokov's fictional young Russian man of letters in exile and Osip Mandelstam who, shortly before going for a semester to Heidelberg on the eve of World War I, wrote to Viacheslav Ivanov, the celebrated *poeta doctus* and a disciple of Mommsen, to ask him who the lyric poets in Germany were at the moment, as he did not know a single one besides Dehmel: "Nemtsy tozhe ne znaiut—a liriki vse-taki dolzhny byt'" (Germans don't know either,—yet there must be *some* lyric poets).[4]

While very critical of contemporary German literature, with the exception of Kafka and, possibly, Rilke, Nabokov proceeded to invent, in his novels, a series of German authors and literary situations. The latter are presented as actual and presumably well-known fictional or poetic episodes, but cannot, in fact, be identified and located. There are at least three such enigmatic instances in *The Gift*: in what book did the ghost of Bonaparte haunt the shop of Stolz the antiquarian in search of his tricorne? who was Albrecht Koch that longed for "golden logic" in the world of the insane? did Goethe indeed point at the starry sky with his stick and say, paraphrasing Kant, "This is my conscience"?

In the original Russian edition of *Camera Obscura* (1932), there appears a German writer residing in the south of France, Dietrich von Segelkranz,[5] "who used to know personally the late Marcel Proust, and imitated him and some other innovators" (the suggestive linkage of names is clandestinely supported by a trilingual verbal association: *Segel* "sail," "*parus*" in Russian, *Proust*). Segelkranz's style, however, judging by a longish excerpt out of the book he is writing, resembles more pointedly Rilke's Proustianism *avant la lettre* in *Die Aufzeichnungen des Malte Laurids Brigge* than *À la recherche du temps perdu.*

In the greatly altered later English version of the novel, *Laughter in the Dark* (1938), the name of the German expatriate who resides in France is "Conrad (not the famous Pole, but Udo Conrad who wrote the *Memoirs of a Forgetful Man* and that other thing about the old conjurer who spirited himself away at his farewell performance)." This Conrad, who would "gladly write in French," his "fatherland not being quite at the right level to appreciate his writings," but is "loath to part with the experience and the riches amassed in the course of" his handling of his native German language, prefigures a protagonist of Nabokov's later novel *Transparent Things* (1972): R., a German who writes in English and who embodies, as we shall see, at least one of the two pairs of cardinal motifs that Nabokov has inherited from Goethe.

Both pairs of motifs, arranged by the principle of complementarity rather than contrast, have to do with the mystery of death as embodied in a specific, strictly delimited range of symbols. The first of these, the twin motif of lepidopteran destiny, appears to have been particularly close to the heart of Nabokov the entomologist and youthful reader of Rozanov.[6] On the one hand, a moth attracted by light manifests a "blissful longing" for fiery death and transformation (*stirb und werde!*); on the other, the pupation of a caterpillar undergoing metamorphosis bespeaks the deceptiveness of bodily corruption and dissolution of flesh, which the uninitiated fear but an inspired poet and a prophet may perceive as a pledge of immortality.

This symbolic thematic context is vital for our understanding of the other pair of motifs that Nabokov has received from Goethe and that recur throughout his literary biography, forming the essential suspense of "Solus Rex" (1940) and "Ultima Thule" (1942), and culminating in *Bend Sinister* (1947) and *Pale Fire* (1962): the motif of the *Erlking*[7] and the motif of the King of Thule, both intimately linked with the Italianate anagram that Gennady Barabtarlo[8] has deciphered in *Invitation to a Beheading*: "mali è trano t'amesti" > **smert' mila, èto taina* (death is sweet, this is a mystery). To arrive at this principal "thetic" solution, the cryptographer has, however, to evaluate also a number of "antithetic," subordinate, corollary, or false, deliberately teasing answers to the anagrammatic riddle of the hereafter, such as **material'no tam est' i …* (materially there is also …) or **a telo temnitsa mira* (and body is a prison of the world), etc. The reading "Death is sweet, this is a mystery" is, of course, the closest to the idea of "Selige Sehnsucht." In conjunction with the mysterious gesture, "indicating size—the length, say, of a babe," that Cincinnatus's mother, a midwife, makes at their last meeting, it evokes Viacheslav Ivanov's interpetation of Goethe's "old truth" in "The Tender Mystery" and in "Sub Rosa" (*Cor ardens*):[9]

> Taina nezhna,—vot slovo moe,—a zhizn' kolybel'na;
> Smert'—povitukha; v zemle—novaia nam kolybel'.
>
> The mystery is tender, this is my word, and life is cradle-like;
> Death is a midwife; in the earth there is a new cradle for us.
>
> Taina, o brat'ia, nezhna: znamenuite zhe tainoe—rozoi,
> Nezhnoi pechat'iu liubvi, miloi ulybkoi mogil.
>
> The mystery, o brethren, is tender; therefore mark what is
> secret with a rose,
> With a tender seal of love, with a sweet smile of graves.

The first approach to the motif of physical dissolution as a promise of redemption is found in an early poem by Nabokov, entitled in the first publication[10] "A Legend (from the Apocrypha)," and in the verse collection *Grozd'* (1922), "On the Anniversary of Dostoevsky's Death":

> Na godovshchinu smerti Dostoevskogo
>
> Sadom shel Khristos s uchenikami ...
> Mezh kustov na solnechnom peske,
> vytkannom pavlin'imi glazkami,
> pesii trup lezhal nevdaleke.
>
> I reztsy beleli iz-pod chernoi
> skladki, i zlovonnym torzhestvom
> smerti – zaglushen byl ladan sladkii
> teplykh mirtov, mleiushchikh krugom.
>
> Trup gniiushchii, treskaias', razdulsia,
> polnyi slizkikh, slipshikhsia chervei ...
> Ioann, kak deva, otvernulsia,
> sgorblennyi pomorshchilsia Matvei.
>
> Govoril apostolu apostol:
> "zloi byl pes; i smert' ego naga,
> merzostna" ...
>   Khristos zhe molvil prosto:
>   "zuby u nego, kak zhemchuga ..."

In a literal English translation, the poem would read as follows: In the garden walked Christ with His disciples ... / Among the bushes on the sunny sand / interlaced with peacock spots / a dog's carcass was lying not far away. // The incisors showed white from underneath the black / fold, and in the fetid triumph / of death was drowned the sweet frankincense / of warm myrtles sweltering around. // The rotting carcass cracked open as it swelled, / teeming with the slithery tangle of worms ... / John, like a maiden, turned his head away; / hunched Matthew winced. // One apostle told another apostle: / "Vicious was the dog; and his death is stark, / abominable" ... Christ, however, said simply: / "His teeth are like pearls."

Nabokov's poem actually is a very free translation, a paraphrase, of Goethe's likewise quite free adaptation of Nisami's poem in *West-Eastern Divan*:

> Herr Jesus, der die Welt durchwandert,
> Ging einst an einem Markt vorbei;
> Ein toter Hund lag auf dem Wege,

Geschleppet vor des Hauses Tor;
Ein Haufe stand ums Aas umher,
Wie Geier sich um Äser sammeln.
Der eine sprach: Mir wird das Hirn
Von dem Gestank ganz ausgelöscht.
Der andre sprach: Was braucht es viel,
Der Gräber Auswurf bringt nur Unglück.
So sang ein jeder seine Weise,
Des toten Hundes Leib zu schmächen.
Als nun an Jesus kam die Reih,
Sprach, ohne Schmähn, er guten Sinns,
Er sprach aus gütiger Natur:
Die Zähne sind wie Perlen weiß.
Dies Wort macht den Umstehenden,
Durchglühten Muscheln ähnlich, heiß.

Nabokov's source may have been, along with Goethe's adaptation of Nisami, some other version of the Muslim *logia agrapha* of Jesus, because, while the original Islamic text does not mention pearls, it refers, unlike Goethe, to Christ's disciples: "What stench," said the disciples passing by a dog's carrion. "How white the teeth," said Isa.[11]

The relevance of the theme of putrefaction to Dostoevsky, especially to the Zosima-Smerdiakov confrontation in *The Brothers Karamazov,* is quite plain, although the exact meaning of the parable in reference to Dostoevsky is ambivalent. It may be a reproach to Dostoevsky for his inability to perceive the beauty of this world, which Nabokov has expressed also in another poem of this period, published in 1923 and entitled "Dostoevsky":

Toskuia v mire, kak v adu,
urodliv, sudorozhno-svetel,
v svoem prorocheskom bredu
on vek nash bedstvennyi nametil.

Uslysha vopl' ego nochnoi,
podumal Bog: uzhel' vozmozhno,
chto vse darovannoe mnoi
tak strashno bylo by i slozhno?

Anguishing in the world as if in hell,
ugly, fitfully lucid,
in his prophetic delirium
he had mapped out our disastrous age.

and coarseness." Similarly, Sergei Bulgakov, in *Ot marksizma k idealizmu* (From Marxism to Idealism, 1903), admitted that Faust, while being a universal type, nevertheless sometimes reeked of a German *Philister*; this passage is quoted in Viktor Zhirmunskii's[14] fundamental monograph on Goethe in Russian literature, which Nabokov may have consulted in connection with his research on Gogol ("Gants Kiukhel'garten" in particular).

Alexander Dolinin, in his commentaries to a Russian edition of *Lolita*,[15] has pointed out an ironic reference to *The Sorrows of Young Werther* when Humbert Humbert attempts to imagine Charlotte as his "life companion." However, the subsequent blending of the names of Goethe's love and Humbert's, *Lottelita, Lolitchen*, mocks not *Werther*, but Thomas Mann's *Lotte in Weimar* (1939, U.S. edition 1940), with its old Charlotte and youthful Lottchen. Moreover, in Charlotte's interrogation of Humbert concerning his origins and his relations with "Our Christian God," and Humbert's "pious platitude" about his belief in "a cosmic spirit," it is easy to notice a spoof of Faust being catechized by Gretchen, with a generous share of Marthe the lonesome widow thrown in for good measure.

There are several other, less immediately apparent onomastic jests at the expense of Goethe's ambience in *Lolita*: the girl eventually marries a destitute (Richard) F. Schiller just as another Lotte, poor Lottchen von Lengefeld did, the joke being that Lolita's husband is, in an allusion to Gogol's famous jest in "Nevsky Prospect," not Schiller the poet, but Schiller the tinker.

On the other hand, the motif of the Erlking is developed with only a slight touch of irony in the scene of Humbert's ride to the Elphinstone hospital with a sick and delirious Lolita.

These parodies are relatively gentle in comparison to the bitterly satirical part that Goethe's presence is made to play in *Pnin* (1957). Here references to Goethe create a compositional ring, linking, as Gennady Barabtarlo[16] has pointed out in his commentaries, the beginning and the end of Chapter 5. At the outset, Pnin's car gingerly making progress ...ng a country road and Pnin's ultimate destination, the house of his ...ds the Cooks, or Kukolnikovs, the modern Philemon and Baucis, n... Mount Ettrick, are evoked as they might have been spotted from th... of a lookout, a "prospect tower," by some presumably absent or no...nt tourist observers ("Mary, Miranda, or Almira," that is, nomina...ntatives of Pnin's guardian angel, his dead love Mira Belochkin... that matter, Wolfgang von Goethe, whose name had been carv... ...ballustrade by some old-fashioned wag."

From the purely literary point of view, an old-fashioned wag might have had in mind Lynceus der Türmer, Lynceus the tower warden, in the last act of *Faust* II, who sees in all things the eternally bright ("die ewige Zier"), but finally is forced to watch the flames consume the innocent and happy hut of Philemon and Baucis, set on fire as part of Faust's quest for the infinite. "Nun lodert's frei," says Mephistopheles, "Als Scheiterhaufen dieser drei," that is, their home is ablaze as a funeral pyre for Philemon, Baucis, and a wanderer, their guest.

The "Wahrheit," not "Dichtung," of this evocation of Goethe is revealed at the close of the chapter, when Pnin is made to remember another funeral pyre, the one that consumed Mira Belochkin on what Pnin imagines to be, by association with the name Buchenwald, "a gasoline-soaked pile of beechwood." According to the information Pnin received from an investigator in Washington, Nabokov writes, "she was selected to die and was cremated only a few days after her arrival in Buchenwald, in the beautifully wooded Grosser Ettersberg, as the region is resoundingly called." Now it becomes clear to the reader that it is the name of Ettersberg that resounds in Mount Ettrick.[17] Nabokov continues:

> It is an hour's stroll from Weimar, where walked Goethe, Herder, Schiller, Wieland, the inimitable Kotzebue and others. "*Aber warum*— but why—" Dr. Hagen, the gentlest of souls alive, would wail, "why had one to put that horrid camp so near!" for indeed, it was near—only five miles from the cultural heart of Germany—"that nation of universities," as the president of Waindell College, renowned for his use of the *mot juste,* had so elegantly phrased it when reviewing the European situation in a recent Commencement speech, along with the compliment he paid another torture house, "Russia—the country of Tolstoy, Stanislavski, Raskolnikov, and other great and good men."

Thus the theme of fire, the fire of "*selige Sehnsucht*" or of Faust's ultimate experimentation with hell, forever accompanies Goethe in the later writings of Nabokov.

In *Transparent Things* (1972), R., that rather Olympian and, in some personal respects, Goethe-like German author of "diabolically evocative" English prose, helps to the other side the unhappy Hugh Person, his copy editor and personage at once, and truly "ein trüber Gast auf der dunklen Erde," obsessed by fear of fires: a dead writer encourages his dying personage in a burning building with an understated English version of "*selige Sehnsucht*" and its ultimate message, to feel "not the crude anguish of physical death but the incomparable pangs of the mysterious mental maneuver needed to pass from one state of being to another.

Easy, you know, does it, son."

One can see, then, that in his later novels Nabokov comes to a certain reconciliation with Goethe, after a long period of revulsion mixed with fascination. For decades Nabokov courageously weighed on his artistic and moral scales the unspeakable iniquity and the fascinating loveliness of his time and, in particular, considered the two monstrous paradoxes of the age, Germany and Russia. *Ada* (1969) was written precisely about the two irreconcilable aspects of Nabokov's age, the twin texture of time, the demonic beauty of memory and the sordid terror of the real "Terra," which could not be undone by creativity. The protagonists of *Ada* discover, together with the author, that nothing can undo what has happened. Yet they "die and become" in the knowledge that, just as Goethe wrote to Karl Graf Reinhard on 31 December 1809, "what has been created takes its place as equal to what has happened."[18]

This appears indeed to be the ultimate message of hope conveyed by Nabokov's last books.

Notes

1. Mary McCarthy, "A Bolt from the Blue," *New Republic* 146 (4 June 1962): 21–27.
2. B.Ia. Geiman, "Peterburg v 'Fauste' Gete: k tvorcheskoi istorii 2-i chasti 'Fausta'" (Petersburg in Goethe's *Faust*: Toward a History of the Genesis of the 2nd Part of *Faust*), *Doklady i soobshcheniia Filologicheskogo instituta Leningradskogo Gos. Universiteta* 2 (1950): 64–96; A. Meynieux, "Puškin et la conclusion du *Second Faust*," *Revue des études slaves* XLVII (1967): 48–108; M. P. Alekseev, "K 'Stsene iz Fausta' Pushkina" (On Pushkin's "Scene from Faust"), *Vremennik Pushkinskoi komissii 1976* (Leningrad,1979), reprinted in M. P. Alekseev, *Pushkin i mirovaia literatura* (Pushkin and World Literature) (Leningrad, 1987), pp. 469–88.
3. *The Garland Companion to Vladimir Nabokov*, ed. V. E. Alexandrov (New York and London, 1995), index.
4. Osip Mandel'shtam, *Kamen'* (Stone), eds. L.Ia. Ginzburg, A.G. Mets, S.V. Vasilenko, and Iu.L. Freidin (Leningrad, 1990), p. 207.
5. Nabokov may have slightly altered here the surname of colonel Sergei Segerkranz, husband of his cousin Tatiana, born Rausch von Traubenberg. See biographical information on him in E. M. Rausch-Gernet's article on Nabokov's kinsmen, the Rausch von Traubenbergs and the Korffs, in *Nabokovskii vestnik* 2 (1998): 91.

6. Vladimir Nabokov, *Lectures on Russian Literature,* ed. Fredson Bowers (New York and London, 1981), p. 101; Gennadi Barabtarlo, "Within and Without Cincinnatus's Cell: Reference Gauges in Nabokov's *Invitation to a Beheading,*" *Slavic Review* 49.3 (1990): 395 n. 12.
7. The motif of Gretchen's child, in its relation to the Erlking, is collated in Nabokov's overall thematic structure with the pervasive implications of Pushkin's "Rusalka" (see: Aleksandr Dolinin, "Zagadka nedopisannogo romana" (The Riddle of an Unfinished Novel), *Zvezda* 12 (1997): 215–24.
8. Gennadi Barabtarlo, Note, *The Vladimir Nabokov Research Newsletter* 9 (1982): 34–35; id., "Within and Without Cincinnatus's Cell," *Slavic Review* 49.3 (1990): 395–6; cf. Vladimir Nabokov, *Rasskazy. Priglashenie na kazn'. Èsse, interv'iu, retsenzii,* ed. A. A. Dolinin and R. D. Timenchik (Moscow, 1989), pp. 509–10.
9. Michael Wachtel, *Russian Symbolism and Literary Tradition: Goethe, Novalis, and the Poetics of Vyacheslav Ivanov* (Madison, 1994), pp. 170–71.
10. *Rul'*, 11 November 1921.
11. E. Besson, *Les logia agrapha* (Rouen, 1923).
12. N. Mikhailovskii, *Literaturnaia kritika* (Literary Criticism), ed. B. Averin (Leningrad, 1989), p. 158.
13. Ludwig Lewisohn, *Goethe: The Story of a Man, Being the Life of Johann Wolfgang Goethe as Told in His Own Words and the Words of His Contemporaries,* vol. 2 (New York, 1949), pp. 222–23.
14. V. M. Zhirmunskii, *Gete v russkoi literature* (Goethe in Russian Literature) (Leningrad, 1937); 2nd ed. (Leningrad, 1981), p. 446.
15. Vladimir Nabokov, *Lolita,* translated from the English by the author, introduction and commentaries by A. Dolinin (Moscow, 1991), p. 373.
16. Gennadi Barabtarlo, *Phantom of Fact: A Guide to Nabokov's* Pnin (Ann Arbor, 1989), pp. 191–92, 215.
17. Ibid., p. 191.
18. Ludwig Lewisohn, ibid., p. 174.

Chapter 17

Pasternak and the Russian Reception of Heine

Karen Evans-Romaine

Boris Pasternak was an ardent and attentive reader of Heinrich Heine. His reception of Heine's work has been explored by a number of scholars. Among these studies, only Iu. M. Lotman's analysis of Pasternak's poem "Zamestitel'nitsa" (The Substitute) examines Pasternak's use of subtexts not from Heine's original verse, but from Heine in Russian translation.[1] Heine was one of the most popular and thus one of the most frequently translated foreign poets published in Russia. There is a vast body of Russian translations of Heine dating back to Tiutchev's first published Heine translation in 1827.[2] By 1870 Russian interest in Heine was so great that he had been elevated and, to some degree, "fossilized" as a classic, an inveterate part not only of German literary history but of the Russian one as well.[3]

Although Pasternak's knowledge of German was excellent and he read German literature primarily in the original, his reception of Heine could not have remained unaffected by the almost century-long history of Russian translations. Indeed, at the beginning of the twentieth century, three multivolume editions of Heine's collected works were published in Russia, with translations representing a variety of interpretations across the aesthetic and political spectrum.[4] This phenomenal boom in Russian publications of Heine's collected works—an endeavor which came to completion a few years after the Heine centenary, perhaps because of the confusion about the year of Heine's birth[5]—reflected a general revival of interest in Heine after a period of relative decline in both the quantity and quality of translations and critical literature on Heine during the last three decades of the nineteenth century. After the virtual sanctification of Heine as a revolutionary hero in the 1860s and the subsequent loss of interest in the enigmatic writer when this revolutionary wave in literature passed, young Russian poets at the turn of the century returned to Heine, this time for his nonpolitical verse and shattering irony. This Heine revival did not occur among modernist contemporaries in Germany, where, according to Heine biographer Jeffrey Sammons, "Heine had been anthologized and liedered and imitated into cliché." He writes that among turn-of-the-century German poets, Heine "was seen as a blight upon poetry because he had created forms that could be easily imitated by anyone who could count four beats and make a rhyme."[6] Perhaps a similar perception of Heine as a cliché of Russian literature—adored equally by the proponents of art for art's sake, the political writers, and indeed the mass reader, all of whom found in Heine texts to support their varied views and interests—led Russian symbolists Annenskii and Blok to examine the German poet through a new lens.[7] I will argue that the post-symbolist Pasternak rejected the new image of Heine developed by his symbolist contemporaries, along with their Heine translations, and turned instead to the translations of their nineteenth-century predecessors. Although Pasternak also interpreted Heine in a new and original way, taking Heine's work, as I have written elsewhere, as a model for the construction of his own literary biography,[8] Pasternak's use of Heine in his own poetry harkens back to the classical "canon" of Heine translations. He quotes in his verse some of the great translations of A. K. Tolstoi, Minaev, Fet, and the greatest of nineteenth-century Heine translators (though himself only a minor poet), Mikhail Larionovich Mikhailov.

In 1909, at about the time when Pasternak began writing, Innokentii Annenskii's essay "Heine Chained" appeared as part of his *Book of Reflections.* Annenskii's interpretation of Heine reflects a profoundly negative symbolist world-view.[9] As in his translations, Annenskii emphasizes in this essay the dark side of Heine's irony. His Heine is sickly and world-weary. In contrast to the more common interest in the early lyrics of Heine's *Buch der Lieder* among proponents of "pure art," Annenskii chooses instead to study the later, darker, more complex verses of *Romanzero.* Of this collection Annenskii writes the following:

> But if you look more intently into this brilliant succession of phantoms, you will see behind it only a sick man's eyes, widened in horror and anguish.

> I love Heine's "Histories," because they once carried away my illusion of the poet as sorcerer and taught me to divine behind his brightest, most festive robes a helpless and pathetic nakedness.[10]

Aleksandr Blok also emphasizes the dark side of Heine's irony.[11] In his 1919 essay "Heine in Russia," Blok criticizes the standard image of Heine as a martyr for the liberal cause:

> The artistic image of Heine, which, it seems, we were ready to adopt, is fading. In its place is appearing a cumbersome, hundred-*pood,* liberal legend of Heine that is taking on hideous forms that are absolutely disgraceful for an artist: Heine is practically turning into a populist who died because he was honorable. This legend lives up to today ...[12]

In contrast to this "legend," Blok proposes that Heine is uniquely appropriate for contemporary readers because of his "antihumanism."[13] Of this he writes:

> Now Heine has become closer than ever to the world ... finally, perhaps, the true voice of Heine has now been heard among that turbulent sea that is the European world, where humanistic civilization is crumbling to pieces. Heine is in his very essence an *anti-humanist* ... and thus always persecuted, always misunderstood, always adaptable.... We can already see clearly the *crisis of humanism.* The world is being cleansed, throwing off the clothing of humanistic civilization.[14] (Blok's emphasis)

For Pasternak, Heine served as an entirely different model, closer to the one reflected in Viktor Zhirmunskii's 1914 essay, "Heine and Romanticism." In that article, Zhirmunskii writes of Heine's interest in this world, as opposed to a distant world of dreams. There is in Zhirmunskii's argument an implicit comparison to the mood of crisis in the symbolist movement at the time:

> Heine is a characteristic representative of late romanticism. In him the process of self-destruction of romantic feeling comes to its conclusion....
>
> We saw in Heine in Germany a representative of the dualistic worldview of late romantics: the faraway, unearthly ideal, more like an illusion, is juxtaposed with coarse earthly reality, so unlike the ideal.... And now that faraway ideal, that vague, unnecessary illusion no longer demands his faith. Already in the era of the *Reisebilder* we note the birth of this new feeling: a kind of striving toward real life ...[15]

The translations by Annenskii and Blok, some of which were published in 1904 and 1911 respectively, bring Heine into the twentieth century by throwing off both the politicized and the sentimental traditions of Heine translation from the nineteenth century.[16] Their translations, if anything, intensify Heine's bitter irony and reflect a profoundly pessimistic world-view. Given the contrast between Pasternak's image of Heine and the one expressed by his symbolist contemporaries, it is not surprising that Pasternak would overlook their Heine translations. While they wish to distance themselves from the overly sentimentalized image of Heine promoted by their predecessors, Pasternak embraces precisely this highly romantic tradition.

A good example of this is Pasternak's poem "Marburg." As I have noted elsewhere, this poem weaves together a number of subtexts from Heine's poetry, particularly that of "Die Heimkehr."[17] Parallel to the Heine originals, a classic Russian translation of one of these poems also serves as a subtext to "Marburg": Heine's poem from "Die Heimkehr," "Nun ist es Zeit …" (Now it is time …). The choice of this Heine poem is logical: both "Marburg" and "Nun ist es Zeit" are ironic self-portrayals of the poet suffering over a failed love affair. One of the best-known Russian translations of this poem was done by A. K. Tolstoi in 1868. This translation appeared in all three of the collected editions of Heine mentioned earlier, as well as in Sasha Chernyi's 1911 edition of the *Buch der Lieder*. It has appeared in Russian anthologies of Heine to this day.[18] Rather than borrowing Heine's meter, which would be rendered in *dol'niki*—an innovation in the translation of Heine introduced by Blok[19]—Pasternak uses a meter that is closer to Tolstoi's amphibrachic meter, with the exception of a complex, polymetric opening to the 1915 version and a few lines of *dol'niki* in both the 1915 and the 1928 versions of "Marburg."[20] He alters Tolstoi's meter, however, by adhering to amphibrachic tetrameter, rather than alternating lines of amphibrachic tetrameter and trimeter, as Tolstoi did; Tolstoi's alternation in line lengths reflects Heine's use of the same alternation in his poem.

The first word of Tolstoi's translation provides a clue to its treatment in Pasternak's poem. Tolstoi's translation opens:

> Dovol'no! Pora mne zabyt' etot vzdor,
> Pora mne vernut'sia k rassudku!
> Dovol'no s toboi, kak iskusnyi akter,
> Ia dramu razygryval v shutku![21]
>
> Enough! It is time for me to forget this nonsense,
> Time for me to return to reason!
> Enough of playing out with you, like a polished actor,
> This drama, as a farce.

Pasternak echoes the opening of Tolstoi's translation, with its repetition of the exclamation "Dovol'no!" in his line "Dostatochno, tiagostno solntse mne dnem" (Enough, burdensome is the daytime sun for me) in the 1915 version of "Marburg."[22] He uses Tolstoi's word "rassudok" not in the 1915 version of "Marburg," but rather in the revised version of 1928:

> Rassudok? No on—kak luna dlia lunatika.
> My v druzhbe, no ia ne ego sosud (1: 108).

> Reason? But it's like the moon for a sleepwalker.
> We are friends, but I am not its vessel.

Pasternak modifies Tolstoi's phrase "iskusnyi akter" to "tragik" in the following stanza:

> V tot den' vsiu tebia, ot grebenok do nog,
> Kak tragik v provintsii dramu Shekspirovu,
> Nosil ia s soboiu i znal nazubok,
> Shatalsia po gorodu i repetiroval (1: 107, 492).

> That day, all of you, from your combs to your feet,
> Like a provincial tragedian a Shakespeare drama,
> I carried with me and knew by heart,
> I wandered about the city and rehearsed.

The last line of the Pasternak stanza, with the verb "repetiroval," echoes the word "deklamiroval" in the second stanza of Tolstoi's translation:

> Raspisany byli kulisy pestro,
> Ia tak deklamiroval strastno,
> I mantii blesk, i na shliape pero,
> I chuvstva—vse bylo prekrasno.[23]

> The wings were decorated garishly,
> I recited so passionately,
> And the sheen of the cloak, and a feather in my hat,
> And emotions—everything was wonderful.

The penultimate stanza of "Marburg" turns to the image of a chess game with the nights:

> Ved' nochi igrat' sadiatsia v shakhmaty
> So mnoi na lunnom parketnom polu.[24]

> After all, the nights sit down to play chess
> With me on the moonlit parquet floor.

This image serves as a kind of pun on Heine's (and Tolstoi's) ironic portrayal of the lyric hero as an actor playing out a melodrama. Tolstoi's translation closes with the following lines:

> O Bozhe! Ia, ranennyi nasmert', igral,
> Glad'iatora smert' predstavliaia![25]
>
> O Lord! Fatally wounded, I played,
> Showing the death of a gladiator!

The game, also appearing in the Heine original, is a crucial image that Pasternak uses for gently ironic polemic with Heine and Tolstoi. He abandons Heine's vision of a fatally wounded fencer (or the even more dramatic gladiator in Tolstoi) for the much milder depiction of a game of chess, a game which, clearly, the poet will survive. Thus, at the conclusion to his poem about survival and artistic rebirth, Pasternak lightens the bitter irony of the Heine original and Tolstoi's translation.

Pasternak's use of a subtext from Russian translations of Heine in his early poem "Pirshestva" (Feasts), revised as "Piry" in 1928, is intriguing in that, in this case, the Heine original does not significantly influence Pasternak. The subtext in this case is Dmitrii Minaev's translation of Heine's poem "Rote Pantoffeln," from among Heine's last verse of 1853 and 1854. Minaev's translation of this poem, entitled "Krasnye tufli," appeared in all three of the collected editions of Heine from the beginning of this century.[26] Ritz notes in Minaev's translations a tendency to remain loyal to content over form.[27] In this case Pasternak takes the meter not of the Heine original, but of the Minaev translation: iambic hexameter with a caesura after the third foot. Pasternak does make one slight modification to Minaev's meter, however: in Minaev's translation, a feminine ending precedes the caesura, while in Pasternak's poem, the ending before the caesura is consistently masculine. This pairing of text and subtext seems puzzling at first. Heine's poem is a satire with an anti-materialistic moral. Pasternak's poem explores the metaphor of poetic inspiration as a random feast. This theme ties the poem to many of Pasternak's other, quite dark, poems on the subject of inspiration, including "Mel'nitsy"[28] and "Ne kak liudi ..." (1: 74). The image of Cinderella, however, serves as a clue to the humorous subtext in "Pirshestva." The last six lines of Pasternak's poem in its 1913 version read:

> I krokhi iastv nochnykh skital'cheskii anapest
> Nautro podberet, kak kroshka Sandril'on.

I Zolushki shagi, eë samoupravstvo
Ne narushaiut grafstva chopornogo sna,
Pokuda v khrustaliakh neubrannye iastva
Vo grudy tuberoz ne prevratit ona (1: 439).

And the crumbs of the nighttime feasts wandering anapest
Picks up in the morning, like little Cinderella.

And Cinderella's steps, her willful ways
Do not disturb the county of prim sleep,
Until she transforms the remains of the feast in crystal dishes
Into bunches of tuberose.

The 1928 revision introduces explicitly the image of poetry as a mouse gathering crumbs. The tone changes at this point in the poem from darkly serious to sardonically humorous:

Nasledstvennost' i smert'—zastol'tsy nashikh trapez.
I tikhoiu zarei—verkhi derev goriat—
V sukharnitse, kak mysh', kopaetsia anapest,
I Zolushka, spesha, meniaet svoi nariad.

Poly podmeteny, na skaterti—ni kroshki,
Kak detskii potselyi, spokoino dyshit stikh,
I Zolushka bezhit—vo dni udach na drozhkakh,
A sdan poslednii grosh,—i na svoikh dvoikh (1: 58).

Heredity and death are the guests at our tables.
And at peaceful dawn—the tops of the trees burn—
In the biscuit bowl anapest rummages, like a mouse,
And Cinderella, in haste, changes her apparel.

The floors are swept, not a crumb on the tablecloth,
Like a child's kiss, verse breathes peacefully,
And Cinderella hurries—on lucky days in a droshky,
But if the last penny is spent, then on her own two feet.

The paired images of Cinderella and the mouse, as well as the meter, link the Pasternak poem to Heine's humorous satire. The heroine of Heine's poem is a mouse who goes to a cat to buy new shoes, in a scene reminiscent of the glass slipper in the fairy tale. From Minaev's translation of the Heine poem:

Odna mysh' molodaia, sem'i dovol'no znatnoi,
Guliala pered lavkoi s trevogoiu priiatnoi,

Vpered i vzad khodila, stoiala u okna
I, nakonets, reshilas' k bashmachnitse ona
S voprosom obratit'sia: "Poslushaite, nel'zia li,
Chtob mne vy paru tufel' puntsovykh pokazali,
I esli by my s vami v tsene za nikh soshlis',
To ia b u vas kupila ikh, gospozha Kis-Kis."

Ei koshka otvechala: "Akh, baryshnia, voidite,
I dom moi oschastlivit' vy soblagovolite!
Proshu, proshu pokorno. Mogu uverit' vas,
Ne brezgaiut so mnoiu znakomit'sia podchas
Krasiveishie damy, grafini; bez zhemanstva
Ko mne zakhodit chasto vse vysshee dvorianstvo;
A tufel'ki prodat' nedorogo mogu.
Vy tol'ko ikh primer'te, a ia vam pomogu.
Voidite zh i prisiad'te; pozvol'te vashu nozhku."[29]

A young mouse, from a rather distinguished family,
Was walking in front of a shoeshop with pleasant anxiety,
Going back and forth, standing by the window,
And, finally, she decided to go in and see the shop-mistress,
To ask her a question: "Tell me,
Could you please show me that pair of crimson shoes,
And if we agreed on the price,
Then I would buy them, Mrs. Kitty-Kitty."

The cat replied: "Come in, Mademoiselle,
And be so kind as to honor my shop!
Please, I humbly beseech you. I can assure you,
Sometimes the loveliest ladies and countesses
Deign to make my acquaintance; without any affectation
The highest nobility often comes to my shop;
And these shoes will not cost you too dearly:
Just try them on, and I'll help you.
Come in and have a seat; do give me your pretty little foot."

The mouse's fate after this is clear. Heine's satire of bourgeois vanity and materialism adds an interesting layer to Pasternak's gentle self-irony on the fate of the poet. Pasternak, drawing on Heine in Minaev's translation, hints at inspiration as a kind of material wealth that can vary in its value. This is clearest particularly in the 1928 version of the poem, with the banality of the line "I Zolushka, spesha, meniaet svoi nariad" (And Cinderella, in haste, changes her apparel) and the poem's closing lines, comparing the days when Cinderella has the money for a carriage and the days she does not.

A similar "measuring" of inspiration in material terms occurs in Pasternak's 1915 poem "Mel'nitsy," in which poetic inspiration falls like coins of copper, and only occasionally, sharply, like silver: "I padaiut kapli mediashkami v kruzhki/ I rezko, i izredka lish'—serebrom...."[30] In both poems, not only does Pasternak liken the poet to a kind of Cinderella or poor relation, but he also creates a self-deprecating portrait of the poet who must pick up crumbs or beg for coins and does not always receive the best for his efforts.[31]

The link between Pasternak's poem and Minaev's translation, and the underlying theme of wealth and materialism, which the two share to varying degrees, is further reinforced by Pasternak's striking use of the word "grafstvo" in the 1913 "Pirshestva" in the line "Ne narushaiut grafstva chopornogo sna" (Do not disturb the county of prim sleep).[32] This word hints at Minaev's use of the word "grafini" in his mocking portrait of materialism and societal vanity in his translation of the Heine poem, in the lines, "Krasiveishie damy, grafini; bez zhemanstva/ Ko mne zakhodit chasto vse vysshee dvorianstvo ..." (Sometimes the loveliest ladies and countesses/ ... without any affectation/ The highest nobility often comes to my shop ...). In the Heine original, the noble patrons of the shop in question are not countesses, but duchesses, "Herzoginnen":

Geruhen Sie mein Haus zu beehren
Mit Dero Gegenwart; es verkehren
Mit mir die allerschönsten Madel
Und Herzoginnen, der höchste Adel.[33]

The lexical and semantic link based on the word "graf" also ties Minaev's translation indirectly, through "Pirshestva," to Pasternak's 1915 poem "Ballada" (revised in 1928), in which inspiration is represented variously, from the image of the count to the metaphor of rains and coins, as in "Mel'nitsy."[34]

Thus through metric, lexical, and semantic links with Minaev's translation of Heine's ironic later poem, Pasternak heightens the irony in his own self-portrait of poetic mediocrity. His mordant self-portrayal in general owes a great deal to Heine; here, however, the more important subtext is not the Heine original, but Minaev's equally sardonic translation.

Pasternak also borrows subtexts from the best nineteenth-century translator of Heine into Russian, Mikhailov. Even Blok, in his relatively harsh criticism of many nineteenth-century translators of Heine, praises Mikhailov's work.[35] Pasternak's quotations from Mikhailov's renderings of Heine tend to be relatively brief and subtle, generally limited to isolated lines and phrases. His selection of excerpts from Mikhailov's translations reflects his general tendency to quote the romantic aspect of Heine translations. As Blok pointed out in his essay on Heine in Russia, Mikhailov's translations, though "pearls of poetry" in themselves, are not Heine. Blok writes that in Mikhailov "there was too much of what we call 'romanticism.'"[36] It is this "romanticism" that Pasternak conveys, despite all his outward statements against this concept.[37]

One example of a subtle quotation from Mikhailov occurs together with a quotation from Tiutchev quoting Heine. In this case, the link between Tiutchev and Pasternak through Heine arises indirectly, through Tynianov's series of essays published in his 1929 collection *Archaists and Innovators*. This series of essays contains both Tynianov's study "Tiutchev and Heine" and his study of contemporary poets, including Pasternak, entitled "The Interval."[38] In "Tiutchev and Heine," Tynianov points out that Tiutchev quotes Heine's poem "Es treibt dich fort von Ort zu Ort …" (You are driven from place to place …) in his poem which begins with the lines:

> Iz kraia v krai, iz grada v grad
> Sud'ba, kak vikhr', liudei metet.[39]
>
> From land to land, from city to city
> Fate, like a whirlwind, tosses people.

A poem from Pasternak's 1931 cycle "Volny," "Shli dni, shli tuchi …," takes the rhythm and construction of Tiutchev's Heine-inspired poem. The relevant lines of Pasternak's poem read:

> Za godom god, za rodom plemia,
> K goram vo mgle, k goram pod stat'
> Goriankam za chadroi v gareme,
> Za rodom rod, za piad'iu piad'.[40]
>
> Year after year, tribe after clan,
> Toward the mountains in haze, toward the mountains
> That are like women in highlands under a veil, in a harem,
> Generation after generation, plat after plat.

Thus Pasternak quotes Heine, through Tiutchev, through Tynianov. In his essay "The Interval," Tynianov implicitly compares Pasternak to Heine, writing that both poets were able to free themselves from the bonds of their own set themes, bonds in which, according to Tynianov, Pasternak's contemporaries suffered.[41] Tynianov calls the bonds from which Heine freed himself the "Heine manner," a phrase which Pasternak echoes in *Safe Conduct*, in discussing his struggle to shed the "romantic manner."[42] Thus Tynianov connects Pasternak and Heine within the theme of artistic rebirth, in an essay dedicated to Pasternak and published two years before "Volny" was written. Pasternak, in turn, quotes poems from Tynianov's study.

Another linked subtext from Heine and Tiutchev emerges from Mikhailov's translation of Heine's poem "Die Jahre kommen und gehen …" (The years come and go …), also from "Die Heimkehr." Here too, as with the Minaev translation, although Pasternak clearly read "Die Heimkehr" very carefully, the Heine original seems not to have influenced Pasternak's work. The first stanza of Mikhailov's translation reads:

> Idet za plemem plemia,
> Idet za godom god;
> A vse liubov' bylaia
> Iz serdtsa von neidet.[43]
>
> Tribe goes after tribe,
> Year goes after year;
> But my past love
> Will not leave my heart.

If in making use of the Heine-Tiutchev subtext, Pasternak referred to his own wandering in the Caucasus as a kind of metaphor for his self-searching at the time of writing "Volny," in this case Pasternak uses the Heine/Mikhailov subtext ironically, in reference to his troubled first marriage, while setting the poem in the Caucasus, where he traveled with his future second wife. While Heine writes of a former love that does not fade, Pasternak includes this reference to the Heine poem in the context of a complex verse cycle about his affair and the pangs of conscience which accompany it.[44]

Pasternak quotes from another Mikhailov translation of Heine in the opening poem from the cycle "Volny." The last six lines of Pasternak's poem read:

> Ko mne begut moi postupki,
> Ispytannogo grebeshki.
>
> Ikh t'ma, im net chisla i smety,
> Ikh smysl dosel' eshche ne poln,
> No vse ikh smenoiu odeto,
> Kak pen'e moria penoi voln.[45]

My deeds are running toward me,
Crests of my ordeals.

They are so many, without count or estimate,
Their meaning is not yet fulfilled,
But everything is clothed in their changing,
Like the singing of the sea in the foam of waves.

The final line of this Pasternak poem draws a phrase from the opening line of Mikhailov's translation of Heine's poem "Wie die Wellenschaumgeborene ...," "Kak iz peny voln rozhdennaia ..." (Like the one born of wave foam ...). As in the previous subtext to "Volny," Pasternak takes the phrase from a subtext about love and uses it in a context that heightens the irony of his own poem. The role of the subtext is different here, however. If the previous Heine/Mikhailov subtext emphasized the theme of eternal love, the subject of this Heine poem is the love and betrayal of an engaged woman and the lyric hero's suffering. In Pasternak's interpretation of this poem, the theme of the woman's betrayal is transformed to reflect the lyric hero's thoughts on his own betrayal. Both subtexts sharpen the themes of conscience and inner conflict in Pasternak's verse cycle.

As is typical of Mikhailov, the translation of this Heine poem is much gentler in tone than Heine's original. Heine's poem, from *Lyrisches Intermezzo*, reads:

Wie die Wellenschaumgeborene
Strahlt mein Lieb im Schönheitsglanz,
Denn sie ist das auserkorene
Bräutchen eines fremden Manns.

Herz, mein Herz, du vielgeduldiges,
Grolle nicht ob dem Verrath;
Trag es, trag es, und entschuldig' es,
Was die holde Thörin that.[46]

Like the one born of wave foam
Gleams my beloved in the sparkle of beauty,
For she is the chosen
Bride of another man.

Heart, my heart, you long-suffering,
Do not be angry about the betrayal;
Bear it, bear it, and forgive it,
What the dear little fool did.

Mikhailov's translation reads:

> Kak iz peny voln rozhdennaia,
> I prekrasna i pyshna,
> Za drugogo obruchennaia,
> Dyshit prelest'iu ona.
>
> Serdtse mnogoterpelivoe!
> Ne ropshchi i ne grusti,
> I bezumstvo toroplivoe
> Bednoi zhenshchine prosti.[47]
>
> Like the one born of wave foam,
> Beautiful and magnificent,
> Betrothed to another,
> She breathes charm.
>
> Long-suffering heart!
> Do not grumble and do not be sad,
> And forgive the poor woman
> For her hasty mindlessness.

Other characteristics of both Heine's poem and Mikhailov's translation appear elsewhere in Pasternak's book of verse *Second Birth*, which begins with the cycle "Volny." The phrasing and intonation of Heine's second stanza and lines 5 and 6 from Mikhailov's translation inspire the opening of another Pasternak poem from this book. The first stanza of the poem reads:

> Ne volnuisia, ne plach', ne trudi
> Sil issiakshikh i serdtsa ne muchai,
> Ty zhiva, ty vo mne, ty v grudi,
> Kak opora, kak drug i kak sluchai.[48]
>
> Don't worry, don't cry, don't wear out
> Your strength, and don't torment your heart,
> You're alive, you're inside me, you're in my breast,
> Like a support, a friend, a chance.

Although the meter of Pasternak's poem differs from Heine's and Mikhailov's, the repetition (Ne …, ne …; Ty …, ty …; Kak …, kak …) echoes similar repetition in lines 5 and 7 of Heine's poem and line 6 of Mikhailov's translation. As in the subtext from Heine/Mikhailov discussed earlier, "Idet za plemem plemia …," the meaning conveyed in

Pasternak's poem is the opposite of Heine's meaning. Pasternak's lyric hero is comforting the object of the poem, in light of his own deceit; the poet in Heine's poem, characteristically, is comforting himself after his beloved's betrayal. Like the previous subtext, this one heightens the poignancy of Pasternak's struggles with his conscience, a theme that pervades many of the poems throughout *Second Birth*.[49]

Thus Mikhailov's translations left many traces in the verse of Pasternak's *Second Birth*. Given Heine's role in Pasternak's literary autobiography, it is particularly significant that Heine's poems would inspire so profound a response in Pasternak's verse at a critical turning point in his private life, reflected prominently in his poems of this period.[50] In the case of Heine's "Wie die Wellenschaumgeborene ...," both the Heine original and Mikhailov's translation are significant to Pasternak, in both form and content.

In another case, Pasternak's treatment of a Mikhailov translation demonstrates his adoption of Mikhailov's sentimental tone. The Heine original in this instance is certainly not without its share of sentimentality, but Mikhailov heightens this quality of the Heine poem. The poem is "Mir träumte einst von wildem Liebesglühn ..." (I once dreamt of the wild glow of love ...), the first poem from Heine's first verse collection, *Junge Leiden*. Mikhailov's translation, entitled "Mne snilis' strastnye vostorgi i stradan'ia ..." (I dreamt of passionate ecstasies and sufferings ...), appeared with tremendous frequency in Heine anthologies, including the three Heine collections published at the beginning of this century.[51] The relevant stanza from Mikhailov's translation (the last stanza of the poem) reads:

Osirotelaia! umchis' i ty skoree!
Leti, o pesn' moia, vosled moikh videnii!
Naidi moi luchshii son, po svetu ptitsei reia,
I moi vozdushnyi vzdokh otdai vozdushnoi teni![52]

Orphaned one! you too fly away quickly!
Fly, oh my song, after my visions!
Find my best dream, soaring about the world like a bird,
And my airy sigh give to the airy shade!

Pasternak uses this translation as a subtext to his poem "Opredelenie dushi" (Definition of the Soul). Just as Mikhailov in the Heine translation, Pasternak compares his poem to a wandering bird and tells it to fly away. Significantly, there is no bird in the Heine original. The relevant portion of Pasternak's poem reads:

> Nashu rodinu buria sozhgla.
> Uznaesh' li gnezdo svoe, ptenchik?
> O moi list, ty puglivei shchegla!
> Chto ty b'esh'sia, o shelk moi zastenchivyi?
>
> O ne boisia, prirosshaia pesn'![53]
>
> Our motherland has been burned by a storm.
> Do you recognize your nest, fledgling?
> Oh my leaf, you are more timid than a goldfinch!
> Why are you thrashing about, my shy silk?
>
> Oh fear not, song grown to the tree!

Although he does not use the word "orphan," as does Mikhailov, Pasternak implies it in the line, "Uznaesh' li gnezdo svoe, ptenchik?" (Do you recognize your nest, fledgling?). Here Pasternak both imitates the intonation of Mikhailov's translation and borrows imagery that is not in the Heine original. As we see in all of these examples, Mikhailov's translations clearly proved to be rich subtextual ground for Pasternak. Individual words and phrases in them inspired a variety of responses in Pasternak's verse.

Perhaps the best known nineteenth-century translator of Heine who was also considered one of the great poets in his own right is Afanasii Fet. The question of Pasternak's reception of Fet's translations of Heine is complicated by the broader issue of the resemblance between the poetics of Pasternak and Fet, noted by many scholars and writers since Tynianov and Mandel'shtam noted Pasternak's apparent kinship with Fet in the 1920s.[54] There are many points of correspondence between Pasternak's verse and Fet's Heine translations in intonation, phrasing, theme, and motifs. Many of these, however, are merely indicative of the general resemblances in the poetics of Pasternak and Fet. I will limit myself here to points at which one can trace more precise links in metrics, lexicon, sound play, or motifs between Pasternak and the Fet translations.[55]

The theme of music provoking nostalgic recollection is in general a common one and constitutes a recurring motif in Heine.[56] Pasternak's poem "Godami kogda-nibud' v zale kontsertnoi …" (Some years hence in a concert hall …) is linked by this theme not so much to Heine's poem from *Lyrisches Intermezzo*, "Hör' ich das Liedchen klingen …" (I hear the song ringing …),[57] as to Fet's translation of this poem. The Heine poem is written in *dol'niks*, while Fet's and Pasternak's poems are in three-beat meters—Fet's translation in dactyllic trimeter, Pasternak's in amphibrachic tetrameter. Both the Pasternak and the Fet texts feature sound play on the vowel *a* throughout. The key word that ties them, "vyplachu" or "vyplachus'," points the way to other lexical resemblances. The relevant portion of Pasternak's poem reads:

> Godami kogda-nibud' v zale kontsertnoi,
> Mne Bramsa sygraiut,—toskoi izoidu.
> ....
> I srazu zhe budu slezami uvlazhen
> I vymoknu ran'she, chem vyplachus' ia.
> Goriuchaia davnost' udarit iz skvazhin,
> Okolitsy, litsa, druz'ia i sem'ia.[58]
>
> Some years hence in a concert hall,
> They'll play Brahms for me,—and I'll be overcome with anguish.
> ....
> And I'll at once be covered with tears
> And will be soaked through before I have sobbed myself out.
> The combustible past will spring from the wells,
> Fences, faces, friends, and family.

And Fet's translation:

> Slyshu li pesenki zvuki,
> Toi, chto pevala ona,—
> Snova tomitel'noi muki
> Grud', kak byvalo, polna.
>
> Tak i potianet nevol'no
> V gory da k temnym lesam,—
> Vse, chto i gor'ko, i bol'no,
> Mozhet byt', vyplachu tam.[59]
>
> If I hear the strains of the song,
> The one she used to sing,
> My breast fills again with agonizing torment,
> As it used to.
>
> Then, against my will, I'll be pulled
> To the mountains and toward dark forests,
> Everything bitter and painful,
> Perhaps, I'll sob out there.

Other lexical and sound similarities include "toskoi" (Pasternak)/ "tomitel'noi muki" (Fet), "goriuchaia" (Pasternak)/ "gor'ko" and "gory" (Fet). Pasternak's word "goriuchaia" links his poem to the Mikhailov translation of this Heine poem, entitled "Tol'ko do slukha kosnetsia …" (As soon as I hear the sound …),[60] written in the same meter as Fet's; however, there are no other significant links between the two texts. Thus the Heine poem and Fet's translation are tied to Pasternak's "Godami kogda-nibud' v zale kontsertnoi …" through the common themes of music, recollection, nostalgia, and resulting tears. The Fet translation, particularly its lexical and sound features, had a stronger impact on Pasternak in this case than the Heine original.

Another example of Pasternak's response to a Fet translation bears a strong resemblance to a similar subtextual relationship with an original Fet poem: the parallel link is between Pasternak's "Zerkalo" (Mirror) and Fet's poem "Zerkalo v zerkalo …" (A mirror in a mirror …), analyzed by Anna Ljunggren.[61] The poems in this case are Pasternak's "Dushistoiu vetkoiu mashuchii," from *My Sister Life*, like "Zerkalo," and Fet's translation of Heine's poem "Lehn' deine Wang' an meine Wang' …,"[62] entitled "Lanitoi k lanite.…" The Fet poem and translation open in a parallel construction ("Zerkalo v zerkalo," "Lanitoi k lanite"). As in both "Mirror" and "The Substitute," Pasternak distances himself from the romantic subtext through futurist techniques of fragmentation and metonymy.[63] Fet's translation reads:

Lanitoi k lanite moei prikosnis',
Togda nashi slezy sol'iutsia,
I serdtsem tesnee mne k serdtsu prizhmis',—
Ognem oni obshchim zazhgutsia.

I esli v tot plamen' prol'iutsia rekoi
Te obshchie slezy muchen'ia,—
Ia, krepko tebia okhvativshii rukoi,
Umru ot toski naslazhden'ia.[64]

Touch my cheek with your cheek,
Then our tears will flow together,
And press your heart closer to my heart,
They will ignite and burn together.

And if those common tears of torment
Flow like a river into that flame,
I, tightly wrapping my arm around you,
Will die from the pangs of delight.

Pasternak's response to this text follows:

Dushistoiu vetkoiu mashuchi,
Vpivaia vpot'makh eto blago,

Bezhala na chashechku s chashechki
  Grozoi odurennaia vlaga.

Na chashechku s chashechki skatyvaias',
  Skol'znula po dvum,—i v obeikh
Ogromnoiu kaplei agatovoiu
  Povisla, sverkaet, robeet.

Pust' veter, po tavolge veiushchii,
  Tu kapel'ku muchit i pliushchit.
Tsela, ne drobitsia,—ikh dve eshche
  Tseluiushchikhsia i p'iushchikh.

Smeiutsia i vyrvat'sia siliatsia
  I vypriamit'sia, kak prezhde,
Da kaple iz rylets ne vylit'sia,
  I ne razluchatsia, khot' rezh'te.[65]

Swaying a fragrant branch,
  Drinking in that blessing in the dark,
From calyx to calyx ran
  A raindrop, stunned by the storm.

Rolling from calyx to calyx,
  It slipped along the two, and in both,
Like an enormous agate drop,
  It hung, sparkling yet shy.

Let the wind, blowing along the meadow-sweet,
  Torment and flatten that droplet.
It is whole, it won't splatter—there are still two of them,
  Kissing and imbibing.

They laugh and try to pull apart
  And straighten themselves out, as before.
But the drop can't spill out of the stigmas,
  And they won't separate, even if you cut them in two.

The underlying trope connecting Pasternak's and Heine's poems is the romantic metaphor of love as two tears fused into one. The secondary metaphor of flame appears directly in Heine and Fet, and indirectly in Pasternak, as lightning. Thunderstorms are among the most important leitmotifs of Pasternak's *My Sister Life*.[66] The lightning in this Pasternak poem is indicated metonymically, in the verb "sverkaet," which appears in the second stanza after the thunderstorm has been mentioned in the first. The Pasternak and Fet texts also have in common an amphibrachic meter—amphibrachic trimeter in Pasternak and alternating lines of amphibrachic tetrameter and trimeter in Fet. Heine's poem is written in *dol'niks*. Both the Pasternak and the Fet texts contain frequent repetitions of the consonants *s* and *sh*; however, Pasternak's sound orchestration is much richer and more elaborate than Fet's. Fet's construction "Lanitoi k lanite" is reflected in Pasternak's phrase "Na chashechku s chashechki," appearing in the first stanza and repeated in the second. Pasternak hints at Fet's closing line, a translation of Heine's line "Sterb' ich vor Liebessehnen!" in his closing remark on eternal love, "khot' rezh'te."[67] Both the Heine poem and the Fet translation as subtexts underline the eroticism clearly implied in Pasternak's poem.[68]

We see in Pasternak's reception of these translations two different modes of interpretation. In the former, Pasternak keeps to Fet's sentimental tone and uses fewer characteristically futurist devices to subvert the romantic message of the Heine poem and Fet's translation. This greater faithfulness to Heine's and Fet's nostalgic mood is typical of the overall tone of *Second Birth*. In the latter case, we see a distancing from the romantic tone of the Heine text and Fet's translation through the elimination of the lyric hero and his beloved, except through their apparent metonymic projection onto the drops of water in the rain-drenched trees. This fragmentation and distancing from the nineteenth-century antecedent text is typical of Pasternak in *My Sister Life*, as Lotman, Nilsson, and Ljunggren have shown.[69]

I have tried to point out that Pasternak draws from the rich tradition of Russian translations of Heine, in some cases as least as much as from Heine originals. He ignores many aspects of this tradition: the vast body of translations of Heine's political verse, the equally extensive repertoire of mediocrities completed by second-rate translators, and, in contrast, the new approach to Heine taken by the symbolist predecessors he so admired. Rather, Pasternak quotes the meter, imagery, intonation, and lexicon of some of the great Heine translators of the nineteenth century, whose work one could call classics of the Russian Heine tradition. In his choice of Heine translations, Pasternak shows his true nature: even while using futurist techniques to distance himself from the original texts and their romantic-era translations, he proves, once again, his allegiance to his nineteenth-century romantic, including late romantic, forebears.

Notes

I am grateful to Gennady Barabtarlo for his suggestions on my translations. The translation of "Dushistoiu vetkoiu mashuchi ..." was done with reference to Katherine Tiernan O'Connor, *Boris Pasternak's* My Sister—Life: *The Illusion of Narrative* (Ann Arbor, 1988), pp. 62–63 and Vladimir Markov and Merrill Sparks, *Modern Russian Poetry: An Anthology with Verse Translations* (Indianapolis, 1967), pp. 592–95.

1. Iu. M. Lotman, "Analiz dvukh stikhotvorenii," *Tret'ia letniaia shkola po vtorichnym modeliruiushchim sistemam. Tezisy. Doklady* (Tartu, 1968), pp. 191–224. A. K. Zholkovskii does, however, mention Lermontov's free translation of Heine's "Ein Fichtenbaum steht einsam ..." (Na severe dikom stoit odinoko ...) as a subtext to Pasternak's poem "Veter": see A. K. Zholkovskii, "Liubovnaia lodka, upriazh' dlia Pegasa i pokhoronnaia kolybel'naia (Tri stikhotvoreniia i tri perioda Pasternaka)," in Zholkovskii and Iu. K. Shcheglov, *Mir avtora i struktura teksta. Stat'i o russkoi literature* (Tenafly, 1986), p. 252. On the role of Russian translations of Heine in establishing a metric tradition continued by Pasternak, see Valentin Belentschikow, *Zur Poetik Boris Pasternaks: Der Berliner Gedichtzyklus 1922–1923* (Berlin, 1998), p. 76. On Pasternak's reception of Heine's verse see, for example, Erika Freiberger Sheikholeslami, "Der deutsche Einfluss im Werke von Boris Pasternak" (Ph.D. diss., University of Pennsylvania, 1973), pp. 130–55, summarized in Erika Freiberger, "Pasternak und die deutsche Literatur," in *Boris Pasternak und Deutschland*, ed. Sergej Dorzweiler, Bernhard Lauer, and Katharina Linke (Kassel, 1992), pp. 42–43; I. P. Smirnov, *Porozhdenie interteksta (Elementy intertekstual'nogo analiza s primerami iz tvorchestva B. L. Pasternaka)* (Vienna, 1985); M. L. Gasparov, "Semantika metra u rannego Pasternaka," *Izvestiia Akademii Nauk SSSR. Seriia literatury i iazyka*, vol. 47, no. 2 (1988), pp. 142–47; O'Connor, *Boris Pasternak's* My Sister—Life, pp. 100–101, 203; V. S. Baevskii, *Boris Pasternak—lirik. Osnovy poeticheskoi sistemy* (Smolensk, 1993), pp. 100–106; Karen Evans-Romaine, *Boris Pasternak and the Tradition of German Romanticism* (Munich, 1997), pp. 229–80.
2. German Ritz, *150 Jahre russische Heine-Übersetzung* (Bern, Frankfurt-am-Main, Las Vegas, 1981).
3. On the history of these translations in Russia and the Soviet Union, see Ritz, *150 Jahre*; Ia. I. Gordon, *Geine v Rossii (1830–1860e gody)* (Dushanbe, 1973); idem, *Geine v Rossii (1870–1917)* (Dushanbe, 1979); idem, *Geine v Rossii. XX vek* (Dushanbe, 1983); Leslie Bouton Peterson, "Heinrich Heine and Aleksandr Blok: Two Episodes in the Development of the Modern Lyric Cycle" (Ph.D. diss., Cornell University, 1988), pp. 309–40. See also references to Heine in Paul Friedrich, *Music in Russian Poetry* (New York, 1998).
4. G. Geine. *Sobranie sochinenii*, ed. P. I. Veinberg, 8 vols. (St. Petersburg, 1898–1902); *Polnoe sobranie sochinenii. V perevode russkikh pisatelei*, ed. P. V. Bykov, 12 vols. (St. Petersburg and Moscow, 1900); *Polnoe sobranie sochinenii*, Izdanie vtoroe, ed. and intro. Veinberg, 6 vols. (St. Petersburg, 1904). On Heine editions in Russia, see Ia. M. Metallov, ed., *Genrikh Geine. Bibliografiia russkikh perevodov i kriticheskoi literatury na russkom iazyke*, compiled by A. G. Levinton (Moscow, 1958).

5. Omry Ronen discusses the confusion related to Heine's birth in the context of the Heine centenary celebrations in Russia in his unpublished paper "Some Poems of 1897," presented at the American Association for the Advancement of Slavic Studies meeting, November 1997. On the history of the complications regarding Heine's year of birth, see Jeffrey L. Sammons, *Heinrich Heine: A Modern Biography* (Princeton, 1979), pp. 11–15.
6. Sammons, *Heinrich Heine*, p. 350.
7. Clara Hollosi, "Views on Heine in Russia in the Beginning of the 20th Century," *Heine-Jahrbuch* 17 (1978), pp. 175–85; Peterson, *Heinrich Heine and Aleksandr Blok*, pp. 21–29.
8. Karen Evans-Romaine, "Zametki o biologicheskom i avtobiograficheskom u Pasternaka," *Sed'mye Tynianovskie chteniia. Materialy dlia obsuzhdeniia* (Riga and Moscow, 1995–96), pp. 151–60; idem, *Boris Pasternak*, pp. 229–80.
9. Hollosi, "Views on Heine in Russia," pp. 181–82.
10. Innokentii Annenskii, "Geine prikovannyi: Geine i ego 'Romantsero,'" in *Izbrannoe* (Moscow, 1987), pp. 366–67.
11. Peterson, *Heinrich Heine and Aleksandr Blok*, pp. 381–97, 407–11.
12. Aleksandr Blok, *Sobranie sochinenii*, ed. V. N. Orlov et al. 8 vols. (Moscow and Leningrad, 1960–63), vol. 6, p. 119.
13. Hollosi, "Views on Heine in Russia," pp. 182–83.
14. Blok, *Sobranie sochinenii*, vol. 6, p. 125.
15. Viktor Zhirmunskii, "Geine i romantizm," *Russkaia mysl'* 5 (1914), section 2, pp. 96, 107.
16. Four of Annenskii's five translations of Heine poems were published with *Tikhie pesni* in 1904 (Nik. T-O. *Tikhie pesni. S prilozheniem sbornika stikhotvornykh perevodov 'Parnasstsy i prokliatye'* [St. Petersburg, 1904]). The fifth, "Ich grolle nicht," was published in 1923 as part of the collection *Posmertnye stikhi Innokentiia Annenskogo* (St. Petersburg). Translations are reprinted in Innokentii Annenskii, *Stikhotvoreniia i tragedii* (Leningrad, 1990), pp. 226–28. Blok translated thirteen poems from "Die Heimkehr" in 1909; twelve of these translations appeared in 1911 with his collection *Nochnye chasy*. He translated eleven other poems in 1920–21, while he was editing Heine's work; they were published posthumously in collections of Heine translations in 1930 and 1931. Blok's translations of Heine poems are reprinted in Blok, *Sobranie sochinenii*, vol. 3, pp. 378–89, commentary pp. 638–39. On Annenskii's and Blok's translations, see Ritz, *150 Jahre*, pp. 315–21, 325–38.
17. Evans-Romaine, "Zametki o biologicheskom," pp. 156–57; idem, *Boris Pasternak*, pp. 262–69. On "subtext," see Omry Ronen, *An Approach to Mandel'shtam* (Jerusalem, 1983); Smirnov, *Porozhdenie interteksta*.
18. For a complete list of the translation's appearances up to 1958, see Metallov, ed., *Genrikh Geine. Bibliografiia*, p. 230, item 2143.
19. On Blok's innovation in rendering Heine's *dol'nik* into Russian, see Ritz, *150 Jahre*, p. 329. He cites a number of studies of meter in Heine and Blok, including James Bailey, "Blok and Heine: An Episode from the History of Russian Dol'niki," *SEEJ*, vol. 13 (1969), pp. 1–22; Robin Kemball, *Alexander Blok: A Study in Rhythm and Metre* (The Hague and Paris, 1965), pp. 255–59;

E. Landa, "Blok i perevody iz Geine," in *Masterstvo perevoda* (Moscow, 1963), pp. 292–328.

20. Olga Raevsky Hughes, "Stikhotvorenie 'Marburg' i tema 'Vtorogo rozhdeniia,'" in Michel Aucouturier, ed., *Boris Pasternak 1890–1960: Colloque de Cérisy-la-Salle* (Paris, 1979), pp. 289–301; Lazar Fleishman, *Boris Pasternak v dvadtsatye gody* (Munich, 1981), p. 107.
21. Genrikh Geine (Heinrich Heine), *Izbrannye sochineniia* (Moscow, 1989), p. 68.
22. Boris Pasternak, *Sobranie sochinenii v piati tomakh* (Moscow, 1989–92), vol. 1, p. 492. All Pasternak texts are cited from this edition, with an indication of volume and page number.
23. Heine, *Izbrannye sochineniia*, p. 68.
24. Pasternak, *Sobranie sochinenii*, vol. 1, pp. 108, 493.
25. Heine, *Izbrannye sochineniia*, p. 68.
26. Metallov, *Genrikh Geine. Bibliografiia*, p. 460, item 4580. In Soviet editions of Heine, Minaev's translation was replaced by Tynianov's.
27. Ritz, *150 Jahre*, p. 286.
28. Pasternak, *Sobranie sochinenii*, vol. 1, pp. 98, 465. On this theme, see also Evans-Romaine, *Boris Pasternak*, pp. 95, 125.
29. Genrikh Geine (Heinrich Heine), *Polnoe sobranie sochinenii*, Izdanie vtoroe, ed. P. I. Veinberg, vol. 6, pp. 140–41.
30. Pasternak, *Sobranie sochinenii*, vol. 1, p. 465.
31. On the theme of inspiration as rain or coins, particularly in regard to "Mel'nitsy" and with reference to "Ballada," see Evans-Romaine, *Boris Pasternak*, pp. 80, 93, 124–25.
32. I am grateful to Omry and Irena Ronen for pointing this out to me when I read this article as a paper at the annual meeting of the American Association of Teachers of Slavic and East European Languages, December 1997.
33. Heinrich Heine, *Historisch-kritische Gesamtausgabe der Werke*, ed. Manfred Windfuhr (Hamburg, 1973–), vol. 3, p. 188.
34. On "Ballada," see Fleishman, *Boris Pasternak v dvadtsatye gody*, pp. 80, 102–6; Per-Arne Bodin, "The Count and His Lackey: An Analysis of Boris Pasternak's Poem 'Ballada,'" in *Studia Filologiczne*, Zeszyt 31/12. Filologia Rosyjska. *Poetica Pasternaka* (Bydgoszcz, 1990), pp. 37–63. On the etymology of the word "graf" and its role in "Pirshestva," see M. L. Gasparov, "'Bliznets v tuchakh' i 'Nachal'naia pora' B. Pasternaka: ot kompozitsii sbornika k kompozitsii tsikla," *Izvestiia Akademii Nauk SSSR. Seriia literatury i iazyka*, vol. 49, no. 3 (1990), pp. 218–22.
35. Blok, *Sobranie sochinenii*, vol. 6, pp. 119, 127.
36. Ibid., p. 119.
37. The literature on Pasternak's definition of and relationship to romanticism is vast. Among the significant contributions to the discussion of this problematic question are Iurii Tynianov, "Promezhutok," in *Arkhaisty i novatory* (Ann Arbor, 1985), pp. 541–80; Victor Terras, "Boris Pasternak and Romantic Aesthetics," *Papers on Language and Literature*, vol. 3, no. 1 (Winter 1967), pp. 42–56; Guy de Mallac, "Pasternak's Critical-Esthetic Views," *Russian Literature Triquarterly* 6 (1973), pp. 502–32; idem, *Boris Pasternak: His Life and Art* (Norman, 1981), pp. 339–56; Johanna Renate Döring, *Die Lyrik Pasternaks in*

den Jahren 1928–1934 (Munich, 1973); Sheikholeslami, "Der deutsche Einfluss," summarized in Freiberger, "Pasternak und die deutsche Literatur," pp. 39–47; Olga Raevsky Hughes, *The Lyric World of Boris Pasternak* (Princeton, 1974); the papers and discussions on romanticism in Aucouturier, ed., *Boris Pasternak 1890–1960*, pp. 55–116; V. S. Baevskii, "Lirika Pasternaka v istoriko-kul'turnom kontekste," *Izvestiia Akademii Nauk SSSR. Seriia literatury i iazyka*, vol. 47, no. 3 (1988), pp. 130–41; idem, *Boris Pasternak—lirik*; Erika Greber, *Intertextualität und Interpretierbarkeit des Texts. Zur frühen Prosa Boris Pasternaks* (Munich, 1989); Lazar Fleishman, *Boris Pasternak: The Poet and His Politics* (Cambridge, MA, 1990), pp. 77–79; Boris Paramonov, "Pasternak protiv romantizma. K ponimaniu problemy," in *Norwich Symposia on Russian Literature and Culture*, vol. 1. *Boris Pasternak 1890–1990*, ed. Lev Loseff (Northfield, 1991), pp. 11–25. For an overview of these and many other works on Pasternak's relationship to romanticism, particularly German romanticism, see Evans-Romaine, *Boris Pasternak*, pp. 1–21.

38. Iurii Tynianov, *Arkhaisty i novatory*, pp. 386–98, 541–80.
39. Tynianov, ibid., 392.
40. Pasternak, *Sobranie sochinenii*, vol. 1, p. 377.
41. Tynianov, *Arkhaisty i novatory*, pp. 550–64.
42. Ibid., p. 550; Pasternak, *Sobranie sochinenii*, vol. 4, p. 227.
43. M. L. Mikhailov, *Sochineniia*, ed. B. P. Koz'min, vol. 1 (Moscow, 1958), p. 301.
44. E. B. Pasternak mentions Pasternak's indebtedness to Heine, particularly to his *Reisebilder*, in writing his own "travel pictures" of the Caucasus in "Volny": see E. B. Pasternak, *Boris Pasternak: Biografiia* (Moscow, 1997), pp. 474–75. On the biographical background to "Volny" and to the book of verse *Vtoroe rozhdenie* in general, see Henry Gifford, *Pasternak: A Critical Study* (Cambridge, 1977), pp. 136–46; E. B. Pasternak, *Boris Pasternak: Biografiia*, pp. 434–91, 498; Christopher Barnes, *Boris Pasternak: A Literary Biography, Vol. 2: 1928–1960* (Cambridge, 1998), pp. 42–60.
45. Pasternak, *Sobranie sochinenii*, vol. 1, p. 374.
46. Heine, *Historisch-kritische Gesamtausgabe*, vol. 1, p. 149.
47. Mikhailov, *Sochineniia*, vol. 1, pp. 297–98.
48. Pasternak, *Sobranie sochinenii*, vol. 1, p. 394.
49. Yet another Pasternak link in this subtextual chain is his poem from *Vtoroe rozhdenie* "Liubit' inykh—tiazhelyi krest ..." (Pasternak, *Sobranie sochinenii*, vol. 1, p. 396), which is connected to Mikhailov's translation by lexical and sound features.
50. On another Heine subtext to "Volny," without reference to translations, see Evans-Romaine, *Boris Pasternak*, pp. 253–57.
51. Mettalov, ed., *Genrikh Geine. Bibliografiia*, pp. 112–13.
52. Mikhailov, *Sochineniia*, vol. 1, p. 258.
53. Pasternak, *Sobranie sochinenii*, vol. 1, p. 135.
54. See, for example, Tynianov, *Arkhaisty i novatory*, pp. 562–68; Osip Mandel'shtam, *Sochineniia v dvukh tomakh* (Moscow, 1990), vol. 2, pp. 207–10; Terras, "Boris Pasternak and Romantic Aesthetics"; Nils Åke Nilsson, "'With Oars at Rest' and the Poetic Tradition," in *Boris Pasternak: Essays*, ed. Nils Åke Nilsson (Stockholm, 1976), pp. 180–202; Kirill Taranovsky, *Essays on Mandel'shtam* (Cambridge, MA,

1976), pp. 21–47; idem, "On the Poetics of Boris Pasternak," *Russian Literature* 9 (1981), p. 350; idem, "Two Notes on Mandel'shtam's 'Hayloft' Poems," in *Text and Context: Essays to Honor Nils Åke Nilsson*, ed. Peter Alberg Jensen et al. (Stockholm, 1987), pp. 122–24; Lazar Fleishman, *Stat'i o Pasternake* (Bremen, 1977), p. 107; Smirnov, *Porozhdenie interteksta*, pp. 122–25; Peter Alberg Jensen, "Boris Pasternak's 'Opredelenie poezii,'" in *Text and Context*, ed. Jensen et al., pp. 96–110; M. L. Gasparov, "Semantika metra"; V. S. Baevskii, "Lirika Pasternaka"; idem, *Boris Pasternak—lirik*; Anna Ljunggren, "'Sam' i 'Ia sam': smysl i kompozitsiia stikhotvoreniia 'Zerkalo,'" in *Boris Pasternak and His Times: Selected Papers from the Second International Symposium on Pasternak*, ed. Lazar Fleishman (Berkeley, 1989), pp. 224–37; V. V. Musatov, "K probleme genezisa liriki Borisa Pasternaka," *Izvestiia Akademii Nauk SSSR. Seriia literatury i iazyka*, vol. 49, no. 5 (1990), pp. 403–13; I. Iu. Podgaetskaia, "Pasternak i Verlen," in *Pasternak-Studien: Beiträge zum Internationalen Pasternak-Kongreß in Marburg 1991*, ed. Sergej Dorzweiler and Hans-Bernd Harder (Munich, 1993), pp. 107–22, reprinted in *De Visu* 1, 1993, pp. 47–56; Konstantin Polivanov, "'Khor' Pasternaka. Opyt kommentariia," *Literaturnoe obozrenie* 4 (1996), pp. 50–55.

55. The texts under examination here were published in the two-volume A. A. Fet, *Stikhotvoreniia* (Moscow, 1863). See commentary in A. A. Fet, *Stikhotvoreniia i poemy*, ed. I. Ia. Bukhshtab (Leningrad, 1986), pp. 629–33, 707–9.
56. On another Heine subtext on music and nostalgic reflection to this Pasternak poem, see Evans-Romaine, "Zametki o biologicheskom," pp. 155–56; idem, *Boris Pasternak*, pp. 258–60. On this poem, see Krystyna Pomorska, "Music as Theme and Structure," in *Themes and Variations in Pasternak's Poetics* (Lisse, 1975), pp. 29–50; on this theme, see also Boris Kats, "'... Muzykoi khlynuv s dugi bytiia.' Zametki k teme 'Boris Pasternak i muzyka,'" *Literaturnoe obozrenie* 2 (1990), pp. 79–84.
57. Heine, *Historisch-kritische Gesamtausgabe*, vol. 1, pp. 172–73.
58. Pasternak, *Sobranie sochinenii*, vol. 1, p. 392.
59. Fet, *Stikhotvoreniia i poemy*, p. 568.
60. Mikhailov, *Sochineniia*, vol. 1, p. 279.
61. Ljunggren, "'Sam' i 'Ia sam.'"
62. Heine, *Historisch-kritische Gesamtausgabe*, vol. 1, p. 139.
63. Nilsson, "'With Oars at Rest,'" analyzes similar futurist techniques with romantic subtexts. In fact, there is a Fet translation of Heine which could be added to Nilsson's list of poems in the tradition that he so thoroughly explores: "Moi drug, my s toboiu sideli" (Fet, *Stikhotvoreniia i poemy*, pp. 568–69), a translation of Heine's "Mein Liebchen, wir saßen beisammen" (Heine, *Historisch-kritische Gesamtausgabe*, vol. 1, pp. 173, 175).
64. Fet, *Stikhotvoreniia i poemy*, p. 566.
65. Pasternak, *Sobranie sochinenii*, vol. 1, p. 128.
66. The most obvious examples of this leitmotif are in Pasternak's "Nasha groza" (*Sobranie sochinenii*, vol. 1, pp. 138–39) and "Groza, momental'naia navek" (ibid., vol. 1, p. 165). The most famous essay on this theme is Marina Tsvetaeva, "Svetovoi liven'," *Epopeia* 3 (1922), reprinted in idem, *Sobranie sochinenii v semi tomakh*, ed. Anna Saakiants and Lev Mnukhin (Moscow, 1994–95), vol. 5, pp. 231–45.

67. The Mikhailov translation of this Heine poem, "Shchekoi k shcheke ty moei prikosnis' ...," is similar in meter and construction to the Fet translation. In contrast to Pasternak's reception of Mikhailov's lofty and somewhat melodramatic tone in "Mne snilis' strastnye vostorgi i stradan'ia," in this case Pasternak apparently ignores the Mikhailov variant, as emotional in tone as the other Mikhailov translation, for the somewhat more distant and reserved Fet variant. Examples of Mikhailov's greater emotionality in his translation include his third-person imperatives "Puskai nashi slezy sol'iutsia!" and "Odnim ognem pust' zazhgutsia!" and his emphatic repetition of "umru": "... Somknu u tebia za spinoi i umru,/ Umru ot blazhenstva i muki" (Mikhailov, *Sochineniia,* vol. 1, p. 276).
68. On this aspect of the Pasternak poem, see Dale Plank, "Readings of 'My Sister Life,'" *Russian Literature Triquarterly* 2 (1972), p. 335; see also O'Connor, *Boris Pasternak's* My Sister—Life, pp. 62–63. Plank analyzes sound orchestration in Pasternak's verse, including some of the poems discussed here, in his study, *Pasternak's Lyric: A Study of Sound and Imagery* (The Hague and Paris, 1966).
69. Lotman, "Analiz dvukh stikhotvorenii"; Nilsson, "'With Oars at Rest'"; Ljunggren, "'Sam' i 'Ia sam.'"

Chapter 18

Kafka's "The Hunger Artist" and Shalamov's "The Artist of the Spade"

The Discourse of Lent

Leona Toker

> It is sad to have to arrive at an assessment of art as cause so different from our assessment of art as effect: how tremendous it is as cause, how paralyzed and hollow as effect.
>
> —Nietzsche, "Schopenhauer as Educator"

Varlam Shalamov's story "The Artist of the Spade"[1] shares the theme of starvation and a number of minor motifs with Franz Kafka's "The Hunger Artist." The case for a conscious allusion or a reminiscence cannot, however, be proved. It is not clear whether Shalamov was familiar with "The Hunger Artist" in 1964, when he wrote "The Artist of the Spade." The similarity of the titles is not sufficient evidence: "the artist of the spade" may have been a camp idiom—indeed, it is used by three different characters of Shalamov's story. On the other hand, it is not impossible that Shalamov might have read "The Hunger Artist": the January 1964 issue of the Soviet journal *Inostrannaia literatura* (Foreign Literature) published Russian translations of Kafka's "Metamorphosis" and "In the Penal Colony"; other Kafka stories in Russian translation circulated in the Samizdat.[2]

The year 1964 was the peak year of Khrushchev's "thaw" in Soviet literary life. Solzhenitsyn's "One Day in the Life of Ivan Denisovich" had already come out, flanked by a few works whose partial critique of the recent Soviet history loyally toed the Party line. By contrast, Shalamov's camp stories were stuck in editorial offices. Their way to the press was blocked: despite their neglect of ideological issues, they were perceived as far more radically subversive than even Solzhenitsyn's narratives. In 1966 the ban on their publication would be practically finalized. One might venture a guess that Kafka's "The Hunger Artist" was not included in the first journal selection for a similar reason: its motifs and symbols were potentially more disconcerting for a Soviet reader than those of "In the Penal Colony." If, for instance, the henchman-turned-victim ending of the latter story could strike one as spectacularly prophetic of the already encapsulated history of the Nazi regime in Germany (the fact that great numbers of Stalin's victims had themselves perpetrated atrocities in his service was still largely downplayed), "The Hunger Artist," with its story of a secular fast turning into lethal starvation, could also trail in the shadows of what had not yet been publicly disclosed about Soviet experience—the near-genocidal forced famine in the Ukraine in 1932–33 and the multimillion death toll due to starvation in Stalin's camps.[3] The politics of monitoring historical consciousness may thus have created a historical alignment between Kafka's Hunger Artist and the starving artists of the spade and wheelbarrow in Shalamov's stories.

Whether or not "Der Hungerkünstler" may have been a biographical pre-text for "The Artist of the Spade," the overlap in the repertoire of the two works can be accounted for by the common semiotic system, such as the discourse of Lent, and the common exposure, direct or mediated, to the ideas of Arthur Schopenhauer and Friedrich Nietzsche in whose work the semiotic system in question finds its termini.

The discourse of Lent is the photonegative of the Bakhtinian carnivalesque. Much as numerous turn-of-the-century Russian intellectuals before him, Bakhtin was to a large extent influenced by the works of Nietzsche.[4] The Dionysian spirit discussed in Nietzsche's *The Birth of Tragedy*—the syndrome of vertiginous transport, of intoxication, of blurring borderlines between personalities and their environment, of vitality that transcends itself in mystical moves, of excess in body language, and of dissolution of the individual in the mass—obviously presides over Bakhtin's carnivalesque, especially in its Rabelaisian version. The system of the noncarnivalesque values and the significances that it pits itself against is associated with the Apollonian dream of discrete shape, individual grandeur, and hermetic finish, as well as with cultural sophistication, social hierarchy, and confident restraint.

Recently, scholars have been critical of Bakhtin's valorization of the carnivalesque. For instance, Michael André Bernstein (1992) has pointed to the abject and the bitter sides of the carnival,[5] and Sergei S. Averintsev to the ambivalence of laughter in the Russian cultural tradition.[6] One of the most important and, I believe, still underrated, contributions to the revisionary Bakhtiniana is Ruth Ginsburg's discussion of Lent as the second self of the carnival. In her essay "Karneval und Fasten" Ginsburg argues that Lent, with its body language of deficiency rather than excess, is not the opposite but the second-self of carnival; like carnival, it is the opposite of the quotidian realities of orderly, hierarchical social life.[7]

Ginsburg's main literary example is Kafka's "The Hunger Artist," in which the person who used to stage widely advertised feats of forty-day fasting loses his hold on the public's interest and eventually starves himself to death in an unmonitored and unlimited fast. Canonical works of fiction contain rather few similar examples of rendering the body language of Lent, and this raises the question whether it is possible to draw meaningful extrapolations from Kafka's extraordinary story. I believe, however, that there exists a whole corpus of narratives that can be characterized as the literature of Lent—namely, the testimonies of concentration camp experience, in particular the narratives of the veterans of the Gulag, including stories by Shalamov, such as "At Night," "The Individual Assignment," "Sententia," "Rain," and, among numerous others, "The Artist of the Spade." Shalamov's work is particularly interesting in this context because this highly self-conscious artist regarded himself as an heir of the Russian modernist writers of the Silver Age (the contemporaries of Kafka, who died in 1924). This positioning of his aesthetic principles implied both a revaluation of the classical, Apollonian tradition and the bracketing away of the intervening developments on the Soviet literary scene.

The type of experience whose representation calls for the discourse of Lent is more closely associated with the Gulag than with the Nazi camps. It is believed that, soon after the Ribbentrop-Molotov pact of 1939, a delegation of Nazi officials visited the Soviet Union in order to study its labor camps—with the view, it seems, to improving the patent;[8] the reciprocal education of the Gestapo and the Soviet penal authorities may have continued for at least another year.[9] In 1942, after the Wannsee conference, the innovations in the Nazi use of camp facilities made a qualitative leap; indeed, a comparison between the Gulag and the Lagers is impeded by the issue of the *Vernichtungslager,* whose function was to effect a speedy rather than a slow extermination of the victims.[10] In terms of the routine of labor camps (that is, slow-death camps as distinct from extermination factories), the direction of the Nazi "improvements" on the Soviet system was towards a much more streamlined and demonstrative exercise of unlimited power over the victims. Accordingly, literary accounts of the Nazi camps testify to the grim carnival in which the SS and the "prominents" feasted on the ordinary prisoners' bodies and souls. The incidence of unrestrained sadistic reveling, open or thinly disguised, was, judging by the testimony of Charlotte Delbo, Arne Brun Lie, David Rousset, Elie Wiesel,[11] and others, much higher in the Lagers than in the Soviet camps; the "prescribed hunger"[12] was not the main scourge of the prisoners' existence in the former. These seem to be the main reasons why the discourse of the Lent is more characteristic of Gulag literature than of the literary reflections of the Nazi camps—with the exception, perhaps, of the works of Primo Levi.

The action of Shalamov's "The Artist of the Spade" takes place in a slow-death camp, a Kolyma gold mine, during one of the worst periods in Gulag history, when the supply of convict labor was ample and, in the eyes of the authorities, individual human lives did not count. In the story, the work team led by young Kostochkin, an intellectual from Kharbin, periodically melts away—obviously, its members, hunger artists despite themselves, die of exhaustion and hunger-related diseases. The protagonist, Krist, is sent to this work team as a reinforcement. Physically weak at the start of the story, he goes on losing his strength, even though, in recognition of his virtuoso handling of the spade, Kostochkin assigns him somewhat higher rations. Then, one day, Kostochkin selects Krist as the recipient of a monetary reward. However, it turns out that all the money is going to Kostochkin himself. Drunk on the alcohol purchased with Krist's prize, Kostochkin displays his boxing skills on Krist, sending him into a knockout.

The same night he himself is humbled by the visit of Minia the Greek, head of the team of professional criminals whose code does not allow them to do any work in the camps. In order to get provisions, the thieves' team has to show production results, and, in the total absence of such, Minia imposes taxation on the other teams. This time, like a month earlier, it is Kostochkin's turn. Kostochkin submits to the racket, even though the falsification of his own team's production reports will lead to a further reduction in his own hungry workers' diet.

Yet the story deals with fasting as much as with hunger. The period of Lent, variously defined at different times in different places, involves a voluntary fast for the sake of physical and spiritual purification. This is a period of temporary asceticism, when the Will, as Schopenhauer might have said, is silenced and subdued. Schopenhauer, indeed, is a philosophical authority on the subject of Lent, just as Nietzsche is the philosophical presence behind Bakhtin's carnivalesque. Elsewhere[13] I have argued that Schopenhauer and Shalamov provide useful indirect comments on each other's work.[14] Indeed, the descriptive part of Schopenhauer's exploration of asceticism is fairly applicable to Shalamov's works, though his prescriptive valorization of asceticism breaks down against them. A Schopenhauerian saint starves himself to death not because, like Kafka's Hunger Artist, he has never found the food that he would like to have, but because he (or, all too often, she) wishes to subdue the Will, to disperse the "illusion" of discrete individuality, to remove "the veil of Maya."[15]

Kafka, like Shalamov later, goes only part of the way with Schopenhauer. Through his feats of fasting his Hunger Artist seeks not to dissolve but to affirm his identity: rejecting physical nourishment, he makes a bid for public admiration.[16] Shalamov's characters do not starve voluntarily: they are *denied* sufficient food—in concentration camps hunger is one of the main instruments of discipline and torture. The reason why it makes sense to talk about the Lenten spirit of Gulag literature—both in terms of body language and in ethical terms—is the suggestion that this corpus repeatedly makes: *the answer to hunger is fasting.*

Indeed, fasting was one of the ways of maintaining personal dignity in the camps. Literal fasting took the shape of hunger strikes (held in the earliest and the latest periods in the history of the Gulag) or of Jewish observances of the Day of Atonement. A related but more pragmatic self-discipline could be observed in the attempts of veteran prisoners to avoid talking or thinking about food. Mainly, however, it was in a figurative sense that the Lenten abstention from carnivorousness was practiced by such prisoners as Shalamov's Krist: they refused to become cannibals—that is, to adopt the "you die today and I tomorrow" maxim and prolong their own lives at the expense of the lives of others, to take jobs that involved bending the will of others, to fawn to the powerful, to sell out. This is why the foreman's job, a carnivorous one of driving the undernourished to their death of exhaustion, is unthinkable for Krist, who does not insist even on a larger share of warmth from the barracks stove, for which his body craves.

In Shalamov's Gulag stories, Lent is a communal experience but not a comprehensive one: alongside the majority of the starved there is an aggressive minority (professional criminals like Minia the Greek, brutal team-leaders like Kostochkin, toadies like Os'ka) who seek power and food at all cost. Lent is acted out in the proximity of a corrupted semblance of carnival. The cage, the oxymoronically public *oubliette* in which Kafka's Hunger Artist is eventually left to die, belongs to a side-show of a circus, that fake carnival island amidst everyday burgher life. Instead of being the center of attention, as in the remote past, the Hunger Artist provides a peripheral extra attraction—or obstacle—on the spectators' way to the menagerie during intermissions. Raw meat its taken to the animals daily past his cage; when he dies, his cage is given to a panther, the predatory untamed cat. Shalamov's carnival is likewise insular: it is the lifestyle of professional criminals and of the camp "prominents."[17] Re-enacting a carnivalesque reversal, the middle-aged Os'ka, a paterfamilias history professor, acts as young Kostochkin's bodyguard and court jester; Kostochkin himself is known as "the Tamer," and meat, metaphorically speaking, is taken past the barracks of political prisoners into the menagerie of the professional predators, the so-called "honest thieves."

Similar to carnivalesque literature, Lenten narratives present pageants in which the borderlines between the actors and the spectators are erased.[18] The artist of the spade is both a spectator and a participant. He first hears of Kostochkin's team, then is sent to join it: the common lot of the doomed that he has heard about is becoming his own lot. He is first an actor, an artist observed by Kostochkin in the toolshed and the pit-face; then, from the dark "orchestra" of the sleeping-planks, he observes Kostochkin, spectator-turned-actor, in the illuminated part of the barrack. Eventually, Krist has to join Kostochkin in that footlighted arena and play the victim in a Roman-type martyr-in-a-lion's-cage show. And after Kostochkin himself joins the tacit chorus of the sleepers at the walls of the barrack, another powerful camp personality, the thug Minia, nicknamed (by chance or choice) "the Greek,"[19] appears on the stage in what looks like an ironic perversion of Nietzsche's account of Greek tragedy—a king-subduing demi-god, evoked, as it were, by Dionysian intoxication[20] (Kostochkin having consumed the alcohol that should have been distributed among the whole chorus). Effecting another carnivalesque reversal, Minia, a would-be magician, intimates that the Tamer may one day be blown to pieces or have his head sawed off:[21] the predator would thus turn into the prey, in the best tradition of the criminals' code of justice.

The Hunger Artist is observed by the public while he also observes and studies the spectators' behavior: on the two sides of the bars of the cage, both freedom and its lack are *relative* (it is not for nothing that Krist defines Kostochkin's education as "knowing who Einstein was"). The bars of the cages give two-way views. The responsibility for the hunger show lies with the public: it is the public that stages an experiment in survival, using the artist as a projection of its hopes and its most intimate fears. At the expiration of the forty days, the period chosen, according to the narrator, for expediency,[22] but also alluding to Christ's forty-day fast in the wilderness, the public watches as its representatives retrieve the hunger artist from his fast—his head lolling, his body hollowed out, his legs clinging to each other at the knees and barely scraping the ground. The smiling ladies who are delegated to lead him to the table[23] grow pale with disgust or else with the fear of the insidiousness of his touch; the one closest to him bursts into tears, "to the great delight of the spectators."[24]

The ritual of carnivalesque laughter partly distances the spectacle from the crowd that has actually financed—and thereby staged—the depletion and the revival of the Hunger Artist's body. Significantly, the Hunger Artist, the crowd's scout to the frontiers of the *Liebestod*,[25] is discarded at about the time when real death looms down from the proud towers of World War I, too large for the public to maintain its need for protracted shows of flirtation with death.[26] What attracts the observers at the end of the story, defining their new desires, is not the hunger artist's endurance but the robust vitality of the panther, the caged beast of prey.

In carnivalesque narratives, identities blend into a crowd when customary partitions, decorum, and self-restraint are voluntarily removed:[27] in the literature of Lent, the discreteness of personality is likewise lost—but through lack instead of excess. From Gulag narratives by Shalamov, Solzhenitsyn, Panin, Kopelev, Margolin, Ginzburg, Herling, and others[28] we learn of the inroads that the environment makes in the convict's famished body (through untreated bruises, scurvy sores, frostbite, osteomyelitis) and in his mind (through, among other things, the total absence of privacy). Body and mind, moreover, are causally related: the chronic starvation that depletes the limbs also contracts the brain, leading—at best—to a loss or a stunting of memory. Shalamov's heroes seek to maintain some sense of their unique moral identity. Yet moral qualities can be generalized, and so it often does not even matter whether separate stories by Shalamov are directly autobiographical or representative, fictionalized: what happens to one ethically staunch former intellectual might also happen, with minor variations, to another. It does not matter much whether the protagonist of separate stories is "Shalamov," or "Andreev," or the anonymous "I," or "Krist"—except that "Krist" (with the inevitable connotations of this name) is the author's avatar mainly in the stories that deal with the periods when the prisoner is sliding down the "vicious spiral"[29] toward exhaustion.

Most of the action in Kafka's story is likewise "iterative":[30] at the beginning, the descriptions of the fasting ritual seem to be representative of the art of show-fasting as it was generally practiced, and the term

"The Hunger Artist" is perceived as standing for any show-fasting performer, a stock character of the contemporary popular-culture side-show, with the Harlequins and Columbinas acting within the circus tent. Then the narrative passes from an account of the general practice to the repeated experience of one particular hunger artist whose name is withheld—if only because some time after one loses track of his fasting days, the placards bearing his name get dirty and are torn down. Yet the story contains one noniterative scene—that of the Hunger Artist's last words and death—as in Kafka's "Before the Law,"[31] the gate that closes in the end is the one that has been kept open for him alone. The sentence "Well, clear this out now!" which the overseer utters in lieu of a funeral oration, is one that could well be used in Shalamov's Gulag world.

In fact, the two Lenten narratives present not the regulated Lent but rather its corruption—the unlimited fast that does not lead to a moral or physical recovery. They also present the corruption of game, of that *ludus*, which, if we believe Huizinga and his followers, lies at the basis of most cultural traditions and religious observances. The anthropological game theory is, in fact, a complementary system of concepts that can be regarded as mediating between the extratextual repertoire of the stories and its textual rendering. The Bakhtinian discussion of the carnival and the game theory of Roger Caillois, whose usefulness for literary studies has been amply demonstrated by Wolfgang Iser,[32] intersect most significantly on the issue of mandatory time limits as well as on the issue of *vertigo*—what Caillois calls the games of *ilinx*.[33] Vertigo can be achieved through carnivalesque intoxication, but it is also one of the results of hunger and of the weakening of the body after prolonged starvation.

The game, like the period of Carnival and Lent, is circumscribed in time. Like an oxymoronic Carnival, it is a licensed subversion of the everyday order of things; it reinvigorates this order by forcing it to lose itself in order to find itself again.[34] The corruption of the game, however, means that loss does not lead to finding. Indeed, so long as the impressario is in charge of the Hunger Artist's forty-day fast and does not allow it to continue indefinitely, the controlled framework of the ludic experiment is maintained. Yet the framework is abandoned when the artist and the impressario part ways: no one updates the billboard stating what day of the fast has been achieved, no public resurrection of the "dry bones" is enacted, and the suffering that has started as an object of unceasing scrutiny ends in lethal neglect.[35]

In Shalamov's story, the corruption of ilinx, the game of vertigo, is given no emphasis, since the foreground is occupied by more immediately evident forms of moral and administrative corruption. At the same time, it is clear that no outer authority places a limit on the trial by hunger endured by Krist and his mates; hence the periodic disintegration of Kostochkin's brigade. As a corpus, Shalamov's stories do include the themes of resurrection and redemption,[36] yet these forms of revival are presented in terms of a more generalized communal rather than individual experience. Shalamov rejected the redemptive view of individual suffering: with Aristotle, he believed that sore trials (such as the experience of the labor camp) could only make one worse. Shalamov's characters are generally driven to an unheroic slow death, at the threshold of which they are consistently denied outward dignity. According to Schopenhauer, each person's life in its particulars is a comedy, but viewed as a whole, in retrospect, it is tragic.[37] It seems, however, that the tragedy of the characters of Lenten narratives is evident only to the narrator and the reader, not to the people around them. The existential predicament of the Hunger Artist lacks communally recognized tragic stature, and Shalamov's characters are blankly denied tragic grandeur in their suffering and death: their claims to attention are erased by communal unconcern, by statistics, and by the emotional depletion of the fellow *morituri* who would have been best qualified to respond to such claims.[38] Shalamov's narratives, with the single and not unproblematic exception of "Major Pugachev's Last Battle," consistently emphasize the unheroic quality of in-camp deaths.

Only occasionally a touch of the sublime is added to this routine martyrdom: like the Biblical Job, Shalamov's 'goners' (*dokhodiagi*) are sometimes shown to be vaguely comforted by being "at peace with the beasts of the field" and "in league with the stones,"[39] that is, by versions of a mystical insight into the life of things. Yet in "The Artist of the Spade," as in "Individual Assignment," "Handwriting," and "Quiet," instead of a mystical insight, the prisoners get belated instruction in the social semiotics of the system in which they had been entrapped: the topos of near-death enlightenment is (symptomatically of the general reduction of the human condition in the camps) recycled as the motif of insights in the mechanics of camp corruption. The discourse of Lent, like the discourse of carnival discussed by Bakhtin, remains eschatological even when it places the topos of existential insight in an ironic perspective.

Kafka's story is a philosophical fable whose structure of meaning involves an autodescriptive allegory. The protagonist is a *Künstler*[40] not because he produces an aesthetic object (his effect on the senses of the audience, whether sight or touch, is of a negative sort) but because he sustains a claim to the moral-intellectual beauty of *difficulty overcome.* And yet he feels that he is a fraud—because the difficulty for him lies not where the public believes it to be, that is, not in the fasting but in the breaking of the fast, not in the period of Lent but in the return to the quotidian. As to the public, its interest is, as so often, misplaced: it wants to make sure that he is not merely pretending to fast, that he is not eating something on the sly—hence the vigils by the Artist's cage, a parody, as it were, on the prolonging of attention that Shklovsky demands for reality-defamiliarizing works of art (indeed, the rhetorical principle behind the story is rather that of Spencer's "economy of effort," which also characterizes documentary prose like Shalamov's[41]). The "suspicion of fraudulence," which can be regarded as an integral part of the aesthetic response to modernist art [42]—and to documentary prose—concerns likewise not the outward aesthetic function of the work but the genuineness of the achievement, of the difficulty that has been overcome in its creation: Marcel Duchamp's framing of objects or a camp survivor's nonfictional account of his experience may be perceived as *too easy*, hence trivial. However, in the case of such works, the Keatsian "pinnacle and steep of godlike hardship" are largely a matter of the complexity of the cultural and biographical experience in their prehistory, as well as in the intensity with which this experience has fermented in the process of artistic shaping.

Shalamov's story is not a philosophical tale ("We were not philosophers. We were workers," he writes in "Sententia"[43]) but rather a bifunctional documentary narrative whose aesthetic and communicative aspects are marked at different periods of reception. As a work of testimony, it has its own urgencies. Nor can it have the self-sufficiency of Kafka's narrative: the metaphysical vision that is implicit in Shalamov's work can only emerge from the interrelations between a great number of his stories in combination with broader intertextual links. Intertextual seepage is, in general, mandatory rather than optional in the literature of testimony, if only because no internal evidence can vouch for a documentary text's historical veracity and only the buttressing presence of other works can confirm what Philippe Lejeune calls the factographic "pact"[44] between a documentary text and its reader. Shalamov's work, however, occupies a special place in the literature of Gulag testimony: the intertext that it claims is not limited to corroborating testimony from similar geographical and social settings. The reading of his stories leaves one with a sense of unfinished business, of "stumbling blocks"[45] that link his texts to precursor battle-grounds of images and ideas, enacting a Lenten-type partial merger with literary and philosophical hypotexts. At the same time, his matter-of-fact insistence on the actuality of the setting and on the testimonial function of the works holds the intertextual seepage in check, imposing a limit on the interpretive license of the reader. In the case of modernist literature, one hardly ever knows how far interpretation can go or which of the far-fetched linkages are, as Dr. Johnson once put it, "worth the carriage."[46] Hesitation of this type is even more acute in the reading of documentary prose: at issue here is not the expense of the carriage but the risk of leaving the load of the literal testimony too far behind.

It may, by contrast, be argued that Kafka's story can well be read without the activation of the historical and philosophical *hors-text* and that its intertextual readings create only surplus significances that affect interpretation rather than explication. Most readers, however, tend to look up from the last sentence of "The Hunger Artist" with a sense that their own intellectual response has only started. Indeterminacy of meaning characterizes practically all modernist literature and art; the *differentia specifica* of the discourse of Lent (if one can extrapolate from the two examples discussed) is that the openness to other systems of thinking and signification lends itself to being interpreted as a translation of the body language of lack, craving, and restraint into a hermeneutical procedure.

Notes

1. The original text of the story can be found in Varlam Shalamov, *Kolymskie rasskazy* [Kolyma Tales], ed. Mikhail Geller (London, 1978), pp. 306–19, or in later publications edited by I. Sirotinskaya, such as *Kolymskie rasskazy v dvukh tomakh* [Kolyma Tales in Two Volumes] (Moscow, 1992) I: 175–83. An English translation is presented as an Appendix to the present collection.
2. This observation, for which I am initially indebted to Omry Ronen, can be supported by, for instance, the references to "Golodar'," the Russian title of Kafka's "The Hunger Artist" in the memoirs of the dissident Anatoly Murzhenko, *Obraz schastlivogo cheloveka, ili pis'ma iz lagerya osobogo rezhima* [An Image of a Happy Man, or Letters from a Special-Regime Camp], ed. Michail Kheifetz (London, 1985), p. 118. Shalamov's acquaintance with Kafka is also suggested by his short story "Zhuk" [The Insect], in *Shalamovskii sbornink I*, ed. V. V. Esipov (Vologda, 1994), pp. 15–16. On the issue of the extent of the Soviet experimental writers' exposure to Kafka's works, see Edith Clowes, "Kafka and Russian Experimental Fiction in the Thaw, 1956–1965," *Modern Language Review* 89 (1994): 149–53.
3. Not many Soviet readers knew, however, that in the Soviet strict regime prisons and camps of the 1960s, hunger was again being used as a means of punishment and control (this was the near-ubiquitous theme of Marchenko's *My Testimony,*

already circulating in the Samizdat). In the camps, the "Thaw" had ended several years earlier than on the cultural scene.

4. See Edith Clowes, *The Revolution of Moral Consciousness: Nietzsche in Russian Literature 1890–1914* (DeKalb, 1988), pp. 114–223; James M. Curtis, "Michael Bakhtin, Nietzsche, and Russian Pre-Revolutionary Thought," in Bernice Glatzer Rosenthal, ed., *Nietzsche in Russia* (Princeton, 1986), pp. 331–54; Bernice Glatzer Rosenthal, ed. *Nietzsche and Soviet Culture: Ally and Adversary* (Cambridge, 1994), pp. 15–21; and Boris Groys, "Nietzsche's Influence on the Non-Official Culture of the 1930s," ibid., pp. 367–89.
5. See Michael André Bernstein, *Bitter Carnival: Ressentiment and the Abject Hero* (Princeton, 1992).
6. See Sergei S. Averintsev, "Bakhtin and the Russian Attitude to Laughter," in *Bakhtin, Carnival, and Other Subjects: Selected Papers from the Fifth International Bakhtin Conference, University of Manchester, July 1991*, ed. David Shepherd; *Critical Studies* 3–4 (1992): 13–19.
7. See Ruth Ginsburg, "Karneval und Fasten: Excess und Mangel in der Sprache des Körpers," *Poetica* 21 (1989): 26–42.
8. The invention was not, in the first place, Russian. In a recognizable form, concentration camps were first introduced in 1896 by the Spanish in Cuba and then in 1900 by the British in the Boer War in South Africa; see Andrzej J. Kaminski, *Konzentrationslager 1896 bis heute: Geschichte, Funktion, Typologie* [Concentration Camps from 1896 to This Day: History, Function, Typology] (Munich, 1990), pp. 34–35. A set of "improvements" different from the Nazi one was eventually introduced in the Chinese labor-camp system, the so-called *laogai.*
9. See Jacque Rossi, *Spravochnik po GULagu: Istoricheskii slovar' sovetskikh penitentsiarnykh institutsii i terminov, sviazannykh s prinuditel'nym trudom* (London, 1987), p. 183. This valuable reference work has also been published in English, in William A. Burbans's translation, *The Gulag Handbook: An Encyclopedic Dictionary of Soviet Penitentiary Institutions and Terms Related to the Forced Labor Camps* (New York, 1989).
10. The Wannsee conference gave the signal for the "Final Solution" of the Jewish question. This ideological denial of the basic human right—the right to live—to whole nationalities was a qualitative shift in both the history of ideas and the history of murder: cf. Geoffrey H. Hartman, *Minor Prophesies: The Literary Essay in the Culture Wars* (Cambridge, MA, 1991), p. 131; and Berel Lang, *Act and Idea in Nazi Genocide* (Chicago, 1990), pp. 3–29. Individual retrospective attempts to understand the experiential significance of the practical consequences, including the conveyor-line production of death (preceded by the policy of trampling down the souls of the victims), usually lead to "a new order of consciousness, a recognizable shift of being" (Alvin H. Rosenfeld, *A Double Dying: Reflections on Holocaust Literature* [Bloomington, 1980], p. 13).
11. See, for instance, Charlotte Delbo, *Auschwitz and After,* trans. Rosette C. Lamont (New Haven, 1995); Arne Brun Lie with Robby Robbinson, *Night and Fog* (New York, 1990); David Rousset, *The Other Kingdom,* trans. Ramon Gurthrie (New York, 1947); and Elie Wiesel, *Night,* trans. Stella Rodway (New York, 1960).

12. Primo Levi, *If This is a Man / The Truce*, trans. Stuart Woolf (London, 1990), p. 42.
13. See Leona Toker, "*Contra* Schopenhauer: Varlam Shalamov and the *Principium Individuationis*," *REAL* (Germany) 13 (1997): 257–69.
14. A similar point on Kafka's relationship to Schopenhauer is made in Martha Satz and Zsuzsanna Ozsvath, "'A Hunger Artist' and 'In the Penal Colony' in the Light of Schopenhauerian Metaphysics," *German Studies Review* 1 (1978): 200–210.
15. Arthur Schopenhauer, *The World as Will and Representation*, trans. E. F. J. Payne (New York, 1969), I: 352, 380.
16. This may be a Nietzschean moment in Kafka; see Patrick Bridgewater, *Kafka and Nietzche* (Bonn, 1974), pp. 134–35.
17. This is also the case in Tadeusz Borowski's Auschwitz stories, which partly accounts for their complex effect.
18. "Carnival is a pageant without footlights and without a division into performers and spectators." Mikhail Bakhtin, *Problems of Dostoevsky's Poetics*, ed. and trans. Caryl Emerson (Minneapolis, 1984), p. 122.
19. There can, of course, be several explanations of Minia's nickname. The topographical deployment of the story is reminiscent of the chorus and the stage in Greek tragedy, but the name "Minia the Greek" also has a plain reality-effect: most of the "honest thieves" were known by their first names and nicknames that reflected either their origin or a feature of their character. For all we know, Shalamov may be describing an actual person who was known as Minia the Greek—he may, indeed, have been of Greek origin, and, like Os'ka, he may have come from Odessa, which had substantial Jewish and Greek populations.
20. In "The Birth of Tragedy" Nietzsche notes that the "choral parts with which tragedy is interlaced are, as it were, the womb that gave birth to the whole of the so-called dialogue, that is, the entire world of the stage, the real drama … the scene, complete with the action, was basically and originally thought of merely as a *vision*; the chorus is the only 'reality' and generates the vision, speaking of it with the entire symbolism of dance, tone, and words." *Basic Writings of Nietzsche*, trans. and ed. Walter Kaufman (New York, 1966), p. 65.
21. Minia's remark also evokes the carnivalesque body language discussed by Bakhtin in *Rabelais and His World*, trans. Helene Iswolsky (Bloomington, 1984), pp. 303–67.
22. Franz Kafka, "The Hunger Artist," in *Complete Stories*, ed. Nahum N. Glatzer (New York, 1976), p. 270.
23. Cf. Ginsburg, "Karneval und Fasten," p. 36.
24. Kafka, *Complete Stories*, p. 272.
25. Cf. Paul Neumarkt, "Kafka's 'A Hunger Artist': The Ego in Isolation," in *American-Imago: A Psychoanalytic Journal for Culture, Science, and the Arts* 27 (1970): 110.
26. It seems, indeed, to be a historical fact that show fasts were largely discontinued at the outbreak of World War I; see Breon Mitchell, "Kafka and the Hunger Artists," in Alan Udoff, ed., *Kafka and the Contemporary Critical Performance: Centenary Readings* (Bloomington, 1987), pp. 236–55.

27. Cf. Bakhtin: "All distance between people is suspended, and a special carnival category goes into effect: free and familiar contact among people.... People who in life are separated by impenetrable hierarchical barriers enter into free familiar contact on the carnival square." *Problems of Dostoevsky's Poetics*, p. 123.
28. See, for instance, Aleksandr Solzhenitsyn, *The Gulag Archipelago: An Experiment in Literary Investigation*, trans. Thomas P. Whitney (Parts I–IV) and Harry Willetts (Parts V–VII) (New York, 1978); Dimitri Panin, *The Notebooks of Sologdin*, trans. John Moore (New York, 1976); Lev Kopelev, *To Be Preserved Forever*, trans. Anthony Austin (Philadelphia, 1977); Julius Margolin, *Puteshestvie v stranu ze/ka* [Journey into the Country of Z/K] (New York, 1952); Evgenia Ginzburg, *Journey into the Whirlwind*, trans. Paul Stevenson and Max Hayward, and *Within the Whirlwind*, trans. Ian Boland (San Diego, 1967 and 1981); and Gustav Herling, *A World Apart*, trans. Joseph Marek (Oxford, 1987).
29. The term comes from Antoni Ekart, *Vanished Without Trace: The Story of Seven Years in Soviet Russia*, trans. Egerton Sykes and E. S. Virpsha (London, 1954), p. 60.
30. An "iterative" scene presents (once) events that are supposed to have happened many times; see Gérard Genette, *Narrative Discourse: An Essay in Method*, trans. Jane E. Lewin (Ithaca, 1980), p. 116.
31. Kafka, *Complete Stories*, pp. 3–4.
32. See Wolfgang Iser, *Lawrence Sterne's Tristram Shandy* (Cambridge, 1988), pp. 94–106; *Prospecting: From Reader Response to Literary Anthropology* (Baltimore, 1989), pp. 249–61; and *The Fictive and the Imaginary: Charting Literary Anthropology* (Baltimore, 1993), pp. 247–80.
33. Roger Caillois distinguishes four types of games: (1) *agôn*, or competition, (2) *alea*, or chance, (3) *mimesis*, or mimicry, and (4) *ilinx*, or vertigo. "Each of the basic categories of play has socialized aspects of this sort ... For *agôn*, the socialized form is essentially sports to which are added contests in which skill and chance are subtly blended as in games and on radio and as part of advertising. In *alea*, there are casinos, racetracks, state lotteries, and pari-mutuels. For *mimicry*, the arts involved are public spectacles, puppet shows, the Grand Guignol, and much more equivocally, carnivals and masked balls which are already oriented toward vertigo. Finally, *ilinx* is revealed in the traveling show and the annual or cyclical occasions for popular merry-making and jollity." Through *ilinx*, a person "gratifies the desire to temporarily destroy his bodily equilibrium, escape the tyranny of his ordinary perception, and provoke the abdication of conscience." *Man, Play, and Games*, trans. Meyer Barash (New York, 1961), pp. 41 and 44.
34. See Johan Huizinga, *Homo Ludens: A Study of the Play Element in Culture* (Boston, 1955), p. 10.
35. This turn of the plot is actually prophetic of the history of the Nazi camps where, toward the end of the war, the inhuman order began to disintegrate in likewise inhuman ways and the break-up of the machine that had been grinding prisoners down speeded up the deaths of immense numbers of people—in death marches, unheated abandoned barracks, sealed railway cars. Margarete Buber-Neumann, a witness of both the Soviet and the Nazi concentration camps, writes that while the prisoners of the Gulag were exposed to the indig-

nities of neglect, the dignity of the inmates of Ravensbrück was assaulted by the hypertrophied demands for orderliness. Toward the end of the war, however, the orderliness of Ravensbrück disintegrated, and the dirt and chaos similar to those of Karaganda camps were added on to the specific horrors of the Nazi Lager. See Buber-Neumann, *Als Gefangene bei Stalin und Hitler* (Stuttgart, 1968); English translation by Edward Fitzgerald, *Under Two Dictators* (London, 1950). Incidentally, in Ravensbrück Buber-Neumann was a close friend of Milena Jesenska, who had been Kafka's intimate friend and Czech translator; she nursed Jesenska in her final illness, mourned her in-camp death shortly before the liberation, and eventually wrote her biography; see *Milena, Kafkas Freundin* (Munich and Vienna, 1977). English translation by Ralph Manheim, *Milena: The Story of a Remarkable Friendship* (New York, 1978).

36. I have discussed these themes in "Varlam Shalamov's Kolyma," in *Between Heaven and Hell: The Myth of Siberia in Russian Culture*, ed. Galya Diment and Yuri Slezkine (New York, 1993), pp. 151–69.
37. See Schopenhauer, *The World as Will and Representation,* I: 324.
38. As Nietzsche has noted in the essay "On the Genealogy of Morals," "Man, the bravest of animals and the one most accustomed to suffering, does *not* repudiate suffering as such; he *desires* it, he even seeks it out, provided he is shown a *meaning* for it, a *purpose* of suffering." *Basic Writings of Nietzsche*, p. 598.
39. Cf. Nietzsche's "The Birth of Tragedy" on the Dionysian reconciliation of man and nature: "Freely, earth proffers her gifts, and peacefully the beasts of prey of the rocks and deserts approach." Ibid., p. 37.
40. Cf. also James Rolleston, "Purification unto Death: 'A Hunger Artist' as Allegory of Modernism," in Richard T. Gray, ed., *Approaches to Teaching Kafka's Short Fiction* (New York, 1995), pp. 135–42.
41. See Herbert Spencer, "Philosophy of Style," in *Philosophy of Style by Herbert Spencer and Philosophy of Composition by Edgar Allan Poe* (New York, 1959), pp. 9–64; discussion in Leona Toker, "Towards a Poetics of Documentary Prose—from the Perspective of Gulag Testimonies," *Poetics Today* 18 (1997): 210–11.
42. Stanley Cavell, *Must We Mean What We Say? A Book of Essays* (Cambridge, 1976), pp. 188–89.
43. Shalamov, *Kolymskie rasskazy*, p. 886.
44. Philippe Lejeune, *On Autobiography*, ed. Paul John Eakin, trans. Katherine Leary (Minneapolis, 1989), p. 4.
45. Bernard Harrison, "Gaps and Stumbling-Blocks in Fielding: A Response to Černy, Hammond and Hudson," *Connotations* 3 (2) (1993/94): 168.
46. Samuel Johnson, *Lives of the English Poets* (London, 1955), I: 15.

APPENDIX

"The Artist of the Spade"

Varlam Shalamov

On Sunday, after work, Krist was told that he was being transferred to the brigade[1] of Kostochkin to reinforce the rapidly melting gold-mining brigade. The news was important. Whether good or bad—for Krist there was no point in thinking about it, for it was irrevocable. But about the man Kostochkin Krist had heard a lot on this rumorless mine, in the deafened, mute barracks. Like any other convict, Krist did not know from where the new people came into his life—some for a short time, others for longer. They always vanished from his life without having said anything about themselves; they went away as if dying, they died as if going away. Chiefs, foremen, cooks, storage-men, neighbors on the plank-beds, brothers at the wheelbarrow, comrades at the hack …

This kaleidoscope, this movement of endless faces, was not tiresome for Krist. He simply gave it no thought. Life left no time for such reflections. "Don't worry, don't think about the new bosses, Krist. You are one, but your bosses shall be many,"—so a certain joker and philosopher would say; who he was Krist had forgotten. Krist could not recall either the name or the face or the voice that had spoken these important jocular phrases. Important precisely because jocular. He was someone who dared to joke, to smile, if only a deeply concealed innermost smile, but nevertheless a smile, doubtless a smile. Such people did exist, but Krist himself was not one of them.

What foremen had Krist had? Either those of one's own kind, the fifty-eighth,[2] who had taken upon themselves too serious a task and were soon demoted, degraded before they had the time to turn into murderers. Or, again, one's own kind, the fifty-eighth, "fraiers,"[3] but experienced—slick customers who could not only give orders at work but also organize the work, and, in addition, get along with the rate-setters, the office, and different bosses, who could bribe, persuade. But these too, one's fellow fifty-eighth, would not admit the thought that ordering people around at this forced labor was the worst camp sin, that where the reckoning is in blood, where man has no rights, to take upon oneself the responsibility to command somebody else's will to live or die was a too big a sin, a deadly sin, not to be forgiven. There had been foremen who died together with the brigade. There had also been those whom this dreadful power over the life of others promptly corrupted, and then the hack, the handle of the spade in their hands began to help them speak to their comrades. And when they recollected this, they would quote, repeating it like a prayer, a grim camp maxim: "You die today, and I tomorrow." Yet it was not the rule that Krist's foremen should be convicts under the fifty-eighth article. Much more often, and during the most terrible years—always, his foremen were common criminals, sentenced for murder, for dereliction of duties. These were normal people, and only the guilt of power and the heavy pressure from above—the stream of deadly instructions—drove them to acts they would not, perhaps, have dared to commit in their former life. The borderline between felonies and "nonpunishable acts" in the regulations of official duties, as well as in most criminal law articles, is very thin, at times indefinable. Frequently, sentences were passed today for something they were not passed for yesterday, to say nothing about the means of suppression—all this judicial gamut of shades from a misdemeanor to a felony.

Common-criminal foremen were brutes by order. Foremen from among professional criminals were also brutes but not just by order. The worst thing that could happen to a brigade was to get a thug for a foreman. But Kostochkin was neither a professional sharper nor a common criminal. Kostochkin was the only son of a big party or administrative boss on the KVZhD, prosecuted and killed in connection with the "KVZhD case."[4] Kostochkin's only son, who studied in Kharbin, and had seen nothing except Kharbin, was at the age of 25 sentenced to fifteen years, as a *ChS*,[5] a member of the family. Brought up in the foreign Kharbinian lifestyle, where the innocently convicted were only known from novels—translated novels mainly, deep down the young Kostochkin was not convinced that his father had been innocent of the charges. His father had inculcated in him a belief in the infallibility of the NKVD;[6] young Kostochkin was totally unprepared for a different judgment. And when his father was arrested, when Kostochkin himself was sentenced and dispatched from the Very Far East to the Very Far

North, Kostochkin grew first and foremost resentful against his father who had spoiled his life by his mysterious crime. What does he, Kostochkin, know about the life of adults? He, who had mastered four languages, two European and two Oriental, the best dancer in Kharbin who had learned all kinds of blues and rumbas from visiting experts, the best boxer in Kharbin, middle-weight to heavy-weight, who had been trained in uppercuts and jabs by a former European champion—what did he know about all this big politics? If the father was shot—there must have been something. Maybe the NKVD had been hasty, maybe he should have been given ten, fifteen years. And he, the young Kostochkin, should have been given five instead of fifteen—if anything at all.

Kostochkin kept repeating the same words, arranging them in a different order, and every time it came out bad, alarming: "Then something must have been there. Then there must have been something."

Having aroused Kostochkin's hatred for his executed father and a passionate wish to rid himself of the stigma of this paternal curse, the investigating body had achieved an important success. But the interrogator did not know that. The interrogator in charge of Kostochkin's case had, in his turn, been shot in connection with one of the "NKVD cases."

It was not only fox-trots and rumbas that the young Kostochkin had studied in Kharbin. He graduated from the Kharbin polytechnical institute, with a degree in mechanical engineering.

When Kostochkin was brought to the mine, his assigned place, he managed to get an appointment with the chief of the mine and asked to be given a job in his speciality, promising to work honestly, condemning his father, pleading with the local authorities. "He'll write labels on tins," said the chief of the mine dryly, but the local police operative, who was present at the meeting, caught faintly familiar notes in the tone of the young Kharbinian engineer. The chiefs talked it over among themselves, then the police agent spoke with Kostochkin, and suddenly the news spread among the pit-face brigades: a newcomer, one of their own, the fifty-eighth, was appointed foreman of one of the brigades. The optimists saw this appointment as a sign of coming change for the better; the pessimists muttered something about a new broom. But all were surprised, except, of course, those who were long past surprise—Krist was not surprised.

Each brigade lives its own life, in its own section, in a barrack with a separate entrance; it meets the other inmates of the barrack only in the dining-hall. Krist often met Kostochkin, who was so conspicuous—red mug, broad shoulders, powerful build. The cuffs of his bell-shaped gloves were of fur. The poorer foremen's cuffs are made of rags sewn together from wadded trousers. And Kostochkin's cap was that of a free man, fur with ear-flaps, and his felt-boots were the real stuff, not rag boots or string-tied bast shoes. All this singled out Kostochkin. He had been working as foreman for the winter month—which meant that he had fulfilled the plan, the percentage; how much that was could be found out on the blackboard by the guard-house. But such a question did not interest an old convict like Krist.

Krist had composed the biography of his future foreman mentally, on the plank-bed. But he was sure, he was not wrong, could not be wrong. The Kharbinian could not have had any other ways of getting a foreman's job.

Kostochkin's brigade was melting away, as was the case with all the brigades working in the gold pit-face. From time to time—this should read from week to week rather than from month to month, a reinforcement was sent to Kostochkin's brigade. Today this reinforcement was—Krist. "Most likely Kostochkin even knows who Einstein is," thought Krist falling asleep in the new place.

As a newcomer, Krist was given a place further away from the stove. Whoever came to the brigade earlier occupied a better place. This was the general rule, and Krist was well aware of it.

The foreman was sitting at the table in the corner, close to the lamp, reading a book. Even though the foreman, as master over the lives and deaths of his sloggers, could for the sake of his convenience put the only lamp on his own little table, depriving all the other inmates of the barrack of light—not that they would read or talk ... One can talk in the dark just as well, but then there were neither subjects not time for conversation. But foreman Kostochkin settled down at the common lamp and went on reading; at times he drew his chubby childish heart-shaped lips into a smile and screwed up his big beautiful gray eyes. Krist was so appreciative of this peaceful picture of the foreman and the brigade at rest, something he had not seen for so long, that he decided to stay in this brigade by all means, to give all his strength to his new foreman.

There was also a deputy foreman in the brigade, the shortish fatigue-man Os'ka, old enough to be Kostochkin's father. His duties were to sweep the barrack, bring the brigade's food, help the foreman—same as everywhere. And on the point of falling asleep, Krist thought for some reason, that, most probably, his new foreman knew who Einstein was. And, made happy by the thought, warmed by the mug of boiling water he had just drunk for the night, Krist fell asleep.

Nor was there any noise in the new brigade when it was taken to work. Krist was shown the tool-shed—everyone received his tool, and Krist adjusted his spade, as he had adjusted it a thousand times before:

knocked off the short handle with the lug fixed to the American scoop-spade, with the butt of the axe unbent this scoop a little wider on a stone, picked out a very long new haft from the many standing in the corner of the shed, passed it into the ring of the spade, fixed it, placed the spade with its bent blade at his feet, took the measure and made a "notch" on the haft against his own chin, then cut it at this mark. With his sharp ax Krist scraped off the rough edge of the new handle, thoroughly smoothing it out. He got up and turned around. In front of him Kostochkin was standing, carefully watching the movements of the newcomer. Not that Krist had not expected this. Kostochkin did not say a word, and Krist understood, that the foreman was deferring judgment till the beginning of work, till the pit-face.

The pit-face was not far away and the work started. The haft broke into motion, a dull pain crept into his back, the palms of both hands fell into their usual position, the fingers clutched the haft. It was a tiny bit too thick, but Krist would repair it in the evening. He would also sharpen the spade with a file. His hands swung the spade time after time, and a quickened rhythm entered into the melodious grating of metal against stone. The spade screeched, rustled; the stone would slide off the spade at each sweep and drop onto the bottom of the wheel-barrow, and the bottom would respond with a wooden thump, and then stone would respond to stone—all this music of the pit-face was well known to Krist. In every place stood the same wheel-barrows, the same spades screeched, the stone, helpfully hewn with the hack, rustled and slid down the landslip, and the spades screeched again.

Krist put down the spade to replace his mate at the "machine OSO[7]—two handles, one wheel" as the convicts in Kolyma called the wheel-barrow (not in thieves' jargon, but something like it). Krist set the wheel-barrow bottom on the trap board, handles turned in the opposite direction of the pit-face, and quickly filled the wheel-barrow. Then he grasped the handles, arched his back straining the stomach, and, catching his balance, wheeled his barrow to the pan, the gold washing appliance. On the way back Krist wheeled the barrow according to all the rules of the wheel-barrowers, inherited from centuries of penal servitude—handles upward, wheel forward; his hands resting on the handles. Then he put down the wheel-barrow and again took up the spade. The spade started screeching.

The Kharbinian engineer, the foreman Kostochkin, stood listening to the pit-face symphony and watching Krist's movements.

"Well, well, I can see that you are an artist of the spade," said Kostochkin and burst out laughing. His laughter was childish, irrepressible.

With his sleeve the foreman wiped his lips. "What category did you get where you have come from?"

What he meant was the categories of nourishment, "the stomach scale," driving on the convict. These categories, Krist knew, were discovered on the White-Sea canal,[8] during the "reforging."[9] The slobbery romanticism of the reforging had a realistic foundation, cruel and sinister, in the shape of this "stomach scale."

"The third," answered Krist, emphasizing with his voice, as clearly as possible, his contempt for his previous foreman who had not appreciated the talent of the artist of the spade. Realizing the benefit, Krist, as usual, lied a little.

"I shall give you the second. Right as of today."

"Thanks," said Krist.

In the new brigade it seemed to be a little quieter than in the other brigades where Krist had to live and work; the barrack was a little cleaner; there was a little less swearing. Out of a long-time habit Krist wanted to fry a little piece of bread that had remained from his supper on the stove, but his neighbor—Krist didn't yet know and would never learn his surname—gave him a nudge and said that the foreman didn't like them to fry bread on the stove.

Krist went up to the lively burning iron stove, spread his fingers wide above the flow of heat, and put his face into the current of hot air. From the nearest plank-bed Os'ka, the deputy foreman, got up and with a strong arm led the newcomer away from the stove: "Go to your place. Don't block the stove. Let everybody be warm." This was, generally speaking, fair, but it is very difficult to restrain one's own body, its longing for the fire. The convicts of Kostochkin's brigade had mastered restraint. And Krist would have to master it too. Krist returned to his place, took off his pea-jacket. He thrust his feet into the sleeves of the pea-jacket, adjusted his cap, curled up, and fell asleep.

On the point of falling asleep, Krist still saw somebody come into the barrack and give an order. Kostochkin swore, without moving away from the lamp or stopping to read his book. Os'ka ran up to the newcomer, and seizing him by the elbows with quick deft movements, pushed him out of the barrack. In his former life Os'ka had been a history teacher in some institute.

For many following days Krist's spade went on screeching; the sand went on rustling. Kostochkin soon understood that behind the refined technique of Krist's movements there had for long been no strength, and no matter how hard Krist tried, his barrows were always filled a little bit less than prescribed. This did not depend on one's own will: measure is dictated by some inner feeling which controls the muscles—all kinds—healthy and powerless, young and worn out, exhausted. Every time the output was measured, it turned out that not as much had been achieved at the pit-face where Krist worked as the foreman had expected from the highly professional movements of the artist of the spade. But Kostochkin did not take it out on Krist, did not scold him more than he did the others, did not abuse him, did not preach to him. Perhaps he understood that Krist was putting all his strength into his work, saving only what could not be spent to please any foreman in any camp of the world. Or perhaps he felt it rather than understood—our feelings are much richer than thoughts; the bloodless tongue of the convict does not reveal all there is in his heart. Feelings also pale, grow weaker, but much later than thoughts, much later than human speech, language. And Krist was indeed working harder than he had worked in a long time, and, although his output was not sufficient for the second category, he was receiving this second category, for his assiduity, for his efforts …

The second category, indeed, was the most that Krist could possibly get. The first was given to the record-setters who fulfilled 120 percent of the quota, and more. There were no record-setters in Kostochkin's brigade. There were those in the brigade who received the third category, that is, who fulfilled the quota, and those who received the fourth category, who did not fulfill the quota but only 70 or 80 percent of it. Still these were not the obvious idlers who deserved penal rations, the fifth category; there were none of those in Kostochkin's brigade.

Day followed day, and Krist grew weaker and weaker, and the submissive silence in the barrack of Kostochkin's brigade pleased him less and less. But then one evening, Os'ka, the history teacher, took Krist aside and said to him in a low voice: "The cashier is coming today. The brigadier ordered money for you, have it in mind …" Krist's heart was pounding. So, Kostochkin did appreciate Krist's diligence, his skill. This Kharbinian engineer, who knows about Einstein, is, after all no stranger to conscience.

In the brigades where Krist had worked before, he had never received any money. Each brigade inevitably appeared to contain more deserving people: either really physically stronger and better workers or just friends of the foreman—Krist never indulged in such futile reflection, accepting every dinner check (the categories were changed every ten days, the percentage being determined for the output of the recent period) as the finger of fate, as good or bad fortune, success or failure, that would pass, change, and would not be eternal.

The news about the money he would be paid that night filled Krist's soul and body with a great, irrepressible joy. It appears that for joy there are enough feeling and strength. How much money could he be paid?... Even five or six rubles would mean five or six kilograms of bread. Krist was ready to worship Kostochkin and couldn't wait for work to end.

The cashier arrived. He was a very ordinary man but wore a good tanned sheepskin coat; he was a civilian. He was accompanied by a guard who was concealing his revolver or pistol or had left his weapon in the guard-house. The cashier sat down at the table, opened his briefcase packed with much used many-colored banknotes that looked like laundered rags. The cashier took out a list, densely lined, covered with all kinds of signatures—made by people either happy or disappointed with the sums they had got. The cashier called Krist and pointed to a ticked-off spot.

Krist, all attention, felt something special in this premium. Nobody, except Krist, went up to the cashier. There was no queue. Maybe the members of the brigade had been so trained by their solicitous foreman. Well, what is there to think about! The money has been ordered, the cashier pays out. That's Krist's luck.

The foreman himself was not in the barrack; he had not yet returned from the office, and the identification of the recipient was taken care of by the foreman's deputy, Os'ka, the history teacher. With his forefinger Os'ka showed Krist the place for his signature.

"And ... and ... how much?" Krist wheezed, choking.

"Fifty rubles. Satisfied?"

Krist's heart began to sing, to pound. Here's luck. Hurriedly, tearing the paper with the sharp pen and almost overturning the non-drip inkstand, Krist signed the payroll.

"Atta boy," said Os'ka approvingly.

The cashier clapped the briefcase shut.

"Nobody else in your brigade?"

"No."

Krist still could not understand what was going on.

"And the money? the money?"

"I gave the money to Kostochkin," said the cashier. "Already in the morning." And the small Os'ka tore Krist away from the table with an iron arm, with a force no hewer in this brigade had ever had, and hurled him into the darkness.

The brigade was silent. Not a single person stood up for Krist or asked a question. No one even cursed Krist for a fool ... This for Krist was more terrible than the brute Os'ka and his clawing iron hand. More terrible than the childish chubby lips of foreman Kostochkin.

The door of the barrack was flung open, and with quick and light steps foreman Kostochkin walked to the lighted table. The floor of the barrack, made of beams and planks, hardly shook under his light, springy gait.

"Here's the foreman himself—speak to him," said Os'ka, stepping back. Pointing at Krist, he explained to Kostochkin: "He wants money!"

But the foreman had already understood everything while crossing the threshold. Kostochkin immediately felt as he had in the ring in Kharbin. He stretched out his hand toward Krist with the habitual handsome boxing movement "from the shoulder," and Krist fell onto the floor, stunned.

"Knockout, knockout," wheezed Os'ka, hopping around the half-dead Krist and acting the referee in the ring, "eight … nine … Knockout."

Krist did not rise from the floor.

"Money? Money he wants?" Kostochkin, was saying, unhurriedly seating himself at the table and taking a spoon out of Os'ka's hands to get down to a bowl of peas.

"These trotskyites," said Kostochkin slowly and instructively, "they are the undoing of me and you, Osia." Kostochkin raised his voice. "They've ruined the country. And are ruining us. He needs money, the artist of the spade, money. Hey, you!" shouted Kostochkin to the brigade. "You fascists! Do you hear! You won't knife *me* down. Dance, Os'ka!"

Krist was still lying on the floor. The huge figures of the foreman and the orderly stood between him and the light. Suddenly Krist saw that Kostochkin was drunk, very drunk. Those same fifty rubles, which had been assigned to Krist … How much spirit could they buy, alcohol rationed out to the whole brigade …

Os'ka, the foreman's deputy, obediently started to dance, repeating all the while:

I have bought two washing-troughs,
And my wife Rosita …

"That's ours, from Odessa, foreman. It is called 'From the bridge to the slaughter-house.'" And Os'ka, history teacher in some capital city institute, father of four children, again went into a dance.

"Stop, fill up."

Fumbling, Os'ka found a bottle under the plank-beds and poured something into a tin. Kostochkin drank it and ate the remains of the peas in the bowl, picking them up with his fingers.

"Where is that artist of the spade?"

Os'ka raised Krist up and pushed him into the light.

"Why, no strength? Aren't you getting your rations? Who is getting the second category? That's not enough for you, you trotskyite swine?"

Krist kept silent. The brigade kept silent.

"I'll smash all of you! Damned fascists," raged Kostochkin.

"Go, go to your place, artist of the spade, or else the foreman may hit you again," advised Os'ka peaceably, holding the drunken Kostochkin and pushing him into the corner, turning him over onto the foreman's splendid single trestle-bed—the only trestle-bed in the barrack where all the plank-beds were double and two-tiered, "the railway type." Os'ka himself, the foreman's deputy and orderly, who slept on the last cot, was taking up his important third and quite official duties, those of the bodyguard, the night watchman of the foreman's sleep, rest, and life. Krist, groping, reached his bed.

But neither Kostochkin nor Krist had a chance to fall asleep. The door of the barrack opened, letting in a current of white steam, and a man came in wearing a fur cap with ear-flaps and a dark winter overcoat with an astrakhan fur collar. The overcoat was very crumpled, the astrakhan worn out; nevertheless it was a real overcoat and real astrakhan.

The man crossed the whole barrack to the table, to the light, to Kostochkin's trestle-bed. Os'ka greeted him respectfully and started shaking the foreman out of his stupor.

"Minia the Greek is calling you." This name was familiar to Krist. He was the foreman of the thugs. "Minia the Greek is calling you." But Kostochkin had already come to and sat up on the trestle-bed with his face to the light.

"Still carousing, Tamer?"

"Well, you see … they've driven me to this, the skunks …"

Minia the Greek uttered a sympathetic grunt.

"One day, Tamer, you'll be blown up into the air. Eh? They'll lay ammonium under your cot, set fire to the cord, and there you go …" The Greek pointed his finger upwards. "Or they'll saw off your head with a saw. Here, your neck is thick, it will take a long time to saw."

Kostochkin, slowly regaining his wits, was waiting to hear what the Greek had to tell him.

"How about a little one? Say, we'll find something in two seconds."

"No. We've got plenty of this liquor in our brigade, you know that. My business is more serious."

"Pleased to be of service."

"'Pleased to be of service,'" laughed Minia the Greek. "This is how they taught you to speak to people in Kharbin."

"Well, I didn't mean anything," Kostochkin hurried on. "I just don't know what you want."

"Here's what." The Greek started speaking rapidly, and Kostochkin nodded in agreement; the Greek drew something on the table, and Kostochkin started nodding understandingly. Os'ka was following the talk with interest. "I went to see the rate-setter," said Minia the Greek. His manner was neither sullen nor lively; he spoke in a most usual voice. "The rate-setter said it is Kostochkin's turn."

"But they took off from me last month as well …"

"What can I do …" And the Greek's voice became merrier. "Our guys, where can they get … the cubes?[10] I told the rate-setter. The rate-setter says—It's Kostochkin's turn."

"But …"

"Look. You yourself know our position …"

"Well, all right," said Kostochkin. "You'll count it in the office, and then tell them to take it off from us."

"No fear, fraier," said Minia the Greek and patted Kostochkin on the shoulder. "Today you helped me out—tomorrow I'll do the same for you. I pay my debts. You do for me today, I do for you tomorrow."

"Tomorrow both of us shall kiss,"[11] said Os'ka starting to dance, overjoyed at the decision finally taken and afraid that the foreman's hesitation could only spoil the matter.

"Well, good-bye, Tamer," said Minia the Greek, rising from the bench. "The rate-setter says: go boldly to Kostochkin, to the Tamer. There is a drop of the rogue's blood in him. Don't be afraid, don't be shy. Your fellows will manage. You have such artists of the spade …"

Translated by N. Strazhas

Notes

1. The action of the story takes place in a Kolyma labor camp; a "brigade" is a convicts' work crew.
2. Convicts sentenced in accordance with the "political" article 58 of the Soviet criminal code.
3. A contemptuous thieves-jargon term for political prisoners, especially intellectuals, which eventually acquired wider currency and was applied to the easily victimized "suckers" inside and outside the camps.
4. KVZhD, the Chinese Eastern Railway, was sold to China in the 1930s, whereupon most of its employees moved to the Soviet Union, only to be persecuted on suspicion of espionage.
5. Acronym of *chlen sem'i* (a member of the family). Wives and children of the victims of Stalin's terror were liable to administrative incarceration.
6. Acronym for People's Commissariat for Internal Affairs, i.e., the Soviet secret police.
7. Acronym of *Osoboe Soveshchanie* (Special Committee) that passed conveyor-line administrative sentences to victims of large waves of arrests. In the original saying "OSO" rhymes with *koleso*, "wheel."
8. The first major Stalinist construction project accomplished through convict labor, from September 1931 to April 1933. The estimates of convict mortality during the digging of the canal range between 50,000 and 200,000 deaths.
9. "Reforging" (*perekovka*) was a much publicized early 1930s campaign of, as it were, re-educating prisoners through labor. It gave birth to the Gulag version of "incentive," which consisted in not giving the prisoners an assured flat subsistence but rather tying their rations to their production output.
10. Cubic meters. Professional criminals would not do any work in the camps, but their rations were still supposed to be calculated according to the output.
11. Allusion to a popular gypsy romance.

CONTRIBUTORS

Gennady Barabtarlo is Professor in Russian and Chairman of the Department of German and Russian Studies at the University of Missouri. He is the author of *Phantom of Fact* (1989), *Aerial View* (1993), and many articles on Russian literature. His most recent one, entitled "Nabokov's Trinity," appeared in *Nabokov and His Fiction: New Perspectives* (1999).

Michael Bernard-Donals is an Associate Professor of English and Jewish Studies at the University of Wisconsin-Madison, where he teaches courses in rhetorical theory, critical theory, and Holocaust representation. He has published books on Bakhtin and on the connections between rhetorical and critical theory, and has co-edited a collection of essays on rhetoric and anti-foundationalism. He is nearing completion of a book entitled *Sublimity and Redemption*, on Holocaust representation, which will be published by the State University of New York Press.

William Craft Brumfield is Professor of Slavic Languages and Literatures at Tulane University. His recent books include the fundamental *History of Russian Architecture* (1993), *Lost Russia* (1995), and *Landmarks of Russian Architecture* (1997).

Steven Cassedy is Professor of Slavic and Comparative Literature at University of California, San Diego. His most recent publications include *To the Other Shore: The Russian Jewish Intellectuals Who Came to America* (1997) and *Building the Future: Jewish Immigrant Intellectuals and the Making of Tsukunft* (1999).

Roger F. Cook is Professor of German at the University of Missouri. He has published several books and essays on German literature and film. His most recent book is *By the Rivers of Babylon: Heinrich Heine's Late Songs and Reflections* (1998).

Anna Lisa Crone is Professor of Slavic Languages and Literatures at the University of Chicago. She has specialized in Russian religious thought, especially the work of Vasily Rozanov, and in Russian poetry from the eighteenth century through the Silver Age. In recent years she has published some twelve essays on Derzhavin, and her book *The Daring of Derzhavin: The Moral and Aesthetic Independence of the Poet in Russia* will be published in 2000 by Slavica Publishers at Indiana University.

Alexander Dolinin is Professor of Russian Literature at the University of Wisconsin, Madison. He is the author of *History Dressed Up as a Novel* (in Russian) and numerous works on Russian, English, and American literature, especially on Pushkin and Nabokov, including notes and introductions to Russian publications of their writings. He is finishing his new book entitled *A Great Unknown: Vladimir Nabokov as a Russian Writer.*

Annelore Engel-Braunschmidt is Professor in Slavonic Philology and Literary Studies at Christian-Albrechts-Universität zu Kiel. Her latest publication is "Modernisierung durch Literatur: Ch. F. Gellerts 'Betschwester' und Katharinas 'O Zeit!'" in *Russland zur Zeit Katharinas II. Absolutismus – Aufklaerung – Pragmatismus* (1998).

Karen Evans-Romaine is Assistant Professor of Russian at Ohio University. She is the author of *Boris Pasternak and the Tradition of German Romanticism* (1997). Her current research has to do with subtexts in musical themes in the works of Pasternak, Mandelshtam, and Tsvetaeva.

John Burt Foster, Jr., is Professor of English and Cultural Studies at George Mason University, and also edits *The Comparatist*, an annual journal of comparative literature. Among his most recent publications are "Working with Nietzsche, Nabokov, and Tolstoy: Cultural Variables in the Literary Reception of Philosophy," in *Real: The Yearbook for Research in English and American Literature* (1997) and "Zarathustrian Millennialism Before the Millennium: From Bely to Yeats to Malraux," in *Why Nietzsche Still? Reflections on Drama, Culture, Politics* (1999).

Michael Holquist is Professor of Slavic and Comparative Literature at Yale University and Chairman of the Comparative Literature Department. He is the author of *Dostoevsky and the Novel* (1977), *Dialogism: Bakhtin and His World* (1990), and other books and articles. His lectures on literary theory are forthcoming under the title *The Double Gaze of Hermes: Philology as History and as Task.*

Marina Kanevskaya is Assistant Professor of Russian Language and Literature at the University of Montana. She graduated from Indiana University in 1997 with a Ph.D. thesis, "F. M. Dostoevsky's Pushkin Speech in Cultural Context," and is currently working on a monograph, *The Cruel Critic: N. K. Mikhailovsky's Rejection of Dostoevsky.*

Priscilla Meyer is Professor of Russian at Wesleyan University. The author of *Find What the Sailor Has Hidden: Nabokov's "Pale Fire"* (1988), she recently published: "The Real Hound, The Real Knight: Stoppard Reads Nabokov," "How the Bronze Horseman Was Made: Pushkin and *'L'inondation a Nantes',*" and "Dolorous Haze, Hazel Shade: Nabokov and the Spirits." She is co-editor with Jane Grayson and Arnold McMillin of *Nabokov at the Crossroads,* to be published by MacMillan.

Anna Muza is a Lecturer at the Department of Slavic Languages and Literatures, University of California at Berkeley. She has written on Russian and European dramatic literature and theater history, and translated plays, biographies, and memoirs. Among her recent articles are "Meyerhold at Rehearsal" (1996) and "Chukovski and the Nabokovs" (1996).

Omry Ronen is Professor of Slavic at the University of Michigan. He is the author of *An Approach to Mandel'stam* (1983), *The Fallacy of the Silver Age* (1997), and many articles on nineteenth- and twentieth-century Russian poetry and poetics.

Bernice Glatzer Rosenthal is Professor of History at Fordham University. She has edited, or contributed to, *Nietzsche in Russia* (1986), *Nietzsche and Soviet Culture: Ally and Adversary* (1994), and *The Occult in Russian and Soviet Culture* (1997).

Wolf Schmid is Professor in Slavic literatures at the University of Hamburg. His latest books are: *Pushkins Prosa in poetischer Lektüre* (1991; Russian edition: 1996; Serbian edition: 1999) and *Proza kak poezija. Pushkin – Dostoevskij – Chexov – avangard* (1998). Since 1993 he has been Chairman of the jury panel for the German-Russian Pushkin Prize for Russian literature.

Alexander von Schönborn is Associate Professor in Philosophy at the University of Missouri. He is the author of *Karl Leonhard Reinhold: Eine annotierte Bibliographie* (1991), and most recently "Karl Leonard Reinhold: '... Endeavoring to keep up the pace *mit unserem zeitalter*'," in *The Emergence of German Idealism* (1999).

Leona Toker is Associate Professor in the English Department of the Hebrew University of Jerusalem. She has written *Nabokov: The Mystery of Literary Structures* (1989); *Eloquent Reticence: Withholding Information in Fictional Narrative* (1993), and *Return from the Archipelago: Narratives of Gulag Survivors,* to be published by the Indiana University Press. She has also edited *Commitment in Reflection: Essays in Literature and Moral Philosophy* (1994).

Index